Advance Praise for

THE CLIMATE CHALLENGE

If you are wondering what to do about climate change, here is the answer.
The Climate Challenge is not only interesting and informative, it is also exciting.

— Lester R. Brown, author of *Plan B 4.0: Mobilizing to Save Civilization*

A wonderfully clear guide to simplify the issues of global warming and
climate change so that anyone can get involved, doing what they can where they are.
Dauncey's 101 solutions — which people can take at every level from personal
to global — provide both the needed information and the inspiration.

— Hazel Henderson, President of Ethical Markets Media,
and author of *Ethical Markets: Growing The Green Economy*

What an amazingly (insanely!) comprehensive and useful book.
Guy Dauncey gets it. He understands all the individual things we must do,
but also why they won't work unless we also commit to working together and
building a movement. This is a joyous, hope-filled manual for facing the greatest
crisis humanity has ever encountered. It's going to do a lot of good!

— Bill McKibben, 350.org

A lot has been written about climate change over the last few years,
but this is a real cracker. Hugely informative, hard-hitting and very upbeat about
the solutions. Get your head around *The Climate Challenge*, and I think you'll find
there's only one answer to Guy Dauncey's own question ('do we believe in
our ability to create a green, sustainable future?'), and that's 'yes!'

— Sir Jonathon Porritt,
past Chair of the UK Sustainable Development Commission

This is a terrific labor. Nowhere will readers find a more exhaustive,
yet accessible, treatment of the climate challenge. *The Climate Challenge* is a terrific
resource for anyone interested in understanding the preeminent issue of our time.
Guy Dauncey's skills as an educator are on full display in this masterful work!

— Gary Gardner, Senior Researcher, Worldwatch Institute

The Climate Challenge is the handbook for the increasing number of people worldwide who understand the stakes: unchecked, global warming threatens a swing in global temperatures of ice age magnitude, only in the opposite direction, within the lifetime of today's young people. Guy Dauncey provides meaningful, effective solutions at the personal, professional and business level. But he also makes it clear that only if local action builds quickly to serious and sustained national political engagement can we really change the future.

— Eban Goodstein, Director,
Bard Center for Environmental Policy, New York

Very timely and persuasive. *The Climate Challenge* is an essential owner's manual for our planet. Guy Dauncey's clear-eyed presentation of the problem is followed by practical solutions that empower each of us to take action now — and if we follow his advice, we will meet the challenge and win.

— Terry Tamminen, New America Foundation,
former Secretary of California EPA

Guy Dauncey has created something unique in the current literature by blending a highly readable narrative on global warming, a rich picture book on climate solutions, and an up-to-date digest of the relevant heaps of climate change information that have steadily grown into electronic Himalayas. If you wish to grasp the mind-boggling complexity of the climate challenge, read this book.

— John Shellnhuber, Chief Sustainability Scientist for the German Government,
Founding Director of the Potsdam Institute for Climate Impact Research

The Climate Challenge is an informative, yet hopeful look at the climate crisis. Based on the latest science, the book includes a wealth of practical steps for citizens, industries, and governments to help avert catastrophic climate change as well other detrimental environmental impacts.

— Rhett A. Butler, founder of mongabay.com

This book is marvelous! Guy Dauncey's new book is an elegant, insightful and comprehensive examination of the dominant global challenge we face. This attractive work belongs on the desk of every investor, entrepreneur, citizen and policy maker.

— Paul R. Epstein, M.D., M.P.H., Associate Director,
Center for Health and the Global Environment, Harvard Medical School

The Climate Challenge is an effective antidote to despair. The solutions abound
and a new spirit of human creativity must be unleashed to embrace them.
Guy Dauncey has harnessed the planet's surest form of renewable energy — hope.

— Elizabeth May, O.C., Leader, Green Party of Canada

The escalating crisis of global warming mandates that for the first time in
our lives, indeed for the first time in our history, we all must take personal responsibility
for the climate. We must all become climate leaders. *The Climate Challenge* offers
compelling and easily accessible ways all of us can rise to this challenge.
There is literally nothing more important in the world.

— Jim Garrison, founder and President, State of the World Forum

The Climate Challenge offers a concise and accessible summary of global warming
and its causes. Dauncey masters the difficult task of describing the most urgent problem
mankind is facing in the 21st century in an uplifting and accessible manner without
confusing his readers with scientific details. The step-by-step instructions that Dauncey
provides range from easy-to-implement improvements of our personal lifestyles to advice on
how to tackle climate change within governments and industries. Truly a remarkable book
that will prove to be an invaluable starting point for those unfamiliar with the topic.

— Dr. Hermann Scheer, Member of the German Parliament,
author of *Energy Autonomy*

Finally here's a book that combines thorough explanations for newcomers
with innovative perspectives for experts and effective answers for people, industries and
institutions. *The Climate Challenge's* imaginative format is up-to-date, comprehensive,
and immensely handy. It's as if *Whole Earth Catalog* had been reborn 40 years later.
Guy Dauncey pulls no punches and is deeply, urgently persuasive.
Anyone who reads this will be moved to act.

— Felix Kramer, founder of CalCars.org

To employ the book's own martial metaphors, Dauncey writes that we are all
soldiers of circumstance placed on the front line of the great battle of our time: the fight
against the forces of climate change. *The Climate Challenge* does more than orient the
reader and set out the tactics; Dauncey rallies the troops for the struggle ahead
by instilling a vision of the better future that will come with victory.

— David Suzuki, author, scientist, environmental champion.

THE CLIMATE CHALLENGE

101 SOLUTIONS TO GLOBAL WARMING

GUY DAUNCEY

NEW SOCIETY PUBLISHERS

Cataloging in Publication Data:
A catalog record for this publication is available from the National Library of Canada.

Cover design by Diane McIntosh.
Cover images: Bicycle – Dreamstime / Gbh007; all others © iStock: Polar bear – Peter Van Wagner; Horizon – Andreas Guskos; Underwater – Dennis Sabo; New Orleans – Joseph Nickischer; Rainforest – Francisco Romero; Train – Karen Grieve; Wave machine – from authors; Electric bus – Vasiliy Dudenko; Solar panels – brainstorm1962; Light bulbs – mathieukor; Electric car – Recycling box – Gaby Jalbert; Wind turbines – Todd Arbini; Clouds – Peter Blottman

Printed in Canada by Friesens.
First printing October 2009.

New Society Publishers acknowledges the support of the Government of Canada through the Book Publishing Industry Development Program (BPIDP) for our publishing activities.

Paperback ISBN: 978-0-86571-589-9

Inquiries regarding requests to reprint all or part of *The Climate Challenge* should be addressed to New Society Publishers at the address below.

To order directly from the publishers, please call toll-free (North America) 1-800-567-6772, or order online at newsociety.com

Any other inquiries can be directed by mail to:

New Society Publishers
P.O. Box 189, Gabriola Island, BC V0R 1X0, Canada
(250) 247-9737

New Society Publishers' mission is to publish books that contribute in fundamental ways to building an ecologically sustainable and just society, and to do so with the least possible impact on the environment, in a manner that models this vision. We are committed to doing this not just through education, but through action. This book is one step toward ending global deforestation and climate change. It is printed on Forest Stewardship Council-certified acid-free paper that is **100% post-consumer recycled (100% old growth forest-free)**, processed chlorine free, and printed with vegetable-based, low-VOC inks, with covers produced using FSC-certified stock. New Society also works to reduce its carbon footprint, and purchases carbon offsets based on an annual audit to ensure a carbon neutral footprint. For further information, or to browse our full list of books and purchase securely, visit our website at: **newsociety.com**

NEW SOCIETY PUBLISHERS

FSC

Mixed Sources
Cert no. SW-COC-001271
© 1996 FSC

Contents

Preface

The choice we face is not between saving our environment and saving our economy. The nation that leads the world in creating new energy sources will be the nation that leads the 21st-century global economy.

— President Barack Obama

Our existence began many billion years ago, when a tiny fireball with an enormous heart sprang out of nowhere. Out of that fireball came everything we know, from the most majestic nebula to the tiniest kitten.

We are all created from this same great unity. When we act as if we can live separately from it, however, we get feedback. Global warming is very unpleasant feedback. By releasing ancient carbon and destroying ancient forests, we are causing Earth's oceans and atmosphere to warm. If the temperature rises by 3°C, many forests will turn to savannah, huge areas of the Earth will become extreme desert, and the sea level will eventually rise by 25 metres. If it rises by 6°C, most of life will be exterminated.

Those are terrible words to write – and yet if we continue on our current path, that's where we're headed. It is with good reason that many climate scientists are tearing their hair out, seeking to get our attention.

But now let's turn this around. For the past 500 years, even while we have experimented with every variety of cruelty and atrocity, Earth's innovators and pioneers have achieved incredible things.

In the 16th century they gave us the Renaissance. In the 17th century they gave us the scientific method and the beginning of global trade. In the 18th century they gave us the Industrial Revolution and the beginning of the anti-slavery movement.

In the 19th century they gave us the birth of modern medicine and public education. In the 20th century they gave us the liberation of women, the beginning of civil rights for all, the first voyages into space, the Internet, and so much more.

What might the 21st century bring? From the residents of the Austrian town of Güssing, who have almost achieved a fully zero-carbon economy, to the citizens of San Francisco, well on the way to a zero-waste economy, our world is full of people who have initiative and the determination to succeed.

If we have the will, we can end our use of fossil fuels, and harvest all the energy we need from the earth, wind, tides and sun.

We can engage Earth's forest stewards in a quest to end all deforestation, and restore the forests' biodiversity and splendor.

We can engage Earth's farmers and ranchers in a quest to restore the soil's fertility.

We can engage Earth's business leaders and bankers in a quest to ensure that every transaction harmonizes with nature, both in the accounting ledger and on the ground.

We can engage Earth's teachers in a quest to ensure that every child understands the wisdom of Nature, and our need to live in harmony with her laws.

If we have the will, and the ability to overcome our ignorance and pride.

My delight in writing this book – once I moved on from the dire perils we face – has been the joy of being immersed in the ocean of solutions, and the surge of creativity that is happening all over the world. It is my deepest wish that after reading it you will share my optimism about the human condition, and the amazing things we can achieve.

Acknowledgments

In writing this book, I have been helped by many people, including Jim Baak, Rhett Butler, Colin Campbell, Dorothy Cutting, Professor Fred Glasser, Paul Gipe, Martin Golder, Mark Harmon, Richard Heinberg, Britt Karlstrom, Drew Kilback, Felix Kramer, Todd Litman, Patrick Mazza, Steve Nadel, Beth Parke, Phil Thornhill, Fionnuala Walravens, and my friends in the BC Sustainable Energy Association.

For the inspiration they provide, I would like to thank Lester Brown, Gregor Czisch, Laurie David, Tim Flannery, Ross Gelbspan, Paul Gipe, Jane Goodall, Eban Goodstein, Al Gore, James Hansen, Jay Inslee, Vinod Khosla, Felix Kramer, Jeremy Leggett, Jaime Lerner, Amory Lovins, Mark Lynas, Wangari Matthai, Ed Mazria, William McDonough, Bill McKibben, George Monbiot, Fred Pearce, Enrique Peñalosa, Jerome Ringo, Joe Romm, Herman Scheer, Andrew Simms, David Suzuki, and Andrew Weaver.

Many people have worked together to produce this book, and it is their commitment, vision and wonderful teamwork that made it possible. I would like to thank everyone at New Society Publishers, including Judith and Chris Plant, Sue Custance, Ingrid Witvoet, EJ Hurst, Greg Green, Ginny Miller, and my editor Audrey Dorsch.

And finally, I would like to thank and cherish my wife and partner, Carolyn Herriot, who supports me in most of the things I get up to, and reminds me to give her a hand in the garden whenever I spend too long at the computer. If you love gardening and growing food, her best-selling book *A Year on the Garden Path: A 52-Week Organic Gardening Guide* is a total treat — and I'm totally unbiased, of course. See www.earthfuture.com/gardenpath/book.htm, or order it through your local bookshop.

And now - on with the Challenge!

— Guy Dauncey, September 2009

Visit TheClimateChallenge.ca for regular updates, free curriculum and classroom materials, and much more.

Dedication

This book is dedicated to everyone
who is working to build a climate-friendly world.

PART 1

Introduction: The Challenge

The Challenge

I say the debate is over. We know
the science, we see the threat, the
time for action is now.

— Arnold Schwarzenegger

As a civilization, we look back on our past with pride and occasional shame. We know we are the most modern, technologically advanced civilization there has ever been on Earth. We can walk on the Moon, replace our ailing hearts and lungs, and phone each other across the continents. We are accustomed to being proud of our successes, while acknowledging our intermittent stupidity.

The origin of our species, if we allow our minds to travel back down the great chain of being, lies 4 billion years in the past. The fact that you live today is testament that all of your ancestors, without exception, lived long enough to pass their accumulated biological wisdom on to you.

We date the origin of the human species to about five million years ago. We started the adventure of modern science, which has taught us so much about nature, 500 years ago — just one ten-thousandth of that time.

We live inside a bubble of time, so it is difficult to ponder the existence of humans 500 years in the future, let alone a million years, but I invite you to ask your friends what they think will be the condition of humanity in 500 years, and then in a million years. I wager that most will respond with pessimism, suggesting that we will be extinct if not by the former, then certainly by the latter date. Yet a million years is only a tiny 0.028% additional fragment of the time since life began.

What has this to do with global warming? It has to do with our state of mind. I argue throughout this book that the single most important factor that will determine whether or not we navigate the rapids of global warming successfully will be whether or not we view what is happening as an inevitable disaster that is some kind of retribution for human greed and ecological ignorance, or an exciting invitation to embark on a new adventure into a climate-friendly, ecologically harmonious world.

Our future is all in our minds. If we allow negativity and pessimism to prevail, then all will be lost. But if we take hold of our optimism, remembering the incredible things that humans have achieved and the amazing promise of what lies ahead, we will have what it takes to succeed.

There is no mistaking the urgency of the matter. We know that carbon dioxide and the other greenhouse gases trap heat, and we know that during the past million years, Earth's atmospheric CO_2 has never risen above 300 parts per million. Yet today, as a result of our burning fossil fuels and destroying the tropical rainforests, by 2009 the concentration had reached 389 parts per million, and it is rising steadily by more than 2 ppm a year. The world's climate scientists are warning us with increasingly desperate and urgent voices that we may be losing our ability to prevent Earth's temperature from rising by 2°C, and that if we pass 2°C we may be unable to prevent further increases of 3°, 4°, 5° or 6°C, placing the entire existence of life on Earth in jeopardy.

We are most certainly in a pickle, for which we need a most urgent response. We have been in a pickle before, however, and as this book will show, we do not lack for solutions. If we put our minds to it, there is no reason to believe we cannot succeed.

Interestingly, the world's oil and gas supplies are also about to start running out, so even if global warming did not exist, as many skeptics would like to believe, we would still have to achieve an organized transition into a world that can flourish without fossil fuels.

Why should we think that we cannot achieve such a transition? Only a hundred years ago, most of us got around by riding around on horses or walking. Look at what we have achieved. As a species, we may be stubborn, stupid and proud, but we are also intelligent, creative and courageous, and we love a challenge. We climb mountains. We cycle across continents to raise millions for causes that strike a chord in our hearts. Blind people sky-dive. Quadriplegic people go sailing.

Global warming presents us with an enormous challenge. If we fail, we condemn future generations to millennia of grief and destruction. They will curse our names, knowing that even when we knew what would result, we chose to continue to indulge our love of fossil fuels rather than stop and change direction. If we succeed, however, a whole new era will begin.

Will the adventurous side of our nature pick up the challenge and steer us into an ecologically sustainable future? Or will the lazy side of our nature win out, making ever more pathetic excuses as the sea level rises and Earth's species become extinct?

Only Earth's future historians will know. Our task is to take up the challenge before it is too late.

I do believe that if we fail to act in time, it will be the single biggest regret any of us has at the end of our lives.
— Joe Romm, author of *Hell and High Water* and the Climate Progress Blog

Wartime poster of Geraldine Doyle, a riveter in the 1940s.

A Gift from the Past

Our children and grandchildren are going to be mad at us for burning all this oil. It took the Earth 500 million years to create the stuff we're burning in 200 years. Renewable energy sources are where we need to be headed.[2]

— Jack Edwards,
Professor of Geology, University of Colorado

The whole universe is made of energy. This has been true ever since the beginning, from the moment God became pregnant. We think we're smart, but when it comes to understanding how all this energy works, we are just beginners.

Long before humans were a gleam in the eye of a dinosaur, the Sun — that giant thermonuclear reactor in the sky — shone down on the Earth, sending us the warmth that life needed to evolve and flourish.

During the period we call carboniferous, 360 to 286 million years ago, giant trees, ferns and mosses used photosynthesis to capture the Sun's energy, creating carbohydrates. When they died and fell into swamps, their carbon was locked away. Over millions of years it turned into coal. In the oceans, ancient plants, plankton and bacteria became oil and gas.

Millions of years passed. The dinosaurs vanished, the mammals evolved and humans emerged. About 600,000 years ago, we discovered how to make fire. Later, we used fire to make copper, bronze, iron and steel. Using metal instead of stone, we were able to cut down the trees that much faster.

By the Middle Ages, parts of Europe were running short of wood. Houses, ships, watermills, windmills and many bridges were all made from wood. The glass industry used wood in its furnaces; the iron industry used charcoal. By 1300, France was almost completely deforested, with forests covering only 13 million of its 54 million hectares. A single iron furnace using charcoal could level a forest for a radius of a kilometer in just 40 days.[1]

In the 13th century, people started using the strange black stuff that they found in the ground called coal: the second great energy revolution had begun. In 1257, Queen Eleanor of England was driven from her castle in Nottingham by foul fumes from the sea coal being burned in the city below. It was coal that powered the Industrial Revolution. There is enough coal for another 200 years, but its impact as a cause of global warming is so great that we must stop using it entirely unless we can capture and safely sequestrate its CO_2, which may or may not be possible. (See p. 60.)

By the mid 1800s, there was renewed hunger for energy. We slaughtered the bowhead whales for their oil, and then in 1858 we struck the first oil in Oil Springs, Ontario, followed in 1859 by the Drake Well in Pennsylvania. During the 1800s, we also discovered natural gas.

Our discovery of fossil fuels has been a journey of wonder and innocence. When we struck

JAN BREUGEL, 1610

Blast Furnace in the Woods.

The waterwheel at the Grist Mill in Keremeos, BC, built in 1877.

The Charcoal Burner's Hut.

the first oil, most people believed the world was created the way the Bible said it was — Darwin's mind-shaking book *On The Origin of the Species* was only published in 1859. We hadn't a clue what fossil fuels were or what their released ancient carbon would do to the atmosphere; most people did not even know Earth had an atmosphere. We burned the fossil fuels in total ignorance of the impact they would have by trapping the Sun's heat in the atmosphere, melting the ice and disrupting the climatic patterns which have given us stability for millennia.

Even if by some strange means the climate crisis did not exist, or someone invented a technology that could miraculously suck the excess carbon out of the atmosphere and store it in massive blocks of calcium carbonate to serve as sea walls for threatened coastlines, there is scarcely a person on Earth who believes the fossil fuels will last forever. Among those who have taken the time to become informed, the debate is whether the oil supply will peak in 2010 or 2020. (See page 36.)

So what happens next? This is the crucial question, and in the darkness of unknowing, our minds go one of two ways. They either go toward hopelessness and despair, convinced that without oil Earth's civilizations will quickly collapse; or they go to a place of wonder and excitement, knowing that we are about to embark on a huge civilizational leap into the Solar Age, replacing fossil fuels with clean, renewable energy, building a future our children and grandchildren will delight to live in.

This choice, this simple inner choice, is the crux on which our future depends.

Negativity lulls creativity. If you allow your mind to switch to the negative, you will activate thoughts of survival and protecting your family during the coming collapse. If you switch to the positive, you will activate creativity, excitement and determination. Across the planet, our inner choice becomes a self-fulfilling prophecy. If we believe we are fated to failure, it is this that will happen. If we believe that we are about to embark on great things, however, and that fossil fuels are a gift from the past that we must now lay down in order to build an amazing future, this will happen instead.

Our Story [1]

Before, if we screwed up, we could move on. But now we don't have an exit option. We don't have another planet.

— Tim Flannery,
author of *The Weathermakers*

13.7 billion years ago: God becomes pregnant. The Universe is born.

5 billion years ago: The Sun is created from a nebula cloud of dust and gas.

4.6 billion years ago: The Sun's solar system forms; the Earth is one of its planets.

4.4–4.1 billion years ago: The Earth forms a thin rocky crust. Volcanoes expel hot magma to the surface and steam condenses into the miracle of rain. The weather cycle begins. Torrential rains fall until rivers flow into great seas.

4 billion years ago: The rich chemical brew produces bacteria, the first living cells.

3.8 billion years ago: Cells invent photosynthesis. Blue-green algae learn to store the Sun's energy chemically, taking hydrogen from the sea and carbon from the atmosphere and releasing oxygen to the air.

2.1 billion years ago: Oxygen-loving cells emerge; oxygen levels rise to near present-day levels.

NASA, HUBBLESITE.ORG

Spiral Galaxy NGC 1309, 100 million light years from Earth, home to millions of stars similar to the Sun.

900 million years ago: Some organisms begin living in colonies, communicating with chemical messages.

600 million years ago: Light-sensitive eyespots evolve into eyesight. The first animals (invertebrates) evolve in the oceans.

500 million years ago: By now, the ozone layer has formed, making it safe for animals to leave the water. Worms, mollusks and crustaceans venture out, along with algae, fungi and insects.

395 million years ago: The first amphibians leave the water.

370–300 million years ago: Carboniferous Period. Great swamps, forests of ferns and early conifers start to store carbon — the origin of coal. Reptiles appear and evolve rapidly.

300–250 million years ago: Permian Period. Glaciations and marine extinctions: reptiles spread; amphibians decline. The world consists of one large continent called Pangaea. The temperature is 8°C warmer than today.

250–200 million years ago: Triassic Period. First dinosaurs. Primitive mammals appear.

200–150 million years ago: Jurassic Period. Zenith of dinosaurs. Flying reptiles, birds and small mammals appear; the first flowering plants appear; continents drift apart.

150–70 million years ago: Cretaceous Period. Extinction of dinosaurs likely caused by intense atmospheric disturbance when a large asteroid hits Earth near the Yucatan peninsula. Flowering plants, marsupials and insectivorous mammals become abundant. The age of mammals begins.

65 million years ago: Something (methane hydrate releases? volcanoes?) causes atmospheric

- Cosmic Walk: rainforestinfo.org.au/deep-eco/cosmic.htm
- *Greenhouse: The 200-Year Story of Global Warming* by Gale E. Christianson, Douglas & McIntyre, 1999
- Greenhouse Timeline: kccesl.tripod.com/greenhousetimeline.html
- NASA's Earth Observatory: earthobservatory.nasa.gov
- North America's first oil well: wandel.ca/oil
- Vital Climate Graphics: grida.no/publications/vg/climate

CO_2 to leap to 3,000–3,500 ppm, 8–9 times the current level. Large-scale burial of surplus carbon in swamps.

23–7 million years ago: Miocene Epoch. Whales, apes, grazing mammals. Spread of grasslands as forests contract.

7 million years ago: Pliocene Epoch begins. Large carnivores, earliest hominids (manlike primates) appear. Temperature is 2.5°C warmer than today.

4 million years ago: Hominids leave the forest and start to spread around the world.

1.6 million years ago: Pleistocene Epoch of ice ages begins. Temperatures 2° to 4°C colder than today.

800,000 years ago: Humans start using fire as a source of energy.

395,000 years ago: Ice age begins.

335,000 years ago: Ice age ends. Brief warm period.

310,000 years ago: Ice age begins.

240,000 years ago: Ice age ends. Brief warm period.

230,000 years ago: Ice age begins.

135,000 years ago: Ice age ends. Eemian interglacial warm period begins.[2]

125,000 years ago: Half of Greenland's ice cap melts, raising sea levels by 4–5 meters.

122,000 years ago: Sudden cooling event in North Atlantic, switching off ocean current, maybe leading to end of warm period.

100,000 years ago: Modern humans leave Africa, develop language, religion and art.

11,500 years ago: Ice age ends. Holocene interglacial warm period begins.

10,000 years ago: Humans develop agriculture and begin to shape our environment.[3]

3,000 years ago: We develop writing, cities and classical religions.

1440: We learn how to print and share knowledge more widely.

1550: We begin serious exploration of the material world (the Scientific Revolution).

1750: We begin burning fossil fuels (the Industrial Revolution).

1824: Jean Fourier suggests that CO_2 emissions from burning fossil fuels might accumulate in Earth's atmosphere, enhancing the planet's greenhouse effect.

1858: First oil well is drilled, in Petrolia, Ontario.

1859: Darwin publishes his book *On the Origin of Species*.

1860: Étienne Lenoir develops the first internal combustion engine in Belgium.

1896: Svante Arrhenius confirms connection between CO_2 and Earth's temperature.

1955: Charles Keeling confirms that CO_2 levels in the atmosphere are rising.

1966: We see Earth from space for the very first time.

1992: Rio Earth Summit. We agree to protect our planet's environment.

1997: Kyoto Treaty —Developed nations agree to reduce their greenhouse gas emissions.

1990-2007: Our greenhouse gas emissions grow by 38%.

2009: Copenhagen Treaty — We make another commitment to reduce our greenhouse gas emissions.

Earth's Miraculous Atmosphere

The atmosphere almost looks like an eggshell on an egg, it's so thin.
— Eileen Collins, Commander of space shuttle *Discovery*

The Sun pours its energy into our solar system, which is by nature rather cold. If you stood on Mars, which has a very thin atmosphere and an average temperature of –63°C (–81°F), it would be bye-bye life, hello deep-frozen astronaut. If you stood on Venus, where the CO_2-rich atmosphere traps so much heat that the average temperature is over 400ºC (752°F), it would be bye-bye life, hello astronaut fricassée.

The difference is caused by Earth's amazing atmosphere, which traps the Sun's heat through the "greenhouse effect." Thanks to our atmosphere, that tiniest skim of greenhouse gases that covers the land and oceans, Earth has enjoyed a stable average temperature of 13.8°C (57°F) for the past 10,000 years, which is just right for farming and growing forests. In an atmosphere that is 77% nitrogen, 21% oxygen and 0.9% argon, it is the miniscule presence of 1% water vapor and 0.3% trace gases (CO_2, methane, and nitrous oxide) that makes the difference. If Earth's atmosphere had no greenhouse effect, the average temperature would be –18°C (0°F).

When the Sun's heat reaches Earth, 30% is reflected back into space, and 70% is trapped by the greenhouse gases. Earth's plants, soils, algae and ocean phytoplankton use photosynthesis to convert the CO_2 into carbon, forming nature's carbon sink.

During the past 650,000 years there have been seven occasions when the CO_2 in Earth's atmosphere fell below 200 ppm, causing the temperature to fall by 8–10°C, plunging Earth into an ice age. Throughout those years, the CO_2 never rose above 300 ppm. By 2010, however, it will have reached 390 ppm, the highest in 20 million years.[1]

Nature's First Carbon Sink: The Oceans

Approximately half of nature's photosynthesis is carried out in the oceans, where every year, on average, phytoplankton absorb 92.4 billion tonnes of carbon and release 90 billion tonnes, storing 2.4 billion tonnes as dissolved organic carbon in the deep ocean.[2] Over the millennia, the oceans have locked away 40,000 billion tonnes of carbon, where it will hopefully remain. A quarter is stored as frozen methane hydrates, which have the troubling potential to thaw as the ocean warms.

The Southern Ocean around Antarctica is the single largest carbon sink, capturing about 15% of it, but there is disturbing evidence that since 1981 it has been absorbing 5% to 30% less CO_2, an outcome that researchers were not expecting until 2050. The reason is increased winds over the past 50 years. The winds normally carry atmospheric CO_2 into the ocean, where it blends with the ocean's existing stock of CO_2, but when the winds increase they bring more carbon to the surface, making it harder for the ocean to absorb the added carbon from fossil fuels and burning rainforests. The winds have increased partly because ozone depletion above the Southern Ocean has created large temperature changes, generating wind, and partly because the uneven nature of global warming has increased temperatures in the north, driving stronger wind activity in the south.[3] Oh, what a complex web we weave.

Nature's Second Sink: The Soil

A quarter of nature's photosynthesis is carried out by soil, a mass of minerals, moisture, bacteria and micro-organisms that absorb carbon from plants

NASA, APOLLO 16, 1972

and trees as they die. Every year the world's soils absorb 50 billion tonnes of carbon from dying vegetation and release 50 billion tonnes through decomposition. Forest destruction and modern farming are weakening the soil, however, losing 1.5 billion tonnes of carbon each year to the atmosphere. In western Canada, when an old-growth Douglas fir forest is clearcut, it can take 150 years for the forest and soil to recover the carbon, which is why it is so important to change our methods of forestry.[4] Over the millennia, the world's soils have accumulated 1,500 billion tonnes of carbon, of which 500–800 billion tonnes are locked up in peat lands, including 500 billion tonnes in the Arctic tundra. Rising temperatures are now unlocking this as the snow cover disappears.

Earth's atmosphere is so thin that from space it seems almost non-existent.

- The Carbon Cycle: grida.no/publications/vg/climate/page/3066.aspx
- The Carbon Cycle (1992–1997): cdiac.esd.ornl.gov/pns/graphics/globcarb.gif
- Carbon in Live Vegetation: cdiac.esd.ornl.gov/ftp/ndp017/table.html
- CO_2 Sinks and Sources: tinyurl.com/38ye9n
- Greenhouse Effect: grida.no/publications/vg/climate/page/3058.aspx

Nature's Third Sink: The Forests and Vegetation

Forests and vegetation store some 550 billion tonnes of carbon, 40% in the tropical forests. Every year, forests release 50 billion tonnes of carbon to the soil and 50 billion tonnes to the atmosphere through respiration, but they absorb 101.5 billion tonnes from the atmosphere, reducing its load by 1.5 billion tonnes. Yes, trees matter. Scientists fear that the tropical forests will soon cease storing their CO_2, however, and start releasing it.[5]

In the balanced natural carbon cycle, half of the atmosphere's CO_2 is exchanged with the soil,

plants and forests and half with the oceans, resulting in no net increase. In the human-disturbed carbon cycle, a quarter is going into the soil and vegetation, a quarter is going into the oceans and half is accumulating in the atmosphere, where it traps the Sun's heat. This is the crux of the crisis we face today.

The Greenhouse Gases

We have a mighty task before us.
The Earth needs our assistance.
— Sir Laurens van der Post,
author, farmer, soldier, philosopher.

The impact of each gas is measured by its radiative forcing — the extent to which it alters the balance of incoming and outgoing energy in the atmosphere. The Sun's natural radiation is about 200 watts per square meter (200 W/m²). The human addition of greenhouse gases and black carbon has increased this by 1.8% (3.6 W/m²).

Each gas has been assigned a global warming potential (GWP), which is the standard used to compare the radiative forcing of each gas, using $CO_2 = 1$ as the baseline. When the world's climate scientists developed GWP to compare different gases and put them in a common basket, they chose to measure their impact over 100 years because this seemed like a reasonable approach. Now that the time frame for an urgent turnaround is understood to be within 10 years, the choice of 100 years may be an error of tremendous

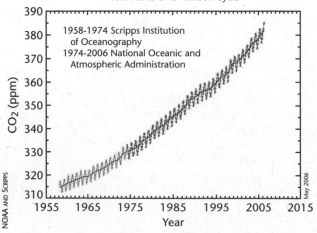

Mauna Loa Monthly Mean Carbon Dioxide
NOAA ESRL GMD Carbon Cycle

1958-1974 Scripps Institution
of Oceanography
1974-2006 National Oceanic and
Atmospheric Administration

NOAA AND SCRIPPS

May 2006

proportions, causing a massive underestimation of the role of methane and HFC-134a and an overestimation of the role of the PFCs and SF6.

Carbon Dioxide. Carbon dioxide is released whenever we burn fossil fuels, whenever oil and gas companies flare their waste gas, by the production of cement, when forests are burned or clearcut and when farmers manage their land unsustainably. CO_2's life in the atmosphere varies from 30 to 3,000 years. Thus much of the CO_2 that was released in the early Industrial Age is still heating the atmosphere, and the CO_2 we release today will be heating it for a long time into the future. About a quarter of the fossil-fuel CO_2 emissions will stay in the air "forever," i.e., more than 500 years.[1] To convert carbon into CO_2, we multiply by 3.667 to include the addition of oxygen. Most global data is expressed as carbon, while most national and local data is expressed as carbon dioxide — just to keep us on our toes. CO_2 is responsible for 44% of the warming.

Methane. Methane (CH_4) is considered on the following page.

Nitrous Oxide. Nitrous oxide (N_2O) has a life in the atmosphere of around 100 years, and traps heat 298 times more effectively than CO_2. It is produced by the use of nitrogen-based chemical fertilizers, transportation, nitric acid production, poor manure management and from some other sources. It is responsible for 4% of the warming.

The F Gases. The chlorofluorocarbons (CFCs) that were used as coolants in fridges and air-conditioning, causing the hole in the ozone layer, trap heat up to 10,000 times more effectively than CO_2. They are being phased out under the

- CO_2 at Mauna Loa, Hawaii:
  esrl.noaa.gov/gmd/ccgg/trends
- Current greenhouse gas concentrations:
  cdiac.esd.ornl.gov/pns/current_ghg.html
- Emission Database for Global Atmospheric Research:
  mnp.nl/edgar
- Global Carbon Project: globalcarbonproject.org
- Global Dimming: pbs.org/wgbh/nova/sun
- Greenhouse gases in 35 developed nations:
  maps.grida.no/go/collection/cop-7-emission-graphics
- IPCC 4th Assessment Report Science:
  ipcc.ch/ipccreports/assessments-reports.htm
- James Hansen: columbia.edu/~jeh1
- Radiative Forcing: tinyurl.com/3cf7sp

Montreal Protocol on Ozone Depletion, but their replacements, the hydrofluorocarbons (HFCs), trap up to 14,800 times more heat, depending on the chemical; so as we solve one problem, we are creating another. This is particularly true with HFC-134a, a fluorinated chemical being used as a replacement for air-conditioning, which is so potent that it may be the equivalent to 17% of all CO_2 emissions. (See Solution #97 for details.) Another powerful greenhouse gas, sulfuryl fluoride (SO_2F_2), is being used as a fumigant in greenhouses in place of methyl bromide, which destroys the ozone layer.

Other Industrial Chemicals. Perfluorocarbons (PFCs) are used in the manufacture of aluminum and semi-conductors; sulfur hexafluoride (SF_6) is produced in magnesium production, as a dielectric in electrical transmission and distribution systems, and in the semi-conductor industry. The GWP error that affects methane also affects the PFCs, which have a very long atmospheric life of 50,000 years. When their GWP is squeezed into 100 years, it makes them seem 7,390 times more powerful than CO_2, exaggerating their impact 500-fold. The same applies to SF_6, exaggerating its impact 320-fold. When a company uses PFC or SF_6 reductions to meet its carbon reduction goal, therefore, the apparent result is extremely misleading.

Tropospheric Ozone. When fossil fuels burn, their pollution creates ozone, which acts as a greenhouse gas in the lower atmosphere. If it was in the upper atmosphere 11 to 50 kilometers (seven to 30 miles) up, it would be a good thing — but it's not. Ozone is also caused by the breakdown of methane and is responsible for 9% of the warming.

Aerosols. Aerosols, also known as dust particles or particulates, are a "negative" greenhouse gas. They result from fossil fuel burning, forest and biomass burning, volcanoes and industrial pollution, and with the exception of black carbon, they reflect the Sun's heat back into space, protecting us from the full impact of global warming. It is for this reason that they have been blamed for "global dimming." There is a high level of scientific uncertainty around most aerosol gases.

Black Carbon. Black carbon is created when biomass and fossil fuels burn. This is a greenhouse particle that should definitely be included in the new Post-Kyoto treaty, as it is responsible for 21% of the warming. Its blackness attracts and holds heat in the atmosphere, and it causes further problems when it falls on ice, because it reduces ice's natural reflectiveness, or albedo.[2] (See p. 14.)

Methane — The forgotten Gas

If we control methane ... we are likely to mitigate global warming more than one would have thought, so that's a very positive outcome.

*— Drew Shindell,
NASA Goddard Institute*

Methane (CH_4) is the source of many jokes about cows and beans, but underlying this is a very serious problem. Methane is the second most impactful greenhouse gas after CO_2, and it has been trapping heat in Earth's atmosphere for 3.46 billion years.[1]

Over the past 650,000 years, methane's presence in the atmosphere has varied from 400 parts per billion (ppb) during the ice ages to 770 ppb in the warm interglacial periods. Based on ice-core drilling in the Antarctic, we know that before the industrial age, it was 715 parts per billion.[2]

Since then, it has increased by 248% to 1775 ppb (in 2005), a far greater proportional rise than CO_2. In 2001, the Intergovernmental Panel on Climate Change (IPCC) increased methane's Global Warming Potential (GWP) from 21 to 23, and in 2007 they increased it again to 25. Its presence in the atmosphere is far less than CO_2, but molecule for molecule, over 100 years it traps 25 times more heat than CO_2.[3]

This is a moot point, however, because we need to achieve a turnaround in our global emissions by 2016, not in 100 years, and methane has

The northern edge of a giant iceberg in the Ross Sea, Antarctica, a fragment of a much larger iceberg that broke away from the Ross Ice Shelf in March 2000. If it melts, the West Antarctic Ice Sheet will raise global sea levels by 5 metres.

JOSH LANDIS, COURTESY NATIONAL SCIENCE FOUNDATION

Human Activity Causes	%	Solutions	
Coal mining and oil wells		Methane capture	#60
Natural gas leaks and venting	26%	Methane capture	#59
Cows, sheep, goats and pigs	35%	Changed diets	#43
Animal manure and slurry	7%	Methane capture	#44
Landfills and sewage treatment	12%	Methane capture	#28
Rice paddies	8%	Changed methods	#72
Some hydroelectric dams	n/a	Prior analysis	#72
Biomass burning	12%	Reduce forest fires	#72
All of the above	100%	Government policy	#72

a very *short* average atmospheric life of only 8.4 years before it breaks down into hydrogen and oxygen. Over 20 years its GWP is 72, and over 8.4 years may be as high as 125, though the IPCC does not list GWPs for less than 20 years. This means that over its actual life span, measures to reduce methane have five times more impact than people think. If GWPs were compared over 20 years, not 100 years, methane would jump to the top of the priority list, along with the reduction of HFC-134a. (See #97.)

After increasing rapidly, methane seemed to stabilize after 2000, leading some scientists to think that organized reductions were having an effect. The increase restarted in 2006, however, and the pause is now thought to have been caused by the drying up of natural wetlands as a result of climate change, which hid the continuing rise from human causes, especially in China.[4]

There is also evidence that methane's impact on global warming may be double what it is thought to be, becoming a full third of the cause, because of the amplified effect that methane has when it mixes with tropospheric ozone caused by local air pollution.[5] As well as reducing methane's heat-trapping impact, a 20% reduction in methane emissions would also prevent 370,000 premature deaths between 2010 and 2030 by reducing air pollution.[6]

Methane's Sources

So where does the methane come from — and how can we reduce it? The dramatic increase in emissions that has been observed over the past 200 years comes from a variety of human activities.

It escapes whenever coal, oil and gas are extracted, and it is produced whenever carbon breaks down in the absence of oxygen — in the stomachs of cows, sheep, goats and pigs, in liquid slurry ponds of animal manure, in rice paddies and in landfills and sewage treatment plants. It is also produced by some hydroelectric reservoirs, where flooded soils and plants break down without oxygen. Methane is also produced by natural wetlands and when biomass burns.

Ancient methane is stored in enormous quantities in the Arctic permafrost, where it has the potential to create a rapid acceleration in global warming as the permafrost melts, which is already happening in western Siberia. It is also stored as frozen methane hydrates under shallow ocean floors, from which its potential release is one of the most catastrophic outcomes that has been foreseen in some future scenarios. A similar release is a leading suspect for the devastating Permian mass extinction, 250 million years ago.[7]

Carbon dioxide has the limelight on the global warming stage, but methane emissions *must* also be addressed, because their reduction has great — and urgently needed — potential to slow global warming in the immediate future.

Black Carbon

Control of black carbon, particularly from fossil-fuel sources, is very likely to be the fastest method of slowing global warming.

— Dr. Mark Jacobson, Stanford University

Black carbon is the surprise gatecrasher. It comes late to the "Let's Trash the Atmosphere" party and then makes more noise than anyone else, except CO_2 — which is why it merits such prominent attention.

Black carbon is soot, produced by incomplete combustion in dirty diesel engines, open wood fires, cooking stoves and burning forests and savannas. It is not a greenhouse "gas," so it was not included under the Kyoto Protocol, although it urgently needs to be. It causes global warming by absorbing solar radiation and releasing it back into the atmosphere, and by falling on snow and ice, reducing their ability to reflect the sun's heat back into space (known as "albedo").

Estimates for black carbon's radiative forcing range from +0.5 to +0.9 watts per square meter.[1] The science seems to be pointing to the upper end of the range; if it is +0.8, it is causing 21% of global warming.[2] Because of its impact on snow and ice, it may be causing a quarter of the observed warming in the Arctic and doing as much harm as CO_2 in the Himalayan Plateau,

where by hastening the melting of the glaciers it is threatening the livelihoods of a billion people. (See p. 18.)

The good news is that it remains in the atmosphere for a only few days or weeks, so when it comes to producing an urgent response that could delay the tipping points, black carbon is a prime candidate, along with methane and HFC-134a.

Black carbon comes from these sources:[3]

- Open forest and savanna fires (42%)
- Traditional biofuel cooking stoves, mostly in India and south Asia (18%)
- Diesel engines used for transportation (14%)
- Diesel engines used for industrial use (10%)
- Industrial fossil-fuelled processes, usually from small, inefficient boilers (10%)
- Small, inefficient coal-burning power plants and residential coal fires (6%)

The reason why it has received so little attention until now is that it is a "particulate," one of many dirty air pollutants, most of which act as a shield against global warming. Their polluting haze masks the sun and lessens the warming, which is why it has been called "global dimming." This is a problem, for if we clean up the way we burn fossil fuels without reducing the fossil fuels themselves, we will reduce their shielding, and make the warming warmer.

Black carbon *increases* the warming, however. If we can reduce it, we will reduce its contribution to global warming and neutralize the elimination of the other air pollutants that are shielding us from global warming.

- Black Carbon Briefing Paper: igsd.org/blackcarbon
- HEDON Household Energy Network: hedon.info
- Project Surya: ramanathan.ucsd.edu/ProjectSurya.html
- *Reducing Black Carbon,* by Dennis Clare: worldwatch.org/node/5983
- Wood Stoves: ashdenawards.org/wood-stoves

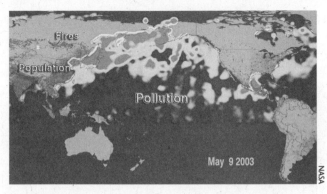

The large shaded areas show heavy aerosol and black carbon concentrations from vehicles, factories and low-tech polluters like coal-burning stoves and forest fires. The tiny dots show intense forest fires.

Fifty years ago, most black carbon came from Europe and North America, part of the industrial pollution that caused tens of thousands of deaths. Today, because the developed world has installed cleaner technologies, most comes from the developing world. The US produces 21% of the world's CO_2, but only 6% of the world's black carbon.

Black carbon from burning forests and savannas

We need to stop the forest burning for a host of reasons, not just reducing black carbon. (See #95.) If the practice of burning savannas and grasslands could change from "slash and burn" to "slash and char," charring the biomass instead of burning it and burying the resulting charcoal/black carbon in the soil, this would create the highly fertile biochar, or *terra preta*. (See #41.) This would require a large initiative by the UN Environment Program and other UN agencies, and it would be encouraged if carbon sequestration in grasslands and farmlands was included in carbon trading, giving farmers around the world a financial incentive to adopt new ways.

Black carbon from traditional biofuel cooking stoves

There are as many as 400 million traditional cooking stoves in the developing world, fueled by firewood or crop residues. Project Surya, founded in rural India by the Scripps Institute atmospheric scientist Professor V. Ramanathan, is developing solar, biogas and renewable alternatives to traditional firewood, cow dung and coal cookers, using

sensors to measure black-carbon reductions. A huge global initiative is needed to provide safe, renewable cookers to three billion rural dwellers around the world, funded by inclusion in global carbon finance mechanisms. If some also produce bury-able biochar, this would be doubly beneficial.

Black carbon from dirty diesel

Highly effective filters already exist to eliminate most diesel pollution, but they require the use of ultra-low sulfur diesel, which needs to become much more widely available in the years before diesel vehicles are phased out entirely. Partial filters can eliminate 40–70% of the pollutants using traditional diesel, so a global initiative is also needed to retrofit many million dirty vehicles around the planet. In the developed world, clean diesel regulations need to be made universal.

Black carbon from dirty coal

China is closing down many of its smaller, dirty, coal-fired power plants. Actions like this need to be supported all over the world, with support from the World Bank and other global agencies.

Greenhouse Gases Chart

A day will come when our children and grandchildren will look back and they'll ask one of two questions. Either they will ask: "What in God's name were they doing?" Or they may look back and say: "How did they find the uncommon moral courage to rise above politics and redeem the promise of American democracy?"

— Al Gore

n/a = not applicable n/k = not known	Pre-Industrial Concentration (1860)	Concentration in 2005[1]	Average Atmospheric Lifetime[2]	Growth Rate per Year
Water vapor	Variable 1–3%	Variable 1–3%	Few days	0.2%[5]
Carbon dioxide CO_2	280 ppm	385 ppm (2008) Rising by 2 ppm per year	Very approx: 50% 30 years 30% 300 yrs, 20% 3,000 yrs	3%
Black carbon	0	n/a	5.5 days	n/k
Methane CH_4[12]	715 ppb	1774 ppb	8.4 years	0.6%
Tropospheric ozone O_3	25 ppb	34 ppb	Hours/days	1%
CFC-12	0	542 ppt	100 years	Decline
HCFC-22	0	174 ppt	12 years	Decline
HFC-23	0	14 ppt[15] (+0.55ppt pa)	270 years	5.1%
PFC-14 — CF_4	0	79 ppt	50,000 years	10%
Sulfur hexafluoride SF_6	0	5.22 ppt	3,200 years	10%
Fluorinated ethers (HFE-125)	0	0.16 ppt	136 years	n/k
Nitrous oxide N_2O	270 ppb	319 ppb	114 years	0.26%
TOTAL				
Aerosols	>0	Variable	Hours/days	n/k[17]
Surface albedo changes	n/a	n/a	n/a	n/a
Solar irradiance	n/a	n/a	10 to100 yrs	n/a

Anthropogenic Sources (from Human Activity)	Global Warming Potential (GWP) over 100 Years[3]	GWP over 20 Years	Radiative Forcing Watts per sq. meter[4]	Share of Cause of Climate Change based on radiative forcing
All of the below[6]	n/a[7]		n/a	n/a
Gas (16%), Oil (33%), Coal (32%) Cement (7.4%)[8], Deforestation (16%)[9]	1	1	+1.66	44%
Fossil fuels (40%) Open biomass burning (42%) Residential traditional biofuel burning (18%)	1,650[10]	4470	+0.80[11]	21%
Fossil fuel extraction (26%) Livestock digestion (35%), Rice paddies (8%) Landfills (12%), Animal manure/slurry (7%) Biomass burning (12%)	25	72	+0.48	13%
Fossil fuel use (50%) Forests/biomass burning (25%), Methane (25%)[13]	n/a	n/a	+0.35	9%
Liquid coolants, foams	10,900	11,000		
Liquid coolants	1,810	5,160		
Liquid coolant CFC & HCFC substitutes	14,800	12,000		
Aluminum manufacture (59%) Solvents and other (26%), Plasma etching (15%)	7,390	5,210	+0.34[14]	9%
Magnesium production Dielectric fluid	22,800	16,300		
Manufacturing of fluoro-chemicals[16]	14,900	13,800		
Nitrogen fertilizers & manures (70%) Transportation (14%), Industrial processes (7%)	298	289	+0.16	4%
TOTAL			**+3.79**	**100%**
Fossil fuels, Biomass fires, Volcanoes	n/a		−0.5[18]	
Deforestation (-0.2), Black carbon on snow (+0.1)	n/a		−0.1[19]	
n/a	n/a		+0.12	

The First and Second Alarm Bells

This is genuinely terrifying. It is a death sentence for many millions of people.

— Andrew Pendleton, Christian Aid, speaking of coming droughts.

When the first climate change warnings came out in the 1980s and 1990s, it was "glaciers will melt," "diseases will spread" and "spring will come earlier." It was bad, but not bad enough to make people pay attention. The threatened sea level rise was "only" 3.5 to 35 inches, and if there were a few more storms and droughts — wasn't that something we were used to? It was easier to change the TV channel.

Then came hurricane Katrina, the flooding of New Orleans, and the destruction of many Gulf Coast communities. Maybe there *was* something wrong with the weather, and we should start worrying about what it would mean. And yet for most people, climate change was still a distant concern that would affect people in 100 years, not now.

The awareness of just how much trouble we're in sinks in only when we stop and pay attention. There is a pivotal moment in Al Gore's movie *An Inconvenient Truth* when he has to use a mechanical lifting device to show how high the CO_2 emissions have risen. That is when people start having sleepless nights.

- Climate Hot Map: climatehotmap.org
- Operational Significant Event Imagery: osei.noaa.gov
- *The Last Generation: How Nature Will Take Her Revenge for Climate Change*, by Fred Pearce, Key Porter, 2007
- US Climate Change Assessment: usgcrp.gov/usgcrp/nacc

There are five particular alarm bells that should make us sit bolt upright and start thinking urgently about what we can do.

The First Alarm Bell: The Century of Drought

The first alarm bell is ringing for the fate of lands fed by glaciers, where hundreds of millions of people live. The Tibetan plateau in the northern Himalayas is the third largest ice field in the world, after Greenland and Antarctica. The region's 46,298 glaciers, covering almost 60,000 square miles, feed seven of Asia's greatest rivers: the Indus in Pakistan; the Ganges and Brahmaputra in India and Bangladesh; the Mekong in Laos, Thailand, Cambodia and Vietnam; the Thanlwin in Burma; and the Yangtze and Yellow in China. Together, these rivers support almost a billion people.

In 2006, the Chinese Academy of Sciences reported that the Tibetan glaciers were shrinking by 7% a year. Four hundred Chinese cities are already running short of water, and 100, including Beijing,[1] face critical shortages. If the warming continues, most of the glaciers will disappear by the end of the century, and except in the monsoon season, the rivers they feed will be reduced to a trickle. How will people survive when their rivers run dry? It is a dauntingly huge concern. The same thing is happening in Peru and Bolivia, where the residents of Lima and La Paz base their existence on water from ancient glaciers that are melting away. Californians face a similar concern, as 40% of the state's water comes from the Sierra Nevada snow pack.[2]

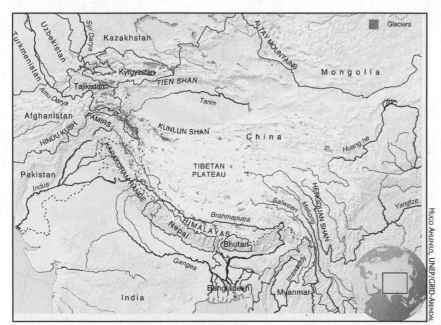

The rivers that drain from the glaciers of the Tibetan Plateau provide water for 40% of the world's population.

In 2006 a study by Britain's Hadley Centre for Climate Prediction showed that by 2100, rising temperatures will cause the area of the Earth's surface that suffers moderate drought to increase from 25% to 50%, and that the extreme drought area will increase from 3% to 30%, becoming essentially uninhabitable.[3] By 2050, 10% of the Earth's land area will suffer from extreme drought, including large areas of Africa, Brazil, China and Australia. Many crops also cease to grow when the temperature rises above 35°C (95°F).[4,5] In 2003, the Ukraine lost 75% of its wheat yield because of the summer heat wave that affected most of Europe.[6]

It is more than alarming. It disturbs many people down to the roots of their being. If we act rapidly, however, we can slow the warming and prevent some of the worst from happening.

The Second Alarm Bell: The Loss of the Amazon

The second alarm bell rings for the Amazon, "the largest living reservoir for carbon dioxide on the land surface of the Earth."[7] Its trees store some 70 billion tonnes of carbon, and its soils perhaps as much again. The Amazon is sustained by rainfall from the Atlantic, but the warming ocean is causing droughts, and the Amazon's trees can sustain only two years of drought before they fall over and die. With drought comes fire, which is already releasing 200 million tonnes of carbon a year. After a point, the forest cannot recover from the fires, and a process of rapid drying begins, which leads first to savannah, then desert. At the Hadley Centre, researchers predict that as temperatures rise, the Amazon will be subject to repeated droughts and fires, and that by 2100 it will be dead. "The region will be able to support only shrubs or grasses at most."[8] In dying, it will release its enormous store of carbon, increasing the expected rate of warming by 50%.

This is another reason that people are having sleepless nights. It is hard to sleep when the alarm bells are ringing so loudly.

The Third and fourth Alarm Bells

> How far can it go? The last time the world was three degrees [C] warmer than today — which is what we expect later this century — sea levels were 25 meters higher. So that is what we can look forward to if we don't act soon.
>
> — Dr. James Hansen, Director of the NASA Goddard Institute for Space Studies.

The Third Alarm Bell:
The Death of Biodiversity

On January 8, 2004, the *Guardian*, a leading British paper, splashed this headline across its front page in huge type: "An unnatural disaster. Global warming to kill off 1 million species. Scientists shocked by results of research. Third of life forms doomed by 2050."

The research was the result of two years of global collaboration led by a team of conservation biologists from the University of Leeds, who examined the impact of climate change in six biologically rich regions of the world, covering 20% of Earth's land surface. The flatter the land, the greater the difficulty species would have because they could not migrate upward to cooler habitats.

In the Brazilian savannah, 70 of 163 tree species would become extinct. In Europe, a quarter of the birds and 11–17% of plant species would disappear. In Mexico, a third of the species they examined would be in trouble.

Two years later, 19 of the world's leading biodiversity specialists warned that life on Earth was facing a catastrophic loss of species that only a global political initiative could maybe stem.[1] Climate change, without including other factors causing loss of habitat, would lead to the extinction of between 15% and 37% of all land-based animals and plants by the end of the century.

As mammals, we are wired to care about the suffering of a kitten or a wild bird. There is nothing in our make-up, however, that tells us how to respond to the loss of a million species. Polar bears. Mountain gorillas. African elephants. Green turtles. Frogs. Saiga antelopes. Sperm whales. One bird in eight. Thirteen of the world's flowering plants. A quarter of all mammals. Sockeye salmon. In Canada, 46% of the country's life-sustaining habitats will be destroyed by 2050 if CO_2 levels are not reduced.[2]

The whole Antarctic food chain, with its whales, seals and penguins, is based on the krill, which have declined by over 80%

Greenland is covered with enough ice to raise global sea levels by 6-7 metres.

EARTH SCIENCES AND IMAGE ANALYSIS LABORATORY, NASA JOHNSON SPACE CENTER

because of the warming water. The krill cannot exist without sea ice, where their young hide from predators while feeding on the algae that live under the ice.[3] Since 1974, the number of Adelie penguins in Antarctica has declined by 70%. Ten of the world's 17 penguin species are listed as endangered or threatened.

We know that gradual climate change likely caused mass extinctions in the past.[4] This time, however, it is not volcanoes pumping out CO_2 that are causing the extinctions. It's us.

The Fourth Alarm Bell: Sea-Level Rise

As humans, we have often settled at the edge of the sea where sailing boats could come and go, carrying trade goods from distant lands. London, Amsterdam, Rome, Venice, Cairo, New York, New Orleans, San Francisco, Shanghai, Bombay, Calcutta — they are all close to sea level.

Most of these cities are less than a few thousand years old. Our folk memory is too poor to recall the drowned civilizations now being revealed off the east and west coasts of India, abandoned when the sea level rose at the end of the last ice age.[5]

In 2007, the IPCC warned that sea levels would rise by 18–59 cm (7–23 inches) by 2100, or up to a meter if uncertainties are factored in.[6] The models did not include the carbon-cycle feedbacks, however; nor did they account for the likelihood that the Greenland and West Antarctic ice sheets will melt dramatically and abruptly in a dynamic, non-linear fashion.

In Greenland, some glaciers are discharging ice into the ocean three times faster than in 1988.

- Climate Change and Sea Level: tinyurl.com/ca73h
- Climate Code Red: climatecodered.com
- Coastal Impact Maps (US): architecture2030.org/current_situation/cutting_edge.html
- Extinction Risk from Climate Change: leeds.ac.uk/media/current/extinction.htm
- *Fighting for Love in the Century of Extinction — How Passion and Politics can Stop Global Warming,* by Eban Goodstein
- Millennium Ecosystem Assessment: millenniumassessment.org
- NASA's Eyes on the Earth: climate.jpl.nasa.gov
- Sea level rise around the world (dynamic maps): flood.firetree.net
- Waterworld: pbs.org/wgbh/nova/warnings/waterworld

In Antarctica, glaciers that were held back by the Larsen B ice shelf are flowing into the ocean eight times faster than they did before the shelf collapsed. If we carry on with business as usual, sea levels may rise not by 59 centimeters but by up to two meters by the end of this century,[7,8] and by much higher levels later. "I find it almost inconceivable that 'business as usual' climate change will not result in a rise in sea level measured in meters within a century," wrote NASA's climate scientist James Hansen in 2007.[9] Twenty million Bangladeshis live within a meter of the sea level.[10] Hundreds of US coastal cities are within a meter of sea level.

Meanwhile, the Arctic sea ice is also melting at an accelerating rate, with the total loss of summer ice possible as early as 2012. As the ice disappears, more heat enters the ocean waters, and the polar bear and walrus face extinction, except in zoos.

The fourth alarm bell rings for all of us, if we want to protect our grandchildren from living in a water world.

The Fifth Alarm Bell: Global Economic Disaster

Climate change presents a unique challenge for economics: it is the greatest and widest-ranging market failure ever seen.

— Sir Nicholas Stern

For many years, the complaint from business leaders around the world was that if we tried to do something about climate change we risked harming the economy and destroying jobs. Yes, climate change might be a concern, and we should do more research, but whatever we did, we must not upset the economy.

Many people, including those in very senior government and corporate positions, do not understand that Earth's economy is a wholly owned subsidiary of Earth's ecology and that we disturb the ecological balance at our peril. The attitude is widespread and stems from our many technological successes. Humans can walk on the moon: can any other species claim as much?

This attitude, combined with the absence of ecological education in our schools, has made people blind to our interconnectedness with nature, and fail to really grasp that the carbon released by burning fossil fuels can accumulate in the atmosphere and destabilize the planet's climate.

In 2006, however, Sir Nicholas Stern, former chief economist at the World Bank, was commissioned by Britain's finance minister (Gordon Brown) to analyze the financial implications of climate change.

The result, for those who had not been woken by the other alarm bells, was a major wake-up call. The cost of climate change, Stern said, would be "on a scale similar to those associated with the great wars and the economic depression of the first half of the 20th century."

If we do not take urgent action to stop the floods, storms, heat waves, declining crop yields, human migrations and other catastrophes that climate change will bring, his report found, they will cost the world's economies up to 20% of their gross domestic product. Worried insurance analysts told Sir Nicholas they feared the insurance claims from global climate change disasters could soon exceed the world's entire GDP.

The *Stern Review* was a massive 700-page analysis that used the best econometric models. "It was hard-headed. It didn't deal in sandals and brown rice. It stuck to the economics," said one UK Treasury source. "It was frighteningly convincing."[1]

A similar report for Germany predicted that climate change would cost the country 800 billion Euros (US$1 trillion) by 2050, including higher energy costs, declining tourism, increased insurance costs and the damage caused by extreme weather. The report also warned that if there was not an appreciable intensification in climate protection, the cost to Germany by 2100 could reach $4 trillion.[2]

On a smaller scale, but typical of studies that could be done for any region of the Earth, staff from the universities of Washington and Oregon analyzed the cost of climate change to the Washington State economy. The list was shocking. By 2020 the cost of fighting wildfires would increase by 50% to more than $75 million a year. The cost of lost timber sales and recreational and tourism opportunities would be many times greater than the cost of fighting the fires. There would be increased costs for heat and disease-related health care costs, shoreline protection, reduced water supplies and the need to relocate

Cost of climate change:
20% of global GDP

**Investment needed to
protect us:**
1% of global GDP

Saving the Earth: *Priceless*

- Climate Change — The Costs of Inaction:
  foe.co.uk/resource/reports/econ_costs_cc.pdf
- Impacts of Climate Change on Washington's Economy:
  ecy.wa.gov/biblio/0701010.html
- Stern Review: sternreview.org.uk

40,000 people from lands that would be inundated by sea-level rise.[3]

The cost of not responding to the looming crisis of peak oil must also be considered. As soon as the world's oil supply gets tight, the price of oil rises rapidly. When oil costs $200 a barrel and most of the money is leaving for the Middle East, the economic impact will be enormous.

The New Green Economy

The good news is that the *Stern Review* estimated that the investments needed to turn things around are no more than 1% per year of the world's GDP by 2050, money that will be invested in technologies that will launch a new energy revolution,

bringing as great a level of positive change as the industrial revolution itself.

As soon as we frame our minds positively, it is possible to see that the world is about to embark on a huge leap of discovery and innovation as we unhook our fossil-fuel dependency and build a climate-friendly world. The investments that we make to reduce our carbon emissions in public transit, zero-energy homes, high-speed trains, renewable energy, electric cars, safe bicycle routes, a new super-grid, zero waste, organic farming, grasslands recovery and much more will build a new green economy. In 2008, a major UN report on the emerging green economy found that tackling climate change would result in the creation of millions of green-collar jobs.[4] (See #83.)

This is not something to fear. It is something to celebrate, especially at a time of financial crisis, when a green stimulus package can be used to reboot the economy in a new green direction.

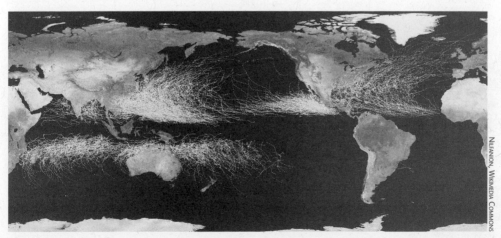

The tracks of all tropical hurricanes/cyclones which formed worldwide from 1985 to 2005.

NILFANION, WIKIMEDIA COMMONS

Six Degrees of Danger

We are still a long way from this hellish six-degree world, but if emissions keep rising over the decades to come ... the nightmare ending will be ever more likely.

— Mark Lynas (marklynas.org)

During 2005, Mark Lynas got up every day and cycled from his home in Oxford, UK, to the Radcliffe Science Library, where he systematically worked his way through tens of thousands of research papers on the impacts of global climate change. The IPCC climate scientists had warned the world that if we continued to burn fossil fuels, Earth's temperature would rise by 1.4°C to 5.8°C (2.5°F to 10.4°F) above the 1990 level, and up to 6.4°C (11.5°F) above the pre-industrial level.

But what did it mean? Step by step, Mark assembled the scientists' findings into six spreadsheets, one for each degree of temperature rise.

These formed the basis of his award-winning book *Six Degrees –Our Future on a Hotter Planet*, made into a National Geographic Channel TV film, *Six Degrees Could Change The World*. If you read just one book about the realities climate change will bring if we do not act, this is it.

Up to + 1°C

- Dust-bowl conditions return to much of the US mid-west.
- Permafrost melt causes buildings to sink across Alaska, Canada and Siberia.
- Mountain glaciers around the world continue to melt.
- Forest fires increase, especially in the Mediterranean.
- Droughts, hurricanes, deluges, and extreme weather events increase.

  Chance of avoiding a 1°C increase: next to zero.

1–2°C

- European summer heatwaves like that of 2003, which killed 35,000 people, become the norm.
- Past 1.2°C, Greenland tips into an irreversible meltdown, leading to an eventual six-metre rise in sea level.
- The Arctic is ice-free in summer months.
- Himalayan glaciers all but disappear, endangering water supplies to 0.5 billion people.
- A third of all land-based animals and plants face extinction as rapid temperature rise makes their habitats disappear.
- Repeated bleaching makes most coral reefs extinct.

Chance of avoiding a 2°C increase: good as long as global emissions peak by 2015 and fall by 80% by 2050.

2–3°C

- The Amazon rainforest passes a 'tipping point' of drought, causing massive fires and an eventual transformation into savannah and desert.
- Tens of millions displaced by spreading deserts in southern Africa and the Mediterranean.
- Farming and food production enter an irreversible decline due to soaring temperatures and droughts in the world's bread-baskets.
- Super-hurricanes, boosted by warming seas, devastate coastal cities.
- Dissolving CO_2 in the oceans makes them increasingly acidic, dissolving the remaining coral reefs and much of the marine food chain.
- Sea levels continue to rise. The last time the temperature was 3°C warmer, more than 3 million years ago in the Pliocene, sea levels were 25 metres higher.

Chance of avoiding a 3°C increase: good as long as emissions peak by 2020, and fall by 60% by mid-century.

3–4°C

- Streams of refugees flee the world's coastlines.
- The world's economy is battered by food shortages and migration.
- China's agricultural production crashes.
- Africa starves.
- Summer temperatures in southern Britain reach up to 45°C.
- More than half the Earth's natural species are wiped out, in the worst mass-extinction since the dinosaurs.
- The Arctic permafrost is in a runaway thaw, releasing billions more tonnes of carbon.
- Preventing a further rise in temperature

becomes next to impossible, due to failure of carbon storage in soils and forests.

Chance of avoiding a 4°C increase: reasonable as long as emissions peak by 2030 and stabilize by 2050.

4–5°C

- Rapid melting of Greenland and Antarctica causes sea level to rise by up to two metres.
- Possible collapse of West Antarctic Ice sheet, leading to further sea level rise.
- Rainforests reduced to tiny refuges as deserts spread through the tropics.
- The Arctic is ice-free all year round.
- Possible destabilization of marine methane hydrates.
- Civilization will likely retreat to a few guarded enclaves. A return to survivalism creates a further disaster for biodiversity.

Chance of avoiding a 5°C increase: good if emissions are stabilized in the second half of the century

5–6°C

- Temperatures on Earth are similar to the Cretaceous period, 144 to 65 million years ago.
- Major mass extinction of life on Earth due to rapid and extreme temperature rises.
- Eventually, over centuries, all ice on the planet melts, turning low-lying countries into archipelagos of islands.
- In the oceans, since less oxygen can dissolve in the warm water, conditions become stagnant and anoxic, further driving the mass marine extinctions.
- Super-hurricanes, floods and other extreme weather make normal life impossible.
- Food production only possible in sub-polar refuges; most of humanity is wiped out.

How To Talk To a Climate Denier

A firm and ever-growing body of evidence points to a clear picture: the world is warming, this warming is due to human activity increasing levels of greenhouse gases in the atmosphere, and if emissions continue unabated the warming will too, with increasingly serious consequences.

— Michael Le Page, *New Scientist*,

It is very discouraging to be speaking at a meeting and have someone stand up and say, "This is all nonsense. There are many climate scientists who dispute what you are saying. There has always been climate change. They were predicting global cooling back in the 1970s, and besides, the hockey stick graph has been proven wrong. Satellite measurements show that the lower atmosphere is cooling, not warming. Volcanoes produce far more CO_2 than humans. The ice-core record from the past shows that the increase in CO_2 follows the rise in temperature; it does not cause it. Why should we trust such flimsy science?"

Fortunately, Britain's *New Scientist* magazine has published a list of 28 climate myths that lay to rest the most common misunderstandings, and "How to Talk to Climate Skeptics" on the *Gristmill* refutes 61 contrary arguments. Here are just a few of the myths that the contrarians like to promote:

Many leading scientists question climate change

Not true. The handful of climate scientists who oppose the consensus stand against tens of thousands who have signed dozens of statements from scientific organizations all around the world supporting the consensus position. In a 2004 review of the abstracts of 928 peer-reviewed papers published from 1993 to 2003 that referenced global climate change, 100% supported the consensus position.[1]

The "hockey stick" graph has been proven wrong

Not true. The hockey stick graph shows that temperatures were basically level during the past 1,000 years and then rose sharply in the late 20th century. In 2006 the US National Academy of Science endorsed its findings and showed that it has been supported by an array of evidence.

Global warming is being caused by the Sun and cosmic rays, not humans

Not true. A 2007 study showed that solar output has been falling since 1985, eliminating also the theory that cosmic rays that create cloud cover, cooling the Earth, are being blocked by the Sun's more intense heat. Most of the 20th century saw a steady decrease in solar output, not an increase.[2]

The cooling after 1940 shows that CO_2 does not cause warming

Not true. The world did cool from 1940 to 1970, largely because the release of aerosols into the atmosphere, resulting from dirty industrial activities and warfare, scattered light from the Sun and reflected its heat back into space. There was also a large volcanic eruption at Mount Agung in 1963 that cooled things down by 0.5 °C.

The lower atmosphere is cooling, not warming

Not true. The apparent cooling was caused by errors in the way satellite data was collected and inaccurate data from weather balloons. More recent data reveals that it is warming as expected.

Ice cores show that past increases in CO_2 lagged behind temperature rises, disproving the link to global warming

The data is correct, but the conclusion is invalid.

Mount St Helens, May 18th, 1980. Volcanoes produce less than 1% of the CO_2 from human activities.

- A Rundown of the Skeptics and Deniers: logicalscience.com/skeptics/skeptics.htm
- Climate Change: A Guide for the Perplexed (New Scientist): tinyurl.com/3bl5e6
- *Climate Cover-up: The Crusade to Deny Global Warming,* by James Hoggan: desmogblog.com/climate-cover-up
- Deep Climate: deepclimate.org
- Desmogblog: desmogblog.com
- Global Temperatures: tinyurl.com/3xmqhm
- Great Global Warming Swindle: realclimate.org/index.php/archives/2007/03/swindled
- Hockey stick graph: tinyurl.com/2f96hb
- How to Talk to a Climate Skeptic: gristmill.grist.org/skeptics
- Naomi Oreskes Study: tinyurl.com/ywtgpj
- NASA/GISS Temperature animation: tinyurl.com/2onur8
- Real Climate Denial Wiki: realclimate.org/wiki
- The Consensus on Global Warming: logicalscience.com/consensus/consensus.htm
- The Heat is Online: heatisonline.org

The initial warming when Earth emerged from an ice age was caused by variations in Earth's orbit, known as the Milankovitch cycles. After a lag of about 800 years, CO_2 emissions from the warmer oceans increased, and CO_2 and temperature rose together for about 4,200 years. The evidence that CO_2 traps heat comes from physics, not from correlations with past temperature.

It was warmer during the medieval period, when there were vineyards in England

Not true. There were some warm periods in Europe from 900 to 1300 AD, but the accumulated evidence shows that the planet has been warmer in the past few decades than at any time during the medieval period, and maybe warmer than it has been for 125,000 years.[3]

Scientists were predicting global cooling in the 1970s

True. "They" were a handful of scientists who were concerned that increased air pollution might outweigh the influence of rising CO_2 emissions, and they called for more research. Subsequent research by thousands of scientists has found that warming caused by greenhouse gases far outweighs the cooling caused by air pollution.

Mars and Pluto are warming too

Maybe true. Our knowledge about these planets is still very sketchy. If it is true, the warming is not being caused by increased solar activity, as the Sun's output has not increased since direct measurements began in 1978.

To many climate skeptics, no amount of debate will change their views. For these people, the alternative framing presented in Solution #65 may be more effective.

The Crucial Data

If it were only a few degrees, that would be serious, but we could adapt to it. But the danger is the warming process might be unstable and run away. We could end up like Venus, covered in clouds and with the surface temperature of 400 degrees. It could be too late if we wait until the bad effects of warming become obvious. We need action now to reduce emission of carbon dioxide.[12]

— Stephen Hawking

Gigatonnes of this, percentages of that — the business of climate change is full of numbers that can easily seem bewildering.

Here are the fundamental numbers that this book uses for its calculations. A gigatonne (Gt) is a billion tonnes. To convert from carbon to carbon dioxide, multiply by 3.667 to account for the added oxygen. The best source of reliable data is the annual *Carbon Trends* from the Global Carbon Project, which publishes new annual data in October each year.

- Before the industrial age began, Earth's atmosphere contained 560 Gt of carbon (280 ppm of CO_2). By 2008 this had risen to over 800 Gt (385 ppm CO_2).

An artist's rendering of NASA's Orbiting Carbon Observatory that crashed on take-off in 2009.

- In 1751, the world released 3 million tonnes (Mt) of carbon by burning fossil fuels. By 1851 this had risen to 54 Mt, and by 1951 to 1700 Mt. By 2008 it had reached 8500 Mt (8.5 Gt).

- If the CO_2 continues to rise at 2ppm a year, it will pass 400 ppm in 2015 and 450 ppm in 2040. If it accelerates to 3 ppm a year, it will pass 400 ppm in 2013 and 450 in 2030.

- Between 1850 and 2007, we released 384 Gt of carbon into the atmosphere by burning fossil fuels and making cement, and 160 Gt from land use change — chiefly deforestation — for a total of 544 GT of added carbon. The world's forests, soils and oceans absorbed 54% of this, leaving an excess of 272 Gt of carbon in the atmosphere.

- The ability of Earth's forests, soils and oceans to absorb our excess carbon is increasing, but slower than the rate at which we are releasing the carbon, causing their relative ability to absorb it to fall by 5% over the past 50 years.

- During 2007 we released a further 8.5 Gt of carbon (31.1 Gt of CO_2) by burning fossil fuels and making cement. Between 2000 and 2007 the emissions grew by 3.5% a year, which is higher than the highest growth rates used by the IPCC climate scientists in their future scenarios.

- During 2007 we also released a further 1.5 Gt of carbon (5.5 Gt of CO_2) through deforestation in the tropical countries — 41% from South and Central America, 43% from South and Southeast Asia, 17% from Africa. Altogether in 2007 we added 10 Gt of carbon

NASA/JPL

CO_2 Emissions/Year (Gt)[2]	
Coal	12
Oil	12
Gas	5.8
Cement	2.7
Deforestation[2]	5.5
Total	36.6[3]

Percentages	CO_2	BC	CH_4	O_3[4]	F	N_2O	Total[5]
Radiative Forcing	44	21	13	9	9	4	100%
Fossil fuels	81	40	26	50	0	14	53%
Farming	-	-	50	-	0	70	32%[6]
Deforestation	15	40	6	12	0	0	17%
F Gases	0	0	0	0	100	0	9%
Cement	7	0	0	0	0	0	3%

	Proven Reserves[8]	CO_2 factor	Gt Carbon
Coal[9]	847 billion tonnes	3 tonnes per tonne	693
Oil[10]	1237 barrels	457 kg per barrel	154
Gas	177 trillion cubic meters	1 tonne per 500 cubic meters	97

- Carbon Dioxide Information Analysis Center: cdiac.ornl.gov
- Global Carbon Project: globalcarbonproject.org
- US CO_2 emissions from energy sources: eia.doe.gov/oiaf/1605/flash/flash.html
- CO_2Now: co2now.org

(36.6 Gt of CO_2) to the atmosphere — 100 million tonnes a day; 1,000 tonnes a second.

- CO_2 produces 44% of the radiative forcing that causes global warming. Eighty-five percent of the CO_2 comes from burning fossil fuels, which are responsible for 53% of the total radiative forcing — the cause of global warming.[1] Because fossil fuels also enable most modern farming, forestry, cement use and industry, our use of fossil fuels is by far the largest cause of global warming.

- When the impact of the other greenhouse gases is included as "carbon dioxide equivalent" (CO_2e), the atmospheric loading rises to 440 ppm CO_2e. The increase is balanced by the effect of the particulates (dirty pollution) that are also released by burning fossil fuels, biomass and forests. With the exception of black carbon (soot), this has a cooling effect, shielding us from some of the warming. Black carbon (BC) is responsible for 21% of the overall problem. (See p. 14.)

- Since 1850, the global temperature has risen by 0.8°C. There is another inevitable 0.5°C rise in the pipeline, because of CO_2's long life in the atmosphere.

- If we burned the world's entire proven reserves of coal, oil and gas, we would release an additional 944 gt of carbon into the atmosphere, 2.5 times more than has been released so far, which would raise global temperatures by 7°C[7] — plus the impact from the associated fugitive methane emissions and black carbon.

- Because CO_2 has such a long atmospheric life, we need to measure each nation's impact by its cumulative carbon emissions since 1850[11]:

USA	29.3%	UK	6.3%
EU-25	26.5%	Japan	4.1%
Russia	8.1%	France	2.9%
China	7.6%	India	2.2%
Germany	7.3%	Canada	2.1%

What Targets Should We Adopt?

We have to stabilize emissions of carbon dioxide within a decade, or temperatures will warm by more than one degree. That will be warmer than it has been for half a million years, and many things could become unstoppable.[5]

— James Hansen, NASA, 2006

If we allow the greenhouse gases to continue to accumulate in the atmosphere, we will be signing the death warrant for most life on Earth. The consequences, spelled out in Mark Lynas's book *Six Degrees,* are so grim that it is inconceivable that we could allow this to happen. (See p. 24.)

Instead of falling below the 1990 level, which the Kyoto Protocol required, by 2008 our greenhouse gas emissions had risen to 38% above that level and were growing by 3% a year, compared to 1.1% a year in the 1990s. As a result of our continued use of fossil fuels and our inability to stop tropical deforestation, we are releasing 100 million tonnes of CO_2 into the atmosphere every day — more than a thousand tonnes every second.

When the IPCC climate scientists looked at the future in 2007, they predicted a possible temperature rise by the end of this century ranging from 2.4°C to 6.3°C, but none of their projections assumed such a rapid rise in emissions. Almost everyone who has studied the data shares a barely suppressed feeling of emergency, for they know the road we are speeding down leads to catastrophe.

- 350: 350.org
- One Hundred Months: onehundredmonths.org
- *The 2° Target: How Far Should Carbon Emissions be Cut?*: carbonequity.info/PDFs/2degree.pdf
- Flaming Methane: alaska.edu/uaf/cem/ine/walter/videopage.xml

The road is not flat, however. As we speed down it, burning ever more fossil fuels and trapping ever more heat, Earth's climate system passes a number of "tipping points" — thresholds of change that are built into the way nature works. As these are passed they cause kick-on consequences, known as "positive feedback." There are many of these tipping points,[1] including:

- The melting of the northern permafrost, which contains enormous quantities of CO_2 and methane. This has already begun in western Siberia — scientists have been able to set a stream of methane on fire as it gushes up through the ice.[2] If the methane all seeps out over 100 years, it will add 700 million tonnes of carbon to the atmosphere each year, leading to a 10–25% increase in global warming, and add 730 Gt (billion tonnes) in total to the Earth's atmospheric burden.

- The loss of the Arctic sea ice, allowing the Sun's heat to be absorbed by the water instead of being reflected back into space. This is already underway.

- The melting of Greenland's ice sheet, which will cause the sea level eventually to rise by 6–7 meters (20 feet). This too is underway.

Some hope that the 2008-2009 financial crisis will cause fossil fuel emissions to slow. In 2008, OPEC forecast that the use of oil would decline by 150,000 barrels a day in 2009. This is only a 0.2% decline, however, and because oil is only 23% of the cause of global warming,[3] the impact of the decline is a 20th of 1%. We are still speeding down the road to catastrophe.

In December 2008, over 200 people worked with activist artist John Quigley to spell out the message in Freedom Square, Poznan, Poland, during the UN COP-14 Climate Conference.

One of the widely discussed goals is the need to avoid a temperature rise of more than 2°C, which will happen if atmospheric CO_2 reaches 450 ppm. Even this creates an unacceptably high level of risk, however, with a 10% chance that we will be unable to prevent the temperature from rising further.[4] Would anyone stand their child in the middle of a road to face the oncoming traffic if they knew there was a 10% chance the child would die? Every instinct says no. The risk we accept for flying is no more than a one-in-a-million chance of an accident.

In London, England, when staff at the New Economics Foundation looked at the tipping points data, they calculated that if we did not achieve a global turnaround in our emissions within 100 months — by January 2017 — we would probably be unable to stop the temperature from passing 2°C. This means we have to begin reducing immediately and aim for a 30–40% reduction below the 1990 level by 2020 — and this applies to the whole world, not just the nations that sign the Copenhagen Treaty.

It also makes no sense to aim for a negative target (450 ppm). We need a target that will guarantee the safety of our planet — in other words, 280 ppm, the level it was before the Industrial Age. This may be achievable if we can eliminate our GHG emissions by 2040, and change the way we manage our forests, farmlands and grasslands so that they suck far more CO_2 out of the atmosphere. (See p. 66 and #74.) On the way to 280 ppm, we should embrace the global movement to adopt the goal of 350 ppm.

Goals for a Climate-Friendly World

30% below the 1990 level by 2020

40% below by 2025

60% below by 2030

80% below by 2035

100% below by 2040

350 ppm of atmospheric CO_2 by 2100

280 ppm by 2150

What Then Must We Do?

The Earth is a living entity with incredible healing powers, and we have much to learn. It is the task of our generation to leave this sacred Earth, in all its wisdom and beauty, to the generations to come. Let the work begin.
— Icefjord Commitment by Religious Leaders, Greenland, 2007

Past civilizations failed when confronted with similar challenges. In the eastern Pacific, when the Polynesians of Rapa Nui (Easter Island) failed to protect their island's ecology, they resorted to warfare after they had cut down the last tree.

In the western Pacific, however, the Polynesians on the small island of Tikopia responded to a similar challenge by preventing population growth, giving up animals for meat and adopting a permaculture style of forest food cultivation, enabling them to maintain both their culture and their ecology intact with pride.

The difference is that the Rapanui had no contact with other humans, and took refuge in religion rather than rationality, building huge stone Moai in the hope of magical relief. The

Solar housing in Vauban, Freiburg, Germany, where 70% of the residents do not own a car.

Tikopians, on the other hand, retained contact with people in the Solomon Islands, only three weeks' sailing away, which enabled them to remain positive and responsive to change.

With the Internet being so widely available, there is no reason for anyone to retreat into hopelessness or religion in response to the climate crisis. If you search on climate + change + solutions you will get 2.7 million hits. To action, then!

We need an emergency crash program, as we did in World War II. We need millions of volunteers who will step forward and offer to organize in their communities, pushing for legislative changes and creating a solid wall of determination, knowing that we must leave the road to disaster and chart a new course to sustainability.

We need to make all our homes and buildings super energy efficient, using government grants and interest-free loans to pay for the retrofits, incorporating green energy for heat and power wherever possible.

We need to phase out all coal-fired power plants, replacing them with renewable energy from the sun, wind, tides, geothermal power, hydro power and other green sources. Algeria alone has enough solar capacity to meet Western Europe's electricity needs 60 times over. The deserts in the southwest US have a similar capacity. The geothermal energy that we could tap into at depths of 5–10 km down is enormous. Developing just 5% of the US's potential wind energy could supply 25% of what's needed — and all this could be financed by adopting Europe's system of feed-in tariffs. (See #69.)

In our cities, we need to shift half of our car-based trips to walking, cycling, transit, light-rail transit and ridesharing. We also need to retrofit our communities to make them more friendly for walking and cycling, building a far stronger sense of community as people meet and greet each other; and we must make public transit far more frequent, comfortable and efficient.

We need to stop producing cars and light trucks that burn oil, replacing them with electric vehicles and plug-in hybrid electric vehicles, and fitting our cities with battery recharging and replacement stations as they are doing in Israel, Denmark, Hawaii, Berlin, Paris, Portland, Seattle and many other places. We need to make biofuels from sewage, waste, algae and other sources that do not compete with food.

We need to make our trucks far more fuel-efficient, electrifying the main trucking routes and enabling them also to run on algae-based biofuels, blue fuel and instant hydrogen (see Solution #47). We need to build a country-wide network of high-speed trains, as they have been doing in Europe for the past 30 years.

We need to treat all waste as energy and recycle all our waste by 2030, becoming a zero-waste world, while extracting heat and energy from sewage, farms and food wastes.

We need to end our farmlands' dependency on fossil fuels and encourage our farmers and ranchers to adopt holistic methods of soil management that suck enormous quantities of carbon out of the atmosphere. We need to do the same for our forests, increasing their ability to store carbon.

To help all this happen, we need to put a price on carbon, rising to $150 a tonne, returning all the money gathered through reduced taxes. We need to accelerate plans for a system of "cap and Dividend" or "cap and trade" (see #79) but also prepare to legislate for direct reductions in carbon emissions, in case these schemes fail.

We need to get everyone engaged in carbon-reduction projects in their homes, schools, colleges, businesses, churches and communities, just as the World War II effort got everyone growing Victory Gardens, raising money to buy Victory Bonds and volunteering to help wherever help was needed. It will mean personal change, but change can be exciting when we remove the fear and uncertainty that comes with unwanted change.

All these solutions are described in detail in the Solutions section of this book. Most are already being put into practice somewhere in the world. Taken together, they show that taking the road to a sustainable, climate-friendly world is not only possible but highly desirable. It will generate many million new jobs and invigorate our economies with innovation and a sense of great adventure. This is not just about ending our dependency on fossil fuels — it is about stepping into a whole new world that will bring new hopes, new dreams and new horizons.

faith in Earth's future

Today, something is happening to the whole structure of human consciousness. A fresh kind of life is starting. Driven by the forces of love, the fragments of the world are seeking each other, so that the world may come into being.

— Pierre Teilhard de Chardin

It can all seem very intimidating. There are many who cannot speak about global warming without crying.

For centuries, people have prayed and hoped that the future would be better than the present. Today, many fear it will be worse — and if we remain on our current path, they will be correct.

We are the binge generation. We wanted the fish — so we took them. At the current rate, all of the world's commercial fish stocks will be gone by 2050. We wanted the fossil fuels — so we took them. We wanted the land — so we turned the forests into farms and subdivisions, driving out the eagles, frogs, deer and birds. All over the Earth, the things we have used lie discarded and scattered in landfills.

We are like two-year-olds who have known nothing but our parents' love but whose parents are now saying "No — that's enough!" Will we respond with a tantrum, demanding the right to be the center of attention regardless of the distress we cause to the Earth and other people? Or will we come to our senses and learn that there is another way of living, where cooperation and respect create a harmony in which all beings may flourish? Will we learn Earth's household rules — the rules of ecology — before it is too late?

The crisis we face involves more than global warming. That is just one symptom of a more wide-ranging problem caused by our cultural belief that Earth has no limits and that we can grow, consume and increase our quarterly returns for ever.

We live at a turning point for our civilization. If we continue down the current road, a disastrous Rapa Nui ending will be certain.

To pause and turn, however, implies the existence of a different road and an ability to visualize where it is going. The key to all success lies in the mind, in our ability to visualize success and then make it happen.

It is like turning a switch from despair to determination, from hopelessness to hope. It happens in sports all the time. You may be losing, but you dig down and find a new determination, grounded in the belief that you can win. It happens in music, politics, warfare, engineering, science, exploration, business and every field of human endeavor.

With this, we come to the core of our problem:

Do we believe in our ability to create a green, sustainable future where we live in harmony with the Earth and each other, instead of at their expense?

- Auroville: auroville.org
- Bioneers: bioneers.org
- Earth Charter: earthcharter.org
- Earthfuture: earthfuture.com
- Four Reasons for Hope: janegoodall.org/jane/essay.asp
- The Great Turning: thegreatturning.net
- Wiser Earth: wiserearth.org
- Worldwatch Institute: worldwatch.org
- Yes! Magazine: yesmagazine.org

Are we giving her reason to have faith in Earth's future?

© ANTONIA | DREAMSTIME.COM

For millions of years, most of Earth's people have lived in planetary babyhood, enjoying the wealth that nature gave us and the dominance that our language and tools gave us. Only those cultures that have survived ecological collapse have learned the deeper wisdom of humility and respect for all species.

There is well-researched evidence that when we choose to live within Earth's limits, there will be enough for all:

- When we choose to farm organically, everyone in the world will receive more calories and more nutritional value.

- When we choose to manage Earth's forests ecologically, we will harvest more timber while sustaining the old-growth character of the forests.

- When we choose to use renewable energy from the Sun, Earth and Moon, there will more than enough for all our needs.

- When we choose to re-use and recycle the things we use, surrendering the idea of "waste," there will be sufficient material resources for all.

Life in a green, sustainable world can provide all of the things that make life wonderful — friendly neighbors, cooperative families, fulfilling work, exciting challenges, peaceful green spaces, beautiful architecture, protected wilderness and wildlife, music, art, inspiration and love.

The age of fossil fuels lasts for a tiny 200 years, in a span of civilization that is tens of thousands of years old and will stretch tens of millions of years into the future. We are at the moment of change, the historical moment when we either continue to act stupidly and let everything collapse, or choose to take a different road that leads to a sustainable civilization.

We first saw our Earth from space in 1966. We embraced the World Wide Web only in the 1990s. The solar revolution has only just begun. Why should we not have faith that we can do this?

The age of fossil fuels gave us the knowledge and skills we needed to build a civilization that no longer depends on them. They have been our training wheels, our launch ramp to the solar age.

We are young. We are just beginning. Everything is possible.

Is the Oil Running Out?

I'd put my money on the sun and solar energy. What a source of power! I hope we don't have to wait 'til oil and coal run out before we tackle that.

— Thomas Edison

Alongside global warming, there is growing concern among many people about the world's oil supply. Most people find it hard to remember that oil is a finite, non-renewable resource, and when it's gone, it's gone. We are like fish in water, who cannot imagine there being no water.

We depend on oil for 90% of our transport, most of our heat and all our plastics. We use it to make the asphalt for our roads, the tires that run on the roads (seven gallons of oil per tire) and the steaks that people like to eat when they get off the road (three quarters of a gallon of oil for every pound of beef). It is only natural to become concerned when credible professionals are warning us that we face "peak oil" very soon.

In 2008, the Swiss school teacher Louis Palmer drove 53,451 km around the world in his Solar Taxi. At 8 kWh/100 km, his car used a total of 4,276 kWh.

Peak oil is the point at which the global production of oil peaks and starts to decline. In a world that wants more oil every year, this will push prices through the roof, as happened in 2008 before the financial crash. It is this that causes people to lose sleep, knowing that our economy, trade, jobs, suburbs, food supply and much else depends on oil. James Howard Kunstler's book *The Long Emergency* captures these concerns with disturbing eloquence.

The Earth's original endowment of conventional oil was between 2,000 and 2,400 billion barrels. By 2008, we had consumed 1,130 billion barrels and were using 31 billion barrels a year (85 million barrels a day) while discovering only 6 billion.

Until 2008, the International Energy Agency (IEA), which governments defer to for data and analysis, had taken the position that there were 1,200 billion barrels of conventional oil left in the ground (39 years' worth at the 2008 rate of demand), that future oilfield decline would be only 3.7% a year, and that conventional oil would not peak until 2030.

That year, however, the IEA undertook its first detailed study of the world's 800 largest oilfields, admitting that their previous numbers had been based mainly on assumptions. In their new analysis, they found that the future rate of decline would be much higher, that conventional oil would probably peak in 2020[1] and that output was declining by 9.1% a year.[2]

Others, including veteran petroleum geologists such as Colin Campbell, founder of the Association for the Study of Peak Oil, think that only 1,000 billion barrels remain, and the peak will happen in 2010. When 150 petroleum geologists were surveyed at a conference in 2008, 60% thought that global oil production would peak within ten years.[3]

There are also 2,000 to 4,000 billion barrels of heavy oil, oil sands and shale oils, which are much more expensive and energy intensive to extract. But because their total production is only 2.8 million barrels a day, even if Alberta expanded its oil-sands output from one to three million barrels a day, it would make little difference to the reality of the peak.

The relationship of peak oil to climate change is critical, and yet it has been almost completely neglected. In 2004, when the Swedish geologist Anders Sivertsson analyzed the IPCC's 40 scenarios for future emissions, he found that their assumptions about future oil and gas consumption were completely out of sync with the proximity of peak oil, which would cause a forced reduction in demand.[4] The IPCC's scenarios assumed the availability of 5,000 to 18,000 billion barrels of oil and gas (measured as oil equivalent), compared to a real world estimate of only 3,500 billion barrels. The IPCC's justification for the high numbers was their assumption that coal would be substituted for any shortfall in oil or gas, even though this would cause a climatic disaster.[5]

What can we conclude from this situation?

- The climate and peak oil communities should come together and cooperate.

- Are We Running Out of Oil? tinyurl.com/c5cnc9
- ASPO USA: aspousa.org
- Association for the Study of Peak Oil and Gas: peakoil.net
- Bridging Climate Change and Peak Oil Activism: richardheinberg.com/museletter/177
- Life After the Oil Crash: lifeaftertheoilcrash.net
- Local Future: localfuture.org
- Oil Depletion Analysis Centre: odac-info.org
- Peak Moment TV Show: apple-nc.org/peakmoment.html
- Peak Oil video: tinyurl.com/yvwalb
- Post-Carbon Institute: postcarbon.org
- Power Switch: powerswitch.org.uk
- The Coming Global Oil Crisis: oilcrisis.com
- The Oil Drum: theoildrum.com
- *The Last Oil Shock*: davidstrahan.com
- Wolf at the Door — The Beginner's Guide to Peak Oil: wolfatthedoor.org.uk
- World Proved Reserves of Oil and Gas: eia.doe.gov/emeu/international/reserves.html

- Governments and climate scientists should integrate their energy and climate analysis so that their scenarios make mutual sense.
- The peak oil community should take climate change much more seriously because the positive attitude among climate activists is delivering a rich supply of solutions, all of which also address peak oil.
- The climate community should take peak oil much more seriously because the reality of peak oil will create an enormous demand for non-oil alternatives, nearly all of which will reduce carbon emissions.

Energy Security

> We could be faced with a Hobson's choice between economic collapse and a global resource war if we don't do something, and fairly quickly.
>
> — Milton Copulos, President NDCF

The word *security* conjures thoughts of Snoopy's friend Linus, with his blanket. For most people, security means a comfortable home, a safe neighborhood, a reliable source of work and income, and the knowledge that you are not about to be attacked. In Google.ca, the word security gets 780 million hits, more than sex (741 million), God (487 million), peace (268 million) and freedom (231 million) — but thankfully less than love (1,640 million).

Energy security means the energy we require is secure — but when US energy security depends on the daily arrival of 12 million barrels of oil from outside the US, three million barrels of which

The US imports eleven million barrels of oil from countries overseas every day, requiring up to six supertankers, each carrying two million barrels. Each tanker enables 21 million cars to drive 100 miles at 25 mpg.

come from countries in the Persian Gulf where many people hate your guts, the only "security" to be found is through the Department of Homeland Security.

Our world's dependence on oil has given us five enormous headaches: global warming; the looming chaos of peak oil; global financial instability; air pollution, asthma, cancer and heart disease; and terrorism, as armed jihadists and suicide bombers fight to bring down American civilization.

With oil at $100 a barrel, Americans send $300 million dollars to the oil-producing countries in the Middle East *every day*, some of which is known to bankroll Islamic terrorist organizations. The money also serves as an enormous source of financial instability, giving Saudi Arabia, with 260 billion barrels of oil in reserves, a $26 trillion stake at the global table, enough to purchasing a controlling interest in huge swaths of the world's banks and industries.[1] The surging price of oil in 2008 was probably one of the factors that helped bring down the world economy.

With US energy security being so vulnerable to political change in the Middle East, in 1945 President Roosevelt made a commitment to King Abdul Aziz ibn Saud that the US would guarantee Saudi Arabia's security in exchange for access to its oil. This led to the Carter Doctrine, which stated that any effort by a hostile power to block the flow of oil from the Persian Gulf would be seen as an attack on the US's vital interests, and to the military establishment of US Central Command, overseeing 20 nations and 550 million people in its area of responsibility.

- Institute for the Analysis of Global Security: iags.org
- Journal of Energy Security: ensec.org
- National Defense Council Foundation: ndcf.org
- Oil Solutions Initiative: move.rmi.org/osi

When the full economic costs of all this energy insecurity are added up, the result is astonishing. The numbers have been crunched by a team led by Milton Copulos, president of the National Defense Council Foundation (NDCF), which spent 18 months undertaking the most comprehensive analysis of the subject ever conducted, which was "rigorously peer-reviewed."[2]

The fixed costs of defending the Persian Gulf oil come to $138 billion a year. The loss of domestic economic activity due to the export of so much cash adds $117 billion. The loss of domestic investment for the same reason removes $394 billion a year from the economy; the loss of government revenues adds $43 billion; and the losses caused by oil supply disruptions cost $82.5 billion. Together with other costs, these oil-related "hidden" costs come to $825 billion a year, adding a hidden $2.60 to the price of every gallon.[3] The outflow of money also causes unemployment, losing 2.25 million jobs.

Almost all the world's transport depends on oil, and the Middle East produces 37% of it. It also sits on 65% of the world's oil reserves, and 45% of the natural gas. Because the rest of the world's oil is running out a lot faster than the Middle East's, this holds the threat of far greater energy insecurity in the years to come. In 2001, Iran, Russia, Qatar and 12 other nations formed the Gas Exporting Countries Forum, promising future insecurity for natural gas as well as oil.

So how can we create *real* energy security? In 2008, the Brookings institute and Rocky Mountain Institute brought together representatives from 15 organizations that have developed plans to break the US's dependence on oil. They met at the Oil Solutions Summit to find ways to work together and formed The Oil Solutions Initiative. Most of their solutions, apart from more offshore drilling, have close to 100% harmony with the solutions to climate change and peak oil that are laid out in this book.

When we consider the future cost of climate solutions, this is worth factoring into the math: that by ending the need for a US military presence in the Middle East, ending the daily bleeding of money and ending the oil-related energy supply disruptions, the US economy will benefit by $825 billion a year, enough to offset the cost of all the building retrofits, plug-in hybrid cars, solar panels, high-speed trains, smart meters and other climate solutions.

US crude oil suppliers 2007	Oil*
Canada	827
Saudi Arabia	541
Venezuela	488
Mexico	457
Nigeria	413
Algeria	242
Angola	185
Iraq	177
Russia	151
Virgin Islands	126
* Million barrels a year	

Earth's Future Buildings

The surest path to safe streets and peaceful communities is not more police and prisons, but ecologically sound economic development. That same path can lift us to a new, green economy, with the power to lift people out of poverty while respecting and repairing the environment.

— Van Jones, founder of Green For All

Everyone needs a home. In the distant past, we built our homes from locally gathered stone, wood, mud, animal skins, grasses, ice — whatever we could lay our hands on. During the ice age in Poland and the Ukraine, 20,000 years ago, we even used mammoth bones.

When we discovered how to make iron, we started cutting the forests for their timber and fuel. When we discovered fossil fuels, we started using coal, oil and gas to heat our buildings and to make new building materials.

We've come a long way but at huge expense to the Earth. Thirty-three percent of our CO_2 emissions come from the energy we use in buildings — far more than all our transport.

Since we know how to build zero net energy buildings, it makes sense that all new buildings should be built this way. In Germany, the *Passivhaus* design uses 95% less energy than the average building, and Germans have built 6,000 of them.

In Britain, where buildings are responsible for 27% of the CO_2 emissions, the government is requiring all new buildings to be zero carbon by 2016. Five new zero-carbon eco-towns are also to be built, with schools and shops within walking distance of the homes. Germany has built 40 such eco-communities.

In the US, where buildings are responsible for 43% of the CO_2 emissions, architect Ed Mazria has issued the 2030 Challenge, calling for an immediate 50% cut in the use of fossil fuels in new and renovated buildings, and for all new buildings and renovations to be carbon neutral by 2030. In Guangzhou, China, the 69-storey Pearl River Tower will gather its energy from wind turbines built into two floors of the tower. The technologies exist for all new buildings to be zero-carbon — so that should be our goal.

In 2008, Architecture 2030 showed that compared to nuclear energy, a $21.6 billion investment in building efficiency would result in a three times greater reduction in CO_2 emissions, create 216,000 new jobs, generate $8.46 billion in annual consumer savings with a payback of less than three years, and produce new energy for a fifth of the cost of doing so with clean coal or nuclear power.[1] It's one of the best deals going.

Zero-Carbon Buildings

- Smaller homes that make better use of space
- Passive solar orientation and design
- Super efficient walls, windows and insulation
- Heat exchange ventilators
- Heat exchange from the air, ground, water and sewage
- Solar heat, hot water and electricity
- Smart metering, efficient appliances
- Bioenergy from locally harvested wastes
- District heating systems

- 2030 Challenge: architecture2030.org
- Findhorn Ecovillage: Findhorn.org
- Passivhaus Institute: passiv.de
- Passivhaus UK: passivhaus.org.uk/
- European Passive Houses: europeanpassivehouses.org

Ecohouses at the Findhorn Ecovillage, in Scotland. Findhorn's residents have the lowest recorded ecological footprint for any permanent community in the developed world, thanks to their eco-buildings, locally grown organic vegetarian diet, wind turbines, community sharing of resources, and very low car mileage (6% of the UK national average).

- Rooftop water harvesting
- Trees for shade and cooling
- White roofs, to reflect the heat
- Non-toxic paints and finishes
- Recycled materials
- Cob, straw bale and rammed-earth buildings

with

- Neighborhood shops
- Walkable, car-free communities
- Safe cycle paths
- Car-sharing and ride-sharing facilities
- Free public transit
- Green space, parks and trees
- Local food gardens and urban farms

Existing buildings

With a good retrofit — increasing the insulation, upgrading the windows, sealing the cracks and upgrading the lighting and equipment — a building's energy use can be reduced by 20–50%. But how can we make this happen? This is the question that has bewildered a thousand minds.

We need zero-interest loans, tax breaks and grants to reduce the financial hurdles that discourage home owners and landlords from making the effort.

We need local energy-saving partnerships that will help us to retrofit all our buildings, paying for the upgrades from the energy savings, as Berlin (Germany) and Cambridge (Massachusetts) are doing. We need green mortgages and energy-saving companies to shift the cost onto the savings so that the upgrades cost nothing to the owners.

We need far higher standards for energy and water efficiency, and we need to make them apply at the point of sale, requiring a home-owner to do an energy upgrade whenever a house is sold, as San Francisco and Berkeley have done since 1990, rolling the cost into the new owner's mortgage.

We need to follow Germany's example, where the government is spending 1.5 billion euros to upgrade all of the country's pre-1978 housing stock to the *Passivhaus* standard by 2025, at 5% a year.

To use a wartime metaphor, leaving our buildings untouched is like leaving a third of our ports unguarded, wasting energy while the climate crisis worsens.

In the future world, every community — including all suburbia — will have community transit, local shops and cafés, local businesses, farmers markets, children's parks, allotment gardens, car-sharing groups, ride-sharing websites, green spaces and wild nature spaces so that people can build a spirit of community wherever they live and cease being so dependent on long-distance travel. This kind of community already exists in many smaller traditional towns, so we know that it works. We just need to make it work everywhere.

Earth's Future Transport

Rather than continuing to base our economy on a finite supply of dead things, we can base it on sources that are practically infinite and eternal: the sun, the moon and the Earth's inner fire.

— Van Jones

Our fossil-fuelled transport causes 20% of our global CO_2 emissions — and when we include the many other ways in which it impacts the climate, about 20% of global warming:[1]

- 20% of the CO_2
- 20% of the black carbon, from diesel exhaust
- 20% of the F gases, from air conditioning
- 25% of the tropospheric ozone, from transport's air pollutants
- 14% of the N_2O from vehicles, ships and planes

- High-level cloud formation from airplane exhaust
- Ground-level ozone pollution, lessening the ability of plants to absorb CO_2.

If we contemplate the development of our civilization over many millennia, and cast our eyes far into the future, the Age of Fossil Fuels will appear as a brief burst of innovation followed by a far longer period when humans derive all their energy from the sun, wind, tides and geothermal sources. Before fossil fuels, human energy gathering was simple and rough. After fossil fuels, it will shine with a brilliance born of technological advance. So how will we travel in a climate-friendly world?

We will live in communities that have been redesigned to make them far more pedestrian friendly so that we can enjoy that most ancient method of transport — our feet.

There will be many safe bike lanes and trails where we can enjoy the delights of cycling. Where there are hills, people will use electric bikes and scooters. In some towns in Holland cyclists are already the majority, reclaiming town centers from the cars and trucks that dominated in the 20th century.

There will be comfortable and timely public transit, bus rapid transit and light rapid transit, powered by biogas and green electricity. Our cities will have learned from the Brazilian city of Curitiba and built comfortable bus shelters in elevated tubes where you can enter the bus easily, on the level. In some cities public transit will be free, paid by city taxes. (See #25.) Every community

CHRISTOPH GRONECK TRAMS-IN-FRANCE.NET

Electric tram in Mulhouse, France.

will have ride-sharing and car-sharing programs, enabling people to get where they need with the minimum of fuss.

In place of long-distance meetings that require so much travel, every community will have state-of-the-art videoconferencing centers. Many people will also choose to do routine grocery shopping on-line, with home delivery the following day.

The cars of the future will be electric and plug-in tribrid multi-fuel vehicles, combining a battery for local trips with zero-carbon biofuel and instant hydrogen for longer distances. When the battery runs low in all-electric vehicles, there will be recharging posts at office-buildings, parking lots and on-street parking meters; if you need to recharge in a hurry there will be battery swap stations where a hydraulic lift replaces the battery with a new one. All this is already happening in cities such as Berlin, Paris and San Francisco, and in Israel, Denmark, Australia, Hawaii and Toronto, where Better Place is taking the initiative. Biofuels will be made from sewage, food wastes, algae and low-water, low-maintenance crops that do not compete with food. (See #76.)

Our towns and cities will be connected with high-speed trains that travel at 200 kph (125 mph) and faster, powered by green electricity, and slower trains for local travel. A cross-country trip from Los Angeles to New York, a journey of 3,000 miles, will take 10–15 hours, enjoyed in leisure without the stress and worry of flying. (See #48.)

Much of what's now transported by truck will be moved by far more efficient rail freight. The trucks of the future will have multi-fuel drives, burning zero-carbon biofuel made from algae or blue fuel made from surplus wind energy, enhanced with instant hydrogen. In the cities, local delivery trucks will be mostly electric. (See #47.)

On the ocean, ships travel more slowly to save fuel and will be equipped with fuel-saving hulls designed to minimize water-friction. Some may burn biofuel made from algae; some may burn blue fuel made from surplus green electricity. Some may be pulled by kites, high above the ships. (See #49.) To the extent that new fuels may cost more, some long-distance goods may become expensive luxuries, encouraging more manufacturing to be done at home where transport costs are lower.

For a long time it seemed that flying might not be possible in a climate-friendly world, but if the technical problems of using biofuels at altitude can be overcome, as seems likely, the airplanes of the future may be powered with algae-based biofuels grown on marginal land where they do not compete with food. (See #50.)

In the 1950s, we believed that highways would pave our way to paradise — usually in the suburbs. Instead, they paved *over* paradise and gave us commuting gridlock. Transport in the climate-friendly future will be far more integrated, giving people access to many travel choices on a single smartcard — and a zero-carbon footprint.

Those Pesky Energy Data

A child born in a wealthy country is likely to consume, waste and pollute more in his lifetime than fifty children born in developing nations.

— George Carey,
Archbishop of Canterbury, UK

Before we move on, a short diversion into the world of numbers will help for what follows.

Energy is measured in a variety of ways, including joules, calories, Btus, watts and tonnes of oil equivalent. The watts, named after the Scottish inventor and engineer James Watt (1736–1819), are the easiest because they are organized as a family. One watt is the transfer of one joule of energy in one second. Here's the easy-to-remember version:

- Watt (Weeny Watt): 1 watt (one old-fashioned Xmas light bulb)

- Kilowatt (Kiddy Watt): 1,000 watts (KW)
- Megawatt (Mama Watt): 1000 kilowatts (MW)
- Gigawatt (Grandpa Watt): 1,000 megawatts (GW)
- Terawatt (Tyrannosaurus Watt): 1,000 gigawatts (TW)

Watts are a measure of capacity. Their cousins the Watt-hours are a measure of delivered energy. If a 500 MW coal-fired power plant produces power 23 hours a day, 365 days a year, it will produce 4.19 million megawatt-hours (MWh) a year. Wind turbines produce power 6–8 hours a day, and solar panels 5 hours a day, averaged over a year.

Napoleon introduced the metric system to eliminate the cheating that was rampant when every town in Europe had its own weights and measures. The metric system is easy, but the British and Americans didn't like Napoleon, so they stuck to the Imperial system, which measures energy in British thermal units (Btus). A Btu is the heat needed to raise the temperature of a pound of water by 1°F. So while the rest of the world measures large quantities of energy in exajoules (= joule x 10^{18}), Americans measure it in quadrillion Btus, or "quads." A joule is the energy generated when a force of one Newton moves an object by one meter. One kilowatt-hour is 3412 Btus. A quad is 293 terawatt-hours.

To complicate things further, oil is measured in barrels and tonnes (1 tonne of oil = 7 to 11 barrels, depending on the fuel), natural gas in cubic feet and coal in short tons. This makes it hard to

The Watt family.

DALE HITCHCOX

- Amazing "kilowatt-hours into anything" calculator: wattsonschools.com/calculator.htm
- Common conversion factors: tinyurl.com/ny58hc
- Energy A-Z: eia.doe.gov/a-z_index/Energya-z_a.html
- Energy Glossary: energy.ca.gov/glossary
- Energy Information Administration: eia.doe.gov
- International Energy Agency: iea.org
- Global and US electricity data: eia.doe.gov/fuelelectric.html
- World Energy Overview: eia.doe.gov/iea/overview.html
- World Energy Review (BP): bp.com/worldenergy

compare fuels, so the OECD describes energy in "Million tonnes of oil equivalent" (Mtoe). One tonne of oil equivalent is 41.868 gigajoules, which is the net heat content of 1 tonne of crude oil. 1 Mtoe = 4.1868 x 10^4 TJ = 10^7 Gcal = 3.968 x 10^7 MBtu = 11,630 GWh. Don't you love this stuff?

More Muddling Math

In 2006, the world used 469 quads of primary energy (137,534 TWh) — a 15% increase since 2000 and a 39% increase since 1980. The breakdown is 36% from oil, 27% from coal, 23% from natural gas, 6% from hydro, 6% from nuclear, and just 1% from solar, wind, biomass and geothermal power combined. That's what we have to change. The USA used 22% of the energy, Russia 7% and China 13%.

Americans use American gallons, but when Canadians talk about fuel-efficiency they assume British gallons, which are 20% larger, making Canadian cars seem more fuel-efficient than American cars. Hey, it all adds to the fun. A metric tonne (1,000 kilos) is also 10% larger than an American ton (2,000 pounds).

At least electricity is always measured in watts. In 2007, the world consumed 18,000 terawatt hours (TWh) of electricity, using 4 TW of installed capacity. Thirty-eight percent of the world's primary energy was used to generate electricity — 68% from fossil fuels, 9% from hydro, 19% from nuclear, and 2.6% from solar, wind, biomass and geothermal.

	Total CO_2 (million tonnes)	% of global CO_2	CO_2 tonnes per person[1]
Australia	417	1.4	20.6
USA	5900	20	19.8
Canada	614	2	18.8
Holland	260	0.9	15.8
Saudi Arabia	424	1.5	15.7
Taiwan	300	1.0	13.2
Russia	1704	5.8	12.0
Denmark	59	0.2	10.8
South Korea	515	1.8	10.5
Germany	857	2.9	10.4
Japan	1247	3.4	9.8
UK	586	2	9.6
Europe	4720	16	8.0
France	417	1.4	6.6
Sweden	57	0.2	6.4
China	6017	20.6	4.6
Mexico	435	1.5	4.0
Brazil	377	1.3	2.0
India	1293	4.4	1.2
Bangladesh	43	0.15	0.3
World	29,195	100	4.5

Earth's Future Electricity

The world's deserts can supply energy for any presently conceivable demand by humankind.
— Dr. Gerhard Knies

In 2007 the world used 18,000 terawatt hours (TWh) of electricity, 68% of which came from burning fossil fuels. Until the 2008 financial crash, we were using 3% more electricity every year, causing predictions that we would need 35,000 TWh a year by 2050.[1] Can we craft a world in which all our electricity comes from clean, sustainable power? The answer is a very solid *yes*, holding out the prospect of a world that can meet all its needs with renewable power, not just for the coming century but forever.

Efficiency

It starts with efficiency. If every inefficient light, appliance and electric motor were phased out by

The Andasol-1 solar thermal power plant in Grenada, Spain. The plant benefits from Spain's feed-in tariff, receiving just under €0.27/kW/h for the next 25 years.

global treaty; if we were all persuaded to upgrade our homes and buildings; and if the world's cities, businesses and industries were all to upgrade their lights and equipment, supported by the very best policies, we could reduce our need for power by 33%. (See #52 and #66.)

When we switch our cars and light trucks to battery electric drive, however, and add electric trains, buses, trucks and heat-pumps for our buildings, this will add 30%, pushing our power needs again toward 35,000 TWh a year — so let's make that our target.

Solar Power

Enough sunshine falls on the Earth's land to produce 15 million TWh of electricity a year, so a mere 1/400th of this would be sufficient to meet all our needs. The price of solar PV (the photovoltaic cells that generate electricity) is falling steadily and may hit the holy grail of price parity by 2015, at $2 per watt. By the end of 2008, the world had installed 15,200 MW of grid-connected solar PV, which was producing 20,000 GWh of electricity a year, 35% of it in Germany.[2] If the solar industry can sustain its 40% annual growth rate, encouraged by the best policies and incentives, it could produce 26,000 TWh a year by 2030, which is 75% of the power we need.

Concentrating Solar Power

Concentrating solar uses the sun to heat water and run steam turbines, and the heat can be stored in rock or liquid molten salt, enabling it to provide power to the grid for up to 17 hours a day. The world's 33 million square kilometers of desert could generate 80 million TWh of electricity a year, if we ever wanted that much. A plant can generate 0.25 TWh a year from each square kilometer, so for 35,000 TWh we'd need 140,000 square kilometers, which is 1.4% of Algeria, or 5% of Texas.[3]

Wind Power

The world could generate 410,000 TWh from land-based wind turbines (12 times more than we need) and a lot more from offshore wind farms. Wind power has been growing by a steady 25–30% a year, and by the end of 2008 the world had installed 120,100 MW of capacity, producing 260 TWh of electricity, with a 25% capacity factor. If the industry can sustain a 25% growth rate, it will be producing 35,000 TWh a year by 2030, which is 100% of the power we need.[4]

Geothermal Power

In 2008 we generated 64 TWh from geothermal projects around the world, drawing on underground heat. Using conventional technology, the potential has been put at 11,000 TWh, or 22,400 TWh with a mix of conventional and binary technology, a heat-conducting fluid that extracts heat at lower temperatures.[5] This would provide 64% of the power we need.

In 2006, however, a team at the Massachusetts Institute of Technology looked at the potential of enhanced geothermal energy, drilling 6 to 10 km (4 to 6 miles) into granite. If $1 billion were to be invested in research over 15 years, they estimated that the US could develop 100 GW of capacity by 2050, producing 584 TWh of power. They also

concluded that the extractable potential of enhanced geothermal power was 55.5 million TWh a year, providing enough power for 1,500 years, and that with technology improvements this could increase tenfold.[6]

Renewable energy can clearly produce the electricity we need — and there is also the potential for tidal and wave energy, run-of-river hydropower, biomass and biogas power, plus the existing hydroelectric dams, which produced 8,000 TWh in 2007. To make the most of this we need a smart grid, good energy storage systems (see #58), and a supergrid to bring the new power to the cities (#70). With the best policies, this is all possible (#69).

The question that matters is "How fast?" Al Gore's Repower America campaign, with two million supporters, wants the US to achieve a 100% transition to renewables by 2018, because of the urgency of global warming (see #69); many other groups are calling for equally bold goals (#80).

We achieved far greater goals than this in World War II, when American industry re-organized its factories to manufacture 300,000 airplanes. We did it then because of the threat, and we can do it now.

- Canadian Renewable Energy Alliance: canrea.ca
- International Renewable Energy Agency: irena.org
- Renewable Energy World: renewableenergyworld.com
- Renewables 2007 Global Status Report: ren21.net/globalstatusreport
- Repower America: repoweramerica.org

Earth's Future Water Power

> Ideology trumps rationality. Most conservatives cannot abide the solutions to global warming - strong government regulations and a government-led effort to accelerate clean energy technologies in the market.
>
> — Joe Romm.

Humans have been using water to generate energy for millennia. As far back as 31 AD, Tu Shih, Prefect of Nanyang in China, invented a system of water-powered bellows to generate the heat needed to cast iron agricultural implements.[1] In 1086, Britain's Domesday Book listed more than 5,000 water mills being used in agriculture.

In 2007, the world's 48,000 dams produced 16% of the world's electricity, mostly in China, Canada, Brazil, the US, Russia, Norway, and India. China and India are eager to build more, but there is huge concern about the social and ecological costs, as homes, farmlands and ancestral villages are flooded. China's controversial 22.5 GW Three Gorges Dam on the Yangtze River will pre-empt the production of 95 million tonnes of CO_2 a year, but it has also forced 1.2 million people to leave their homes. Globally, technically feasible hydropower potential is 14,000 TWh, representing 40% of our future electricity needs, but most of this will hopefully not be used.[2]

Hydropower does not always have a low carbon footprint. In 2000, the World Commission on Dams reported that organic matter washed into reservoirs breaks down as methane, which is also released by rotting vegetation in the bottom of shallow reservoirs. Globally, the methane from dams could be producing 800 million tonnes of CO_2e a year. In Brazil, the 250 MW Balbina reservoir, only four meters deep but covering 4,400 sq. km. of once tropical forest, will produce 3.3 million tonnes of CO_2e a year over its first 20 years, almost twice as much a 250 MW coal-fired power plant.[3] Different reservoirs produce different quantities of methane, and the reservoirs studied in Brazil varied by a factor of 500. In response, Brazil's National Space Research Institute is researching a technique to extract the methane as a source of power.[4]

Green power can also be generated from small hydro, using run-of-the-river technology to create an impoundment, divert some of the water down an elevation drop, and run it through a Pelton wheel before returning it to the river.

Tidal and Wave Energy

The world's tides allow us to harvest lunar energy, caused by the gravitational pull of the moon. Tidal energy has been gathered since the Middle Ages, when millers used to catch the tide in millponds to drive their water-wheels.

At St. Malo in northern France, a 240 MW tidal barrage at the mouth of the Rance river has been generating power since 1966, and there is a 23 MW project in the Bay of Fundy, Nova Scotia,

PELAMIS WAVE POWER

The Pelamis wave energy converter. The 2.25 MW Aguçadoura wave farm, 5km off the Atlantic coast of Portugal, is supported by a 32 cents/kWh feed-in tariff.

- Canadian Hydropower Association: canhydropower.org
- IEA Hydropower: ieahydro.org
- International Hydropower Association: hydropower.org
- International Network on Small Hydro Power: inshp.org
- Low Impact Hydropower Institute: lowimpacthydro.org
- Microhydro web portal: microhydropower.net
- World Commission on Dams: dams.org

- Aquamarine Power: aquamarinepower.com
- European Marine Energy Centre: emec.org.uk
- European Ocean Energy Association: eu-oea.com
- Hammarfest Strøm: tidevannsenergi.com
- Marine Current Turbines: marineturbines.com
- Ocean Energy Council: oceanenergycouncil.com
- Ocean Renewable Energy Coalition: oceanrenewable.com
- Ocean Renewable Energy Group: oreg.ca
- Open Hydro: openhydro.com
- Pelamis Wave Power: pelamiswave.com
- Puget Sound Tidal Energy: pstidalenergy.org
- Sea Raser: dartmouthwaveenergy.com
- Sea Solar Power International: seasolarpower.com
- Severn Estuary Tidal: severnestuary.net/sep/resource.html
- Tidal Electric: tidalelectric.com
- Tidal Stream: tidalstream.co.uk
- Verdant Power: verdantpower.com
- Wave Energy Centre: wavec.org
- Wave energy in Denmark: waveenergy.dk
- Wavegen: wavegen.co.uk
- World Energy Council: tinyurl.com/agzb3q

where the tidal flux is so high that it has 30,000 MW of potential capacity. Estuary impoundment disrupts estuarine ecology, however, which is why Britain's proposed 8.6 GW tidal barrage on the Severn River is generating heated discussion.

Energy from tidal flow raises far fewer concerns. Europe's Marine Energy Centre in the Orkney Islands, which provides multi-berth, subsea cabled ocean test sites, lists 55 tidal and 95 wave companies that are developing various technologies. In the world's largest project, off the coast of South Korea, 300 ocean-floor turbines will be powering 200,000 homes by 2015.

In 2007, The Carbon Trust estimated that marine energy could provide 20% of Britain's electricity needs. Portugal could meet 30% of its needs with wave energy. For the world as a whole, there may be a realizable potential of 1,000 TWh a year of tidal and 2,000 TWh of wave energy, representing 8.5% of the world's future electricity needs.

Wild Cards

- Ocean thermal energy, generated where the difference between the warm surface and the colder deep waters in the tropics is more than 20°C, using an evaporator to drive a turbine. otecnews.org
- Energy from slow oceans currents such as the Gulf Stream. vortexhydroenergy.com

Marine Current's 1.2 MW SeaGen tidal turbine installed in Strangford Lough, Northern Ireland.

- Salt power energy, released when salt and fresh water mix. wctsus.nl
- Hydrothermal power from ocean vents along the Ring of Fire. marshallsystem.com.

Renewable Energy – the Small Print

I think we have a very brief window of opportunity to deal with climate change… no longer than a decade, at the most.

— James hansen, NASA, 2006

What does wind energy cost? How long does it take solar PV to generate the energy needed to make it? How much energy can renewables provide? The table on the right provides answers to these important questions, and I am indebted to Professor Mark Jacobson of Stanford University for most of the data, except where otherwise endnoted. [1]

In 2008 the world's total global primary energy use for all purposes was about 145,000 terawatt hours (TWh), and electricity use was 18,500 TWh. The stunning efficiency of electric vehicles means that if we shift land transport to a predominantly electrical mode, we will need at most a 33% increase in electrical power, which is how much we lose through the inefficiency of our buildings, electrical motors and equipment.

Column 1 shows that if we take 35,000 TWh as our goal to allow for growing demand (see p. 47), wind offers 12 times more energy than we need, and solar PV offers 85 times more. The world's deserts could provide 2,285 times more power. So if we used just 1/20th of 1% of their area for solar thermal, we could have all the energy we needed. The numbers for hot rocks geothermal are equally huge, with the US potential being 2,000 to 20,000 greater than the demand.

Columns 2 asks how much CO_2 is produced per kWh for each technology, from manufacturing to deconstruction, and shows that wind produces 28 times less CO_2 than coal that is burned using carbon capture and storage technology (CCS).

Column 3 shows the length of time needed to plan, approve and construct each technology, with

	1 Abundance (TWh/year)	2 CO_2 (grams/kWh)	3 Construction time needed (years)	4 CO_2 Opportunity Cost (grams/kW)	5 Energy payback (months)	6 Cost in 2008 (cents/kWh)	7 Cost in 2020 (cents/kWh)	8 Growth rate (%/year)
Efficiency	48,000[3]	-600[4]	1	–600[5]	0	1–3[6]	2–4[7]	—[8]
Wind	410,000	9–15	2–5	9–15	6–9	7–8[9]	5–6	24%[10]
Solar PV	3,000,000	19–59	2–5	19–59	12–42	50–80[11]	10–20	40%[12]
Solar thermal	80,000,000[13]	9–11	2–5	9–11	5–7	13–17[14]	5–6[15]	75%[16]
Geothermal	Enormous[17]	15–55	3–6	16–61		5–7[18]	5–7	3%[19]
Tidal	200[20]	20–41	2–5	34–55	12	8–26[21]	?	—
Wave	2,000[22]	20–41	2–5	42–63	12	10–30[23]	?	—
Hydroelectric	16,500	17–22	8–16	48–71		7–12	7–12	3%[24]
Nuclear	4,000–122,000[25]	9–70	10–19	68–180		15–21[26]	Rising	0.4%[27]
Coal + CCS	11,000	255–440	20[28]	308–571[29]		15–23[30]	?	—

CCS coal being the slowest, as it will not be ready until 2030 (see p. 61), followed by nuclear, which can take up to 21 years before it produces any power (see p. 56). This is relevant, because each year's delay means another year of emissions from the conventional power sources — this is the "opportunity cost" of the delay.

Column 4 includes this data to arrive at a more accurate measure of CO_2 per kWh. Column 5 asks how much energy is needed to manufacture and deconstruct each technology (measured in the months of energy needed to produce it).

Column 5 addresses the important question of energy payback, "net energy", or "energy return on energy invested (EROIE), to address the belief that some forms of renewable energy take more energy to produce than they return. Not true. A 2 MW wind turbine will recover the energy needed to make the steel, ship it around and recycle it at the end of its life in 6-9 months. A solar panel will recover the energy invested in 12-42 months — but a lot of this is the aluminum frame, which direct solar shingles will eliminate. For nuclear power, some estimates suggest 60 months. For clean coal, it is impossible to obtain any such figure yet, since no-one has done it yet.

Columns 6 and 7 compare the cost, both now

Marine Current's 1.2 MW SeaGen tidal turbine installed in Strangford Lough, Northern Ireland.

and in the future. Energy efficiency, which is by far the cheapest way to make new power available, increases in cost as the low hanging fruit is taken. For wind and solar, the efficiencies of scale that come from mass production drive the price down. In the case of solar PV, the price looks set to fall significantly; for nuclear energy, by contrast, the price looks set to rise dramatically. An economic analysis by Mark Cooper at the Vermont Law School suggested that adding 100 new nuclear reactors to the US power grid would cost their customers between $1.9 and $4.1 trillion over the reactors' lifetime. Since the "nuclear renaissance" began, projected costs have risen four-fold.

Earth's future Bioenergy

> There is no science on how we are
> going to adapt to 4 degrees warm-
> ing. It is actually pretty alarming.
> — Prof. Neil Adger, Tyndall Centre
> for Climate Change Research

Bioenergy is stored solar energy, and our world has been gathering and recycling it ever since life began 3.8 billion years ago in the form of seaweed, fish, trees, plants, animals, manure, sewage and every kind of biological debris. Plants absorb the sun's energy through photosynthesis, and humans have been using it by burning wood since we learned how to make fire. Can we also gather it in other ways?

The small Austrian town of Güssing is already doing so. In their efforts to become energy independent using green energy, they are generating 95% of their heat, electricity and fuel from sawdust, maize, waste cooking oil and other means, in more than 30 projects. (See #27.)

Some forms of bioenergy are clearly not so smart. In Malaysia and Indonesia, where ancient rainforests have been cleared and ancient peat lands drained to grow palm oil to make biodiesel, their destruction causes so much loss of carbon that for every tonne of palm oil produced, 33 tonnes of CO_2 are released.[1]

Similarly, growing corn to make ethanol, which is happening on almost every acre in Iowa, produces a very low net reduction of CO_2,[2] and if we used the world's entire grain supply it would only meet 15% of our oil needs.[3] Growing corn ethanol to power all US vehicles would require 90–161 million hectares of land (10–18% of the US land area); doing so with cellulosic ethanol made from woody crops grown on marginal land would require 43–324 million hectares (5–35% of the US land-area).[4]

If those same vehicles were electric, however, powered by the wind, the land needed for the turbine footprints would be less than 300 hectares (three square kilometers), and the land needed for the full spacing of the turbines would be only 3 to 6 million hectares (0.35–0.7% of the US land area), while still being available for crops and wildlife.[5]

Brazilian ethanol made from sugar cane seems to make more sense, producing 6,000 liters of ethanol per hectare. Brazil produced 19 billion liters in 2007, enough to power 38 million efficient cars traveling 10,000 km a year. Close up,

The LEED Platinum-targeted Phase II 'Balance' buildings at Dockside Green, Victoria, Canada, where 2,500 people will live within a sustainable community with heat generated from bioenergy from the gasification of local wood wastes.

Biofuel Source	Liters per hectare[10]	Efficient car kilometers per hectare[11]
Algae	33,000	660,000
Sugar cane	4,000–6,000	80,000–120,000
Biomass to liquid	5,000	100,000
Palm oil	3,750–4,700	75,000–94,000
Corn	3,000	60,000
Sugar beet	2,000–5,000	40,000–100,000
Jatropha	2,000	40,000
Canola (rapeseed)	1,200	24,000
Wheat	1,000–2,500	20,000–50,000

however, the process is distressing. Working 12 hours a day, six days a week, the average cane cutters last only 12 years on the job before "they are so worn out that they have to be replaced."[6] This is a situation that biofuel importers could change if they demanded that all biofuels be certified, including a requirement that the cane workers be allowed to join a union and have guaranteed improved working conditions and a far better wage.

Given these complexities, what makes sense for Earth's future bioenergy? (See also #77.)

- Biogas, electricity, heat and biofuel from sewage, manure, food wastes, farm wastes and municipal wastes
- Biodiesel from farmed algae, for the airline, trucking and shipping industries, which could be by far the most land-efficient way to generate biofuel, if the technology can be proven on a large scale. (See #50 & #77.)
- Biodiesel from the oily seeds of shrubs like jatropha. In India, the government has identified 400,000 square kilometers of land where jatropha can be grown, and aims to replace 20% of India's diesel consumption by 2011. The Indian State Railway has planted 7.5 million jatropha plants along the rail lines, using the oil in its diesel-powered railways. In North America, jatropha can be grown in warm climates such as Texas, Florida and Mexico.
- Biodiesel made from crop trees, forest thinnings and wood wastes, transforming the wood into a synthetic gas and then into a liquid fuel, a technique being developed by Choren in Freiberg, Germany.[7]

- Cellulosic ethanol made from crops such as switchgrass and trees such as poplar, willow and black cottonwood that can be grown on marginal land, using sustainability criteria to protect biodiversity.
- Biofuel from native perennial prairie plants grown on degraded soil, which also store enormous quantities of carbon in their roots.[8] (See #41 and #43.)
- Biofuel from marine algae and seaweeds, which have minimal land-based impacts.
- "Mycodiesel," using Patagonian tree fungus (*Gliocladium roseum*) to break down woody biomass. The fungus has a chemical structure remarkably similar to fossil fuels, enabling the resulting biofuel to be put almost directly into the tank.[9]
- Carbon negative biofuels, produced by making bioethanol and then burning it to produce electricity for cars, capturing and sequestrating the CO_2 underground as is being proposed for "clean coal." If this happened, every kilometer driven would reduce the atmosphere's burden of carbon.

Earth's future Wastes

The human response it calls for is truly heroic, requiring nothing short of rewiring the entire planet with a new generation of clean-energy technologies — and doing that very soon.... Are we, as a species, capable of that kind of deliberate global response?
— **Eban Goodstein, author of** *Fighting for Love in the Century of Extinction — How Passion and Politic can Stop Global Warming*

In nature, there is no such thing as "waste." Everything is recycled in a constant exchange of form, and it had been this way forever until we started creating products that did not break down. The average carpet, made from Earth's resources, will enjoy up to ten years of human use and then spend 20,000 years in a landfill before it finally biodegrades and returns to the earth.

In 2005 the US produced 246 million tons of municipal waste, a quarter of the world's total of a billion tons.[1] Of that, 32% was recycled or composted, 14% was incinerated, and 54% ended up in a landfill where it produces methane, which, during its short 12-year life, warms our planet's atmosphere more than 100 times more effectively than CO_2.

Recycling reduces CO_2 emissions

- Making recycled steel requires 39% less energy than making virgin steel. 98% of automobile steel and 58% of the steel in tin cans is recycled.

- Making recycled paper requires 26–45% less energy than making virgin paper; 40% of paper and paperboard is recycled. For every tonne of paper that's recycled, 15 trees can remain standing, gathering carbon.

- Making recycled aluminum requires only 5% of the energy required to make virgin aluminum. Recycled aluminum accounts for 52% of aluminum packaging and 64% of aluminum cans.

- Making recycled plastic requires 25–43% less energy than making virgin plastic, as well as releasing far fewer toxic chemicals. Only 5% of plastics are recycled.

Source: Natural Resources Defense Council

Landfills produce 12% of the world's methane emissions, and because methane causes 13% of the overall radiative forcing, our collective garbage is causing 1.5% of the overall problem.

The materials we call "waste" also contain embodied energy, and whenever we don't recycle the steel, aluminum, paper or plastic, we have to mine it or make it afresh to make something new.

In a future world where we have learned to live sustainably we will have realized the goal of "zero waste" and moved beyond it, because the whole concept of "waste" will have vanished. All organic wastes will either be composted into topsoil or converted into biofuel, and all mineral wastes will be recycled as minerals.

Incineration is not the answer — this reduces the energy to its lowest form (heat), while releasing chemicals such as dioxins into the air, which are known to cause cancer.

Our sewage will be seen as a source of natural wealth rather than waste. Sewage contains ten times the energy needed to treat it,[2] and can be a source of heat, electricity, biodiesel, biogas, minerals and potentially hydrogen.[3] In Zurich, buildings obtain their heat directly from the city's sewer pipes. In Stockholm, 80,000 apartments are heated with biogas from the city's sewage works. In Kristianstad, Sweden, cars and buses run on biogas generated from 100,000 tonnes a year of sewage mixed with solid organic wastes. In King County, Washington, 1000 homes are powered by fuel cells that obtain their hydrogen from sewage. In Kelowna, British Columbia, Okanagan College uses heat pumps to extract the heat it needs from the Kelowna Wastewater Treatment Plant — a

STEPHEN SALTER

technology that is also being used in the Olympic ski resort of Whistler and Vancouver's Southeast False Creek housing project.

In Victoria, British Columbia, which is preparing to treat its sewage instead of screening it and dumping it in the ocean, engineer Stephen Salter has calculated that the city's sewage contains enough energy to provide pure biodiesel for 200 buses and 5,000 cars, heat for 3,500 homes, or electricity for 2,500 homes.[4] In the same city, the LEED[5] platinum Dockside Green development obtains its heat and hot water from a biomass gasification plant run by Nexterra, using wood wastes that would otherwise have gone to the landfill. The same happens in London's Beddington Zero Energy Development (BedZED).

There is waste in food, too. In 2005 US landfills received 26 million tonnes of food wastes, which, if thermally processed in biorefineries, could produce 50 million barrels of oil a year. As a fraction of the 20 million barrels that Americans use each day, it is only 2.5 days' supply, but if our use of oil was reduced by 90% thanks to bikes,

Hammarby Sjöstad, in Stockholm, Sweden, where 11,000 apartments will be heated with recovered heat from the local sewage treatment plant and the combustion of pre-sorted wastes.

- Beddington Zero Energy Development (BedZED): bioregional.com
- Cradle to Cradle Design: mbdc.com/c2c_home.htm
- Dockside Green: docksidegreen.ca
- Fossil-Fuel Free Kristianstad, Sweden (video): tinyurl.com/dcwy3n
- Inspiration from Sweden: georgiastrait.org/?q=node/359
- Municipal Solid Waste in the US, 2005: tinyurl.com/3daont
- Nexterra: nexterra.ca
- Resources from Waste (Sweden): tinyurl.com/cmmjpg

electric vehicles, ridesharing, public transit and high-speed trains, that becomes 25 days' supply, and a useful contribution. As architect Bill McDonough likes to say, "waste equals food."

Nuclear — Hope or Hype?

Nuclear power has died of an incurable attack of market forces and is way beyond any hope of revival, because the competitors are several-fold cheaper and are getting rapidly more so.

— Amory Lovins

Faced with the climate emergency, some people look to nuclear power as a clean, safe, cost-effective solution. Unfortunately, it is not.

It makes no economic sense

The nuclear industry argues that nuclear power is cheaper than coal-fired power or wind power — but an assessment in Ontario based on the actual performance of existing reactors and the required return of capital, suggested that the realistic lifetime price of new nuclear power would be 20 cents/kWh, more than two and a half times the price of wind or microhydro.[1]

The nuclear industry wants to build 1,000 new plants — but nuclear plants almost always come in over budget and over time, and we need solutions *now*. The average construction time is 15 years (8 years in France); and the last US plant took 23 years to complete. Ontario's five nuclear plants had cost overruns ranging from 40% to 270%. The Olkiluoto nuclear reactor in Finland, showcase of a nuclear renaissance, is costing twice its estimate and taking far longer to build than promised.

There are also big hidden costs the taxpayer has to cover, such as the cost of handling a nuclear disaster, which could cost $600 billion in damages and claims. The nuclear industry's liability is limited to $9.1 billion in the US, $700 million in Europe, and just $75 million in Canada, with the taxpayer picking up the rest. If the nuclear industry had to buy insurance for all its liability costs, it would never get financed.

There's also the problem of radioactive wastes, which have to be stored for up to 250,000 years, twice as long as since modern humans left Africa. In Canada, taxpayers are on the hook for $24 billion to cover just the first 300 years.[2] At the end of a reactor's life, it has to be decommissioned, at $325 million per reactor.[3] In Britain the total estimated cost is £70 billion, 70% of which will be covered by the taxpayer.

As a result of these realities, the private sector won't touch nuclear power unless there is firm government support. In the US since 1948, the nuclear industry has received $74 billion in subsidies, with an additional $13 billion in 2007 and $50 billion in loan guarantees, and an average subsidy of $13 billion per new nuclear plant — roughly its entire cost.[4] In Canada it has received $20 billion in subsidies since 1952.

If we are going to use public money, we should do so intelligently. In 2008, Architecture 2030 reported that compared to nuclear energy, a $21.6 billion investment to increase building efficiency would produce three times more CO_2 reductions, create 216,000 new jobs, and produce new energy for a fifth of the cost of clean coal or nuclear power.[5] Because such options exist, public investments in nuclear power, as opposed to building efficiency, will actually slow down the path to a climate-friendly world.[6]

- Millions Against Nuclear: million-against-nuclear.net
- Nuclear Power: Climate Fix or Folly?: inyurl.com/bdqs5y
- Nuclear Power: The Energy Balance: stormsmith.nl
- World Information Service on Nuclear: 0.antenna.nl/wise

Nuclear power may seem like a good low-carbon option, but in reality it's a very expensive distraction.

© Shevs Dreamstime.com

The world has a limited supply of uranium

The world's 440 nuclear reactors, with a combined capacity of 363 GW, use 67,000 tonnes of uranium a year, averaging 146 tonnes per reactor. At this rate, the world's uranium reserves of about 4.5 million tonnes will last for 70 years. If we build 1,000 new nuclear plants with a combined capacity of 1500 GW, as widely proposed, they will need an additional 277,000 tonnes a year. By 2025, when the new reactors might begin to operate, reserves will have fallen to 3.5 million tonnes, and the demand will now be 344,000 tonnes a year.[7] At this rate, the uranium will be exhausted in 10 years. Some propose that we build fast breeder reactors fueled by their own fissile wastes, but these have been a technical and economic failure. Others argue that higher uranium prices will cause new reserves to be found — but even if reserves doubled, they would be exhausted in 20 years.

More problems

Nuclear power plants need cooling, but in the summer of 2003 France had to close a quarter of its 58 plants because the river water used to cool them was too warm — because of global warming. If we locate them by the sea, they will be vulnerable to sea-level rise. They are also vulnerable to earthquakes, as the 2007 earthquake in northwestern Japan demonstrated

There is also the concern that nuclear technology allows nuclear weapons proliferation, and there is the risk of the use of stolen plutonium to make dirty bombs or that terrorists might fly a hijacked plane into a nuclear reactor.

Nuclear power also poses grave health risks. Uranium mining and nuclear power plants contribute to greater rates of breast cancer, lung cancer and childhood leukemia. Following the Chernobyl nuclear accident, there has been a 90-fold increase in thyroid cancer and thousands of people have died.[8]

Nuclear power produces less CO_2 than coal-fired power but 24 times more than wind[9] — so why take such risks when there are far cheaper, safer ways to generate the energy we need? Those who promote nuclear power as a solution to climate change have simply not done their homework.

Hydrogen — Hope Or Hype?

Fuel-cell cars are expected on about the same schedule as NASA's manned trip to Mars, and have about the same level of likelihood.

— Matthew Wald,
Scientific American, May 2004

In the beginning, there was the Big Bang, and then there was hydrogen, the most abundant element in the Universe. Like nuclear energy and biofuel, hydrogen seems to offer such promise.

Hydrogen exists everywhere — in plant matter, natural gas, water, sewage — and when separated from its partner atoms it can be burned as fuel or used in a fuel cell to produce electricity. It sounds good, but we need to expose it to a good dose of cold water and see what's left.

First, hydrogen is not a source of energy, like solar or wind. It is an energy *carrier.* You have to use energy to make it — and if you use electricity to split water (H_2O), you will need 1.3 kWh of electricity to get 1 kWh of hydrogen. Its only value is that you can store it, which is why some people (but increasingly few) hope it could replace oil in transport.

In reality, 95% of today's hydrogen is derived from natural gas (CH_4), and when the hydrogen and carbon separate, the carbon immediately forms carbon dioxide, the very problem we want to avoid. When such hydrogen is used in a fuel cell car, replacing oil, CO_2 emissions are hardly reduced at all. In 2005 the US annual US production of about 10 Mt of hydrogen used 5% of the US's natural gas and released 100 million tonnes of CO_2.[1] Much of North America's future gas will be shipped in as liquefied natural gas from countries such as Russia and Iran — those paragons of democracy.

The second option is the one that has many people excited: separating hydrogen from water

The first combined hydrogen and gasoline station in North America, in Washington D.C., a joint project of Shell Hydrogen and GM.

SHELL HYDROGEN

- California Hydrogen Highway: hydrogenhighway.ca.gov
- Hydrogen Cars: hydrogencarsnow.com
- Hydrogen Now: hydrogennow.org
- *The Hype about Hydrogen,* by Joseph Romm: cool-companies.org

(H_2O). Back in 1870, in his novel *The Mysterious Island,* Jules Verne wrote:

> Yes, my friends, I believe that water will one day be employed as fuel, that hydrogen and oxygen which constitute it, used singly or together, will furnish an inexhaustible source of heat and light, of an intensity of which coal is not capable.... When the deposits of coal are exhausted we shall heat and warm ourselves with water. Water will be the coal of the future.

The hope of the solar hydrogen economy is that energy from the sun, wind and waves can be used to split water, generating hydrogen to power our cars, trucks, ships and planes. It is technically possible, and there are prototype hydrogen cars on the roads, as well as buses, vans, trucks and forklifts. The cars are rather expensive — Intergalactic Hydrogen will sell you a converted hydrogen Hummer for $60,000 plus the cost of the Hummer, and the Hydrogen Car Company will sell you a hydrogen Shelby Cobra sports car for $150,000 — but the price would come down if mass production ever happened.

The problem is fourfold. The first is the cost of the cars. The second involves finding a leak-proof system for storing hydrogen. The third is the expense of delivery and the lack of a refueling system. In 2008 California had 26 hydrogen fueling stations for its 250 hydrogen vehicles, but a continent-wide refueling network would cost billions and need millions of hydrogen cars to justify the investment. In the hydrogen economy's Catch-22, hydrogen cars will not be mass-produced until a refueling network is in place, and vice-versa, which is one reason why Jim Woolsey, former Director of the CIA, said, "Forget hydrogen, forget hydrogen, forget hydrogen." Right now, most hydrogen is delivered in diesel trucks.

The final problem concerns energy loss. A kilowatt hour of electricity from wind or solar will power an electric vehicle for five miles, but if you use the electricity to make hydrogen for a fuel cell, 70% of the electricity will be lost in the conversion, and the car will not travel much more than one mile.[2] Why put perfectly good renewable electricity in a hydrogen car when the same power in an electric or plug-in hybrid EV that costs four times less will send it four times further? The sequel to *Who Killed the Electric Car?* will be *Who Killed the Hydrogen Car?* The answer will be the electric car.

Should we stop all investment in hydrogen research? Probably not. Because of its portability, hydrogen may have a role in shipping — Iceland is planning to fuel its fishing fleet with hydrogen made with geothermal electricity, replacing imported oil. Hydrogen can also be used to store electricity. On the island of Utsira, off Norway, Norsk Hydro has connected two 600-kW wind turbines to a hydrogen generator and fuel cells, which provide firm power 24 hours a day to ten of the island's households as an alternative to diesel. Researchers are also exploring ways to develop hydrogen directly from algae and sugary carbohydrates.

Overall, however, the hope has turned to hype, and most of the hype has died.

Clean Coal — Hope Or Hype?

In the battle against global warming, carbon capture and sequestration (CCS) will arrive on the battlefield far too late to help the world avoid dangerous climate change.

— United Nations Development Program, 2007

Coal is an amazing legacy from Earth's past, which we have used to transform the planet. Before we started using fossil fuels, we were limited by the supply of firewood, and the tiny amount of power we could squeeze out of wooden waterwheels and windmills. If they had not discovered coal, Europeans would probably have cut down the remaining forests and then reverted to tribal warfare.

It was coal that made the Industrial Revolution possible, and it is still an enormous presence on our planet. We mine and burn 5.5 billion tonnes a year, causing 33% of the world's CO_2 emissions, and there are plans for many more traditional coal-fired power plants in the US, China, India and elsewhere. In 2006 China built 110 GW of new coal-fired power plants — almost two plants a week. When it comes to the climate, coal-fired power is a global disaster.

Mining coal is enormously destructive of the land, especially when mountaintops are blasted to pieces. In the Appalachians of West Virginia an estimated 300,000 to 405,000 hectares of hardwood forest, 1,500 km of waterways, and more than 470 mountains — and their surrounding communities — have been erased from existence since 1990.[2] It is also a major cause of death, both for the miners who die from black lung disease and for the thousands who die prematurely each year because of air pollution from coal-fired power plants.[3]

Share of electricity from coal[1]	
Poland	93%
South Africa	93%
Australia	80%
China	78%
Israel	71%
USA	50%
Germany	47%
Canada	17%
Rest of World	39%

As a result of coal mining, coal seam fires are also burning in China, India, USA and other countries. Most fires in the developed world are under control, but not in the developing world — in China they consume up to 200 million tonnes of coal a year. Globally, the fires may be causing 2–3% of the world's CO_2 emissions, and an urgent international initiative is needed to tackle them.[4]

The coal industry has a solution, however: "clean coal." In Norway, each year Statoil injects a million tonnes a year of unwanted CO_2 that contaminates its natural gas into the Sleipner field under the North Sea, because it's cheaper than paying Norway's carbon tax. Using the same carbon capture and sequestration (CCS) technology, the idea is that future coal-fired power plants would capture their CO_2 and bury it underground in depleted oil and gas fields. The idea has won praise and money from politicians of every stripe, and the American Coalition for Clean Coal Electricity is investing $3.5 billion in CCS projects. When you look more closely, however, CCS loses its shine:

Firstly, CCS requires 10–40% more coal to be burned per kWh of electricity, so more coal has to be mined and burned.[5]

Secondly, because carbon capture itself requires energy, and there are no reductions in coal mining's fugitive methane emissions or coal's transport-related CO_2, only 60–85% of the CO_2e can be captured by CCS.[6] The resulting electricity will produce 225–442 grams of CO_2 per kWh, compared to wind (9–15g) and solar thermal (9–11g).[7]

Each railway hopper carrying 108 tonnes of coal will produce 198 tonnes of CO_2 when the coal is burnt. In 2007, the world consumed 5.5 billion tonnes of coal, producing 12 billion tonnes of CO_2 — a third of the world's CO_2 emissions.

Thirdly, CCS will cost up to 75% more than conventional coal-fired power, requiring a carbon price of $30 a tonne to make it competitive. The coal industry has been fighting any suggestion of carbon pricing because it will make conventional coal uncompetitive.

Fourthly, once buried, the CO_2 may leak out at 0.1–1% a year, raising difficult questions around ownership, liability and permits. If a rush of stored CO_2 escaped in an earthquake, it would kill everyone living nearby.

Fifthly, there's the sheer scale of what's needed. The IEA has estimated that for a 60% reduction in coal's CO_2 emissions, 6,000 Sleipner-sized projects would be needed, along with millions of miles of pipeline to carry the CO_2 to the underground storage sites.

The sixth concern is that CCS will not be technically feasible at a utility scale until 2030,[8] by which time solar and solar thermal power will be cheaper (wind already is), and not commercially viable until after 2050, so the market may simply take a pass, as it is doing for nuclear energy.

Given these realities, CCS is a very expensive distraction, designed to make coal look good, but in reality taking scarce taxpayers' dollars away from solutions that can make a real difference.[9]

What should we do? There is one good hope for CCS technology, which is that by burning biomass instead of coal for power and applying CCS, we could create carbon-negative electricity that would actually suck CO_2 out of the atmosphere. For this reason alone, biomass CCS research should continue.

- American Coalition for Clean Coal Electricity: cleancoalusa.org
- Carbon Capture and Storage: co2capture.org.uk
- CCS Technologies: sequestration.mit.edu
- Coal Fires, by Anupma Prakash: gi.alaska.edu/~prakash/coalfires
- Coal Fires: itc.nl/personal/coalfire
- *False Hope:* greenpeace.org/CCSreport
- Just Transition Alliance: jtalliance.org
- The Future of Coal (MIT): web.mit.edu/coal
- World Coal Institute: worldcoal.org

Governments should place an immediate ban on new non-CCS coal-fired projects, as New Zealand and British Columbia have done, and set a date by which power companies will be required to stop buying coal-fired power, as California has done. We should also begin to phase out all coal mining, compensating the investors and helping the workers to train for new jobs. (See #71.)

Earth's Future Farms

A lot of people go straight from denial to despair without pausing in the middle and doing something about it.

— Al Gore

Deep down, most people want to believe that farming is a peaceful, bucolic activity in which humans, their animals and the land work together in ecological harmony. This is true on some farms, but sadly, on most farms it is not.

The impact of farming on the global climate is particularly harmful. The United Nations report *Livestock's Long Shadow* found that the global livestock industry produces 14–18% of the world's greenhouse gas emissions — more than all the world's emissions from driving, shipping and flying.[1] This is less surprising when you consider that 70% of the world's farm and pastureland is used to raise animals. A meat-eater on a bicycle is responsible for more emissions than a vegan in a small car — unless it is local, organic, pasture-raised meat, which has less impact.

Farming as a whole may be responsible for as much as a third of global warming, because of the accumulation of greenhouse gases.[2] (See box.) The greatest impacts come from the clearance of rainforests to raise cows and grow soybeans to feed to cows; the huge quantities of methane from the ruminant belching of 3.5 billion cow, sheep and goats; the nitrous oxide that results from the use of nitrogen fertilizers and from animal manure; and all the oil and gas that is needed to run the industrial food economy.

Whether farming is causing 25% or 33% of global warming, farming in a climate-friendly world will look and feel very different and will follow principles very different from today's money-driven industrial operations.

Firstly, the farms of the future will be organic, restoring the rural tranquility many people love so much. They will declare ecological peace, no longer using pesticides to make war on the birds and insects, and no longer using fertilizers to artificially

Farming's Heavy Footprint on the Global Climate	MT of CO_2e/yr
CO_2 from the use of natural gas to make nitrogen fertilizer	410[3]
Nitrous oxide from the poor application of nitrogen fertilizers	2128
Nitrous oxide from animal manure	506[4]
Methane from livestock burping	2150[5]
Methane from poor manure management	450[6]
Methane from rice production	616
CO_2 from on-farm energy use	527
CO_2 from food processing, transport, packaging and sales	1500[7]
CO_2 from the clearance of rainforests for cattle and cattle feed	2400
Methane from food wastes rotting in landfills	150[8]
CO_2 from the loss of soil carbon because of poor farming and grassland practices	1100[9]
Total	12,000 = 32% of 36.6 Gt

- Cool Farming: greenpeace.org/international/ press/reports/cool-farming
- Institute for Science in Society: i-sis.org.uk
- *Livestock's Long Shadow:* tinyurl.com/7yzdoy

Bringing the harvest home in British Columbia's Similkameen Valley, home to many organic farms.

drug the soil. In doing so, they will increase the nutritional value of their food, increase the drought-resistance of their crops and reduce their energy needs, while suffering no loss in productivity. A comprehensive review of 293 studies worldwide found that there was no reduction in yields on organic farms in the developed world, while yields in the developing world increased up to four-fold.[10] (See #42.)

Secondly, they will recycle all their wastes — eliminating their methane and nitrous oxide emissions — and produce all their own energy with no need for fossil fuels, primarily by using biogas digesters to turn animal and crop wastes into biogas to generate heat, fuel and electricity. A study in 2007 by the German Green Party found that Germany could produce more biogas by 2020 than all of Europe's imports of natural gas from Russia. They will also use solar and wind energy, solar greenhouses and other renewable energies, and some farmers will use their marginal lands to grow fuel crops under global conditions of sustainability. (See #44.)

Thirdly, they will make a conscious effort to farm carbon. They will draw it down from the atmosphere and store it in the soil by adopting holistic approaches to soil management; by changing the way they graze their animals to mimic the way herds behaved before their predators were hunted to extinction (see #43); and by burning their crop-wastes at very low-oxygen levels to turn them into charcoal and burying it as biochar, locking its carbon away for centuries. Globally, the world's farms and pasturelands could store 6.5 Gt of carbon a year if they changed their soil management practices. (See #41.)

Fourthly, they will raise fewer animals, and not in intensive feedlots but on organic pastures rich in native grasses and medicinal herbs. The need to feed them a continuous flow of drugs and antibiotics will cease because of the return of natural conditions and a grass-fed diet, and their methane emissions will fall because of the more traditional diet. Their changed grazing patterns, meanwhile, will rebuild the carbon in the world's grasslands.

Many farms will establish closer links with local communities and farmers' markets, reducing the distance that food travels, but whether they change or not, they will have to cope with the troubled weather patterns of the warming world, including prolonged droughts and torrential downpours. By going organic and storing more soil carbon, they will increase the resilience of their crops against drought.

How we will untangle the complex web of corporate control, genetically modified crops and subsidies that have made farming so unsustainable is a problem for another book.

Earth's future forests

We're like a two-year old playing with fire.... We're messing around with something really dangerous and don't really understand what will happen.
— William Laurance, Smithsonian Tropical Research Institute, Panama

Whether they are in Indonesia or the Congo, Canada or Surinam, forests are filled with magic and an astonishing wealth of biodiversity. Inhabited by forest people who have lived there for thousands of years, they are a source of wonder to anyone has spent time in them.

They are also immensely important for the world's climate balance. Since the start of the human adventure, we have cut down 80% of the Earth's forests, but 20% remain, covering 12% of Earth's land area, storing 40% of the world's terrestrial carbon — 283 billion tonnes of carbon in their biomass, 1000 billion tonnes when the soil and forest vegetation are included.[1]

As a result of our assault by bulldozer, chainsaw and fire, we are losing 40,000 hectares of tropical rainforest every day, 14.6 million hectares a year, an area the size of Florida. From the Amazon to Indonesia, the Congo to Nigeria, people and businesses are destroying rainforests to raise cattle, grow soybeans for Europe's cattle, plant monoculture trees for pulpwood, plant palm oil plantations for biodiesel, cut the valuable tropical hardwoods — often illegally — and make space for human settlements.

The rainforests host more than half of Earth's known plant and animal species, but at the current rate of loss, even without climate change, most will no longer be functioning ecosystems in 100 years time. Add climate change, and the future becomes very troubling indeed. Most trees can only migrate a few meters a century, but with rising temperatures they will need to move several hundred miles north or south to get away from the heat.

The trouble also affects temperate forests. In 2003 Siberia's boreal forest lost 4 million hectares to forest fires. In Canada 2.6 million hectares are being lost to fire each year, compared to only a million hectares a year in the 1970s. Researchers at the Russian Academy of Sciences Forest Research Institute fear that Russia's boreal forests will be so dry by 2090 that they will turn into grassland. As temperatures rise, a new kind of megafire has been appearing that cannot be controlled by the known methods of fire suppression.[2]

The Amazon rainforest, which stores 120 billion tonnes of carbon, is particularly affected. Research by Britain's Hadley Centre shows that if the temperature rises as predicted, the Amazon could cease being a net store of carbon by 2040

RHETT A. BUTLER MONGABAY.COM

A monkey treefrog (Phyllomedusa bicolor) in Peru's Tambopata rainforest.

because of drought. The Amazon can survive two consecutive years of drought, but a third year causes the trees to fall over and die. In 2005 and 2006 parts of the Amazon suffered extreme drought, causing major tributaries to dry up, but luckily the rains returned in 2007.

The world's forests are losing 1.5 billion tonnes of CO_2 a year, producing 15% of our total CO_2 emissions.[3] Because forest fires also produce black carbon and contribute to the formation of tropospheric ozone, the overall impact of deforestation is around 17% of the cause of global warming.

What should we do?

Costa Rica has made it illegal to convert forest into farmland. The Paraguay government placed a moratorium on deforestation in the eastern half of the country in 2004, using satellites to keep check and sending in forestry officials and police when they spotted a problem, reducing deforestation by 85%. Where the political will is strong, deforestation can be stopped.

In Peru, the government has reduced the loss of forest in protected areas to less than 0.2% a year, using a combination of protected parks and indigenous reserves, the titling of native territories to the forest people who live there, the sanctioning of long-term commercial timber production in chosen areas and satellite monitoring.[4]

Giving forest communities locally enforceable rights over forest management is an important part of the solution. India, which recently brought in a law returning the bulk of its forests to communities for management, is one of the few countries where there is a net increase in

forest cover, along with China and the USA. Forest Stewardship Council certification is also important to provide assurance of sound ecological management.

Globally, we need either a large global forest protection fund financed by a global carbon tax that can be used to purchase threatened forests and give them permanent protection, or a system of "avoided deforestation" credits that countries can trade as an incentive to protect their forests.

Whichever method we choose, we need it urgently. Andrew Mitchell of the Global Canopy Program says, "Tropical forests are the elephant in the living room of climate change," because we have not given this aspect of the problem anything like the attention it needs.

As well as stopping the carbon hemorrhaging, we need to maximize the forests' ability to store carbon by adopting holistic, ecosystem-based forest management. As this becomes the reality worldwide, the world's forests will begin to regain their former glory and recover their biodiversity.

For more details, see #38, #74 and #95.

- Earth's Tree News: olyecology.livejournal.com
- Forest Peoples Programme: forestpeoples.org
- Global Canopy Programme: globalcanopy.org
- Global Forest Alliance: worldwildlife.org/alliance
- Mongabay (a great resource): mongabay.com
- Rainforest Action Network: ran.org
- World Rainforest Movement: wrm.org.uy

Earth's Future Atmosphere

In the years to come, this issue will dwarf all the others combined. It will become the only issue.

— Tim Flannery, Australian zoologist, author of *The Weather Makers*

The science of Earth's atmospheric is complex and full of uncertainties. A host of interacting factors are at play, including the role of carbon cycle feedbacks, aerosol forcings, ocean heat uptake and climate sensitivity behavior. As such, it is far beyond the scope of this book.[1]

It is clear, that reducing our greenhouse gas emissions to zero will not be enough. Almost all scenarios that try to find a "safe landing" require negative CO_2 emissions in the second half of the century — which means capturing the surplus carbon. The current atmospheric carbon (800 Gt) is 260 Gt above the pre-industrial level (560 Gt), and rising by 5 Gt a year. So we will need to draw 400 Gt of carbon out of the atmosphere as quickly as possible.

There are various proposals to do this, including seeding ocean phytoplankton with iron dust to speed CO_2 absorption, which has marine ecologists alarmed; reducing the sun's heat by increasing the albedo of Earth's surface by launching 16 trillion tiny mirrors into space; using hundreds of rockets to release sulfate particles; creating a cloud layer of reflective aluminum;[2] and covering the deserts with white plastic sheeting.

In 2009 Tim Lenton and Naomi Vaughan from Britain's University of East Anglia published the first comprehensive assessment of the various options, comparing their ability to reduce radiative forcing.[3] Aside from proposals that were deemed ineffective or too risky, they found that if there was a strong commitment to reduce emissions, a combination of afforestation, reforestation, bio-char burial and global-scale air capture and storage might be able to reduce the atmospheric CO_2 back to the pre-industrial level by 2100.

Earth's Forests

Trees absorb carbon dioxide, storing it in their timber, roots and forest soil. The older a forest, the greater its capacity to store carbon. (See #38.) We need a global treaty in which nations agree to protect their ancient forests, adopt holistic practices in their working forests and engage in major tree planting. If successful, Earth's forests could draw down 50 to 100 Gt by 2100.[4] At the same time, rising temperatures will be causing more droughts, insect outbreaks and forest fires, so while the strategy is essential, we can't depend on it for our salvation.

Earth's Farmlands and Grasslands

Farmland soils store carbon too — but organic farms store 15–28% more, for an additional 2.2 tonnes per hectare per year (one ton per acre). If all the world's farmers were to go organic on 1.5

IKEA is using The Atlas of Russia's Intact Forest Landscapes *to avoid buying old-growth timber from Russia's 289 million hectares.*

WRI FEATURES

- "A Safe Landing for the Climate," by Bill Hare: worldwatch.org/node/5984
- Bioenergy with Carbon Sequestration Interview: tinyurl.com/dhwtpo
- CO_2 Solution: co2solution.com
- Lenton and Vaughan Study: tinyurl.com/cg24wa
- *The Weather Makers* by Tim Flannery: theweathermakers.org
- Virgin Earth Challenge: virginearth.com

billion hectares of cropland, this could draw down up to 30 Gt of carbon a year, or 200 Gt by 2100.

The same applies to the world's 3.4 billion hectares of grasslands. If 50% of the world's grassland farmers were to graze their animals rotationally, mimicking the way they used to graze when predators were around, they could store an additional one tonne per hectare per year. (See #41 and #43.) Globally, this would capture 1.7 Gt of carbon a year, or up to 150 Gt by 2100. Both farmlands and grasslands will be subject to a saturation ceiling when the soil can absorb no more, but the numbers are encouraging.

Biochar Production

It has been estimated that if farmers could be persuaded — financially or otherwise — to turn their agricultural wastes into charcoal by burning them in very low-oxygen chambers and plowing the charcoal into the soil, they could store 0.5 Gt of carbon a year, rising to 1.75 Gt a year by 2060.[5] By 2100, biochar could store 50 Gt of carbon. Biochar advocates calculate that if biochar additions were applied on 10% of the world's cropland, it could store 8 Gt a year of carbon.[6] (See #41.)

Air Capture with Storage

Bioenergy Capture and Storage (BECS) is a spin-off from clean coal with carbon capture. Instead of burning coal, you burn biomass, capturing and burying its CO_2 to create carbon-negative energy. Peter Read, from Massey University, New Zealand, has estimated that if BECS were adopted as the world's main form of generating energy, up to 298

Carbon Capture Method	Gt of carbon captured by 2100 (280 Gt needed)
Holistic forestry	50–100
Organic farming	200
Carbon farming on grasslands	150
Biochar production	50
Total	340–440

Gt of carbon could be drawn down by 2060, but many of the same doubts arise as for CCS coal.[7] (See p. 60.)

Chemical air capture with storage, building scrubbing towers or artificial trees that use sodium hydroxide and lime to capture CO_2 from the ambient air is another option, but it is subject to numerous doubts and constraints, including the concern that the energy and cement needed to construct the devices and to transport and bury the captured CO_2 could produce more CO_2 than it captures.

William Calvin, at the University of Washington, wants to see trials for the ocean sequestration of carbon at key places in the world's oceans by using wave-powered pumps to drive the carbon rich surface waters down where they will be stored in the cold deep ocean, while driving nutrient-rich waters to the surface where they can absorb more carbon. He believes this system could store up to 30 Gt of carbon a year, which is the level of sequestration he believes to be needed.

Earth's Future Economy

Do we want to be remembered as the generation that saved the banks and let the biosphere collapse?

— George Monbiot

Throughout "the lost years," when politicians around the world were bewitched by the coal, oil and auto industries, and US leaders were obsessed with the dream of military global dominance, the one-liner that old-world conservatives used to kill off discussions about climate solutions was, "It will destroy jobs and kill the economy."

They could not have been more wrong. It is *not* implementing climate solutions that will kill the economy. In California, a study put the future economic damage caused by climate change as high as $47 billion a year. A study for Illinois put the figure at $43 billion a year. In Germany, the cost has been put at $1.23 trillion by 2050, and $5 trillion by 2100. Sir Nicholas Stern, former chief economist at the World Bank, put the figure at up to 20% of the world's gross domestic product — and he has since said that this probably under estimates the likely damage. (See #61.)

There is also the looming reality of peak oil and its impact on the world's economies. We had a glimpse of this in 2008 when speculators drove the price to $147 a barrel ($5.40 a gallon at the pump, $1.50 a liter in Canada). As soon as the economy recovers, oil will take off again, bringing financial grief to households, businesses, cities and nations around the world. The solutions, of course, are the same as those for climate change. Both crises result from our dependence on fossil fuels; both threaten to destroy our economies; and both can be solved by a rapid transition to zero-carbon economy.

Now let us flip to the future, and imagine the world in 2030 when most of the world's nations have embraced the transition and are living in a climate-friendly economy. First, the US will have no more need to maintain a global military to protect its oil supplies, saving billions of dollars. (See p. 38.)

The cost of heating our buildings will have fallen dramatically, thanks to the efficiency of their insulation, windows, heat pumps and district heating systems. The cost of electricity will be falling because of the falling price of solar, wind and geothermal energy. The cost of driving an efficient electric vehicle will be no more than $2 a week — so cheap that governments will introduce road tolls to prevent road congestion, using the income to support cycling, public transit and railways. The cost of food will have fallen because of the productivity of organic farms and the renaissance in backyard and community gardening — with the exception of meat, most of which will cost more because of the carbon and methane taxes it carries.

This will be an economy with no more waste. The landfills will have been closed; products that cannot be dismantled and recycled will have been eliminated; and everything we buy will come with a recycling fee attached. Every sewage treatment plant will have been retrofitted to reclaim its energy, and every community will have a local biogas plant, extracting heat and fuel from organic wastes to supplement the energy from renewables.

It will be an economy with plenty of jobs, especially during the transition when we move away from fossil fuels. Once the buildings have been retrofitted, the renewable energy systems installed and the cycle paths and high-speed

Denmark has been found to be the happiest place on Earth — is it coincidence that Denmark is such a cycling-friendly country?

MIKAEL COLVILLE-ANDERSEN / COPENHAGENCYCLECHIC.COM

trains built, we may enter a very different world in which the low cost of living encourages people to work less and spend more time with their families and communities.

Flying and global shipping will almost certainly be more expensive, encouraging more local manufacturing and trade; and local economies may experience a thriving renaissance, with local currencies co-existing alongside national currencies.

All this will be happening twenty years after the global financial meltdown that started in 2008, when many of the principles that had steered the world's economy since the 1980s were found to be faulty. Following that disaster, the world's nations will almost certainly have re-embraced the importance of government intervention and global cooperation on a wide range of matters, including the regulation of the world's oceans and fishing fleets and protection of the world's forests.

The drive for economic growth will not have stopped, however. It will have been transformed through the progressive dematerialization of the economy. As nations reach a certain level of material sufficiency, people will prefer to spend their hard-earned income on bringing more quality to their lives, rather than stuff. As this happens, Gross National Happiness will become more important in many people's eyes than Gross National Product, and politicians will find themselves promoting measures to strengthen family and community life, restore local ecosystems and help people find fulfillment through meaningful work.

This is the possibility of a world beyond fossil fuels. Looking back, it will seem to have been as natural a part of history as embracing fossil fuels was two hundred years ago. We are on the brink of an economic and technological renaissance — and the most rational response to it — is to embrace it whole-heartedly.

Pathways to a Climate-friendly future

At [the current] rate, it will take almost 200 years to get the tenfold improvement we need. The deep changes required across the global economy will not happen without new incentives and policies at the national and international level.

— McKinsey & Company

From the future we return to the present, and the realists' question: "How can we build the pathway that will lead to the promised land described in the previous sections?"

At the time of writing, our engagement with climate change has been through three stages, out of many:

Stage One: Denial, heads planted firmly in the sand.

Stage Two: Debate between simple solutions. Which is best — nuclear, biofuels or cap-and-trade?

Stage Three: Recognition of the need for a full, multi-sectoral portfolio of solutions.

Stage Four: Construction of the first organized pathways to a climate-friendly future.

Stage Five: Integration of the pathways with community and business strategies.

Stage Six: Broad public engagement, allowing governments to move down the pathways with more intent.

Stage Seven: Financial cooperation to provide the necessary capital flows.

Stages Eight to …: To be announced.

- McKinsey & Company: mckinsey.com
- McKinsey Solutions Climate Desk: solutions.mckinsey.com/climatedesk

Our concern here is with Stage Four, in which some nations are beginning to lay down pathways based on a costed analysis of a wide range of solutions.

One institution that is helping governments with this task is the management consulting firm McKinsey & Company. By 2009 they had created greenhouse-gas-abatement cost curves for the US, UK, Germany, Australia, Netherlands, Sweden, Switzerland and the Czech Republic and were engaged with a dozen other countries, including India, China and Brazil.

For their global study "Pathways to low carbon economy," they worked with ten leading global companies and organizations, including The Carbon Trust, Shell and WWF, to assess more than 200 possible solutions. They found that to get on a pathway to keep the temperature rise below 2°C (as defined by IPCC authors) would require an annual investment of about 530 billion ($700 bn) a year between 2010 and 2020, growing to about 810 billion ($1 trillion) a year between 2021 and 2030. This investment represents 5–6% of the expected capital investment into fixed assets during that period, most of which would be financed through borrowing, based on payback. The total cost to the world economy would be 200–350 billion annually by 2030, ($260–$460 bn) which is less than 1% of the expected global GDP. Fluctuations in the price of oil from 2004–2008, by contrast, cost 5% of the world's GDP. They also found that a delay of even 10 years would mean missing the 2°C target.

Another McKinsey study emphasized the importance of investing in energy efficiency, which could reduce emissions while earning an average 17% annual return, based on oil at $50 a barrel. The higher the price of oil, the greater the return — hence the importance of putting a price on carbon.

With all studies of this kind, the motto is "data in, data out." If the US model were changed to reflect Repower America's goal of 100% renewables by 2018, for instance, it would show different results. What is important is to do the analysis, as this is how we will lay the foundation for our pathways to a climate friendly future.

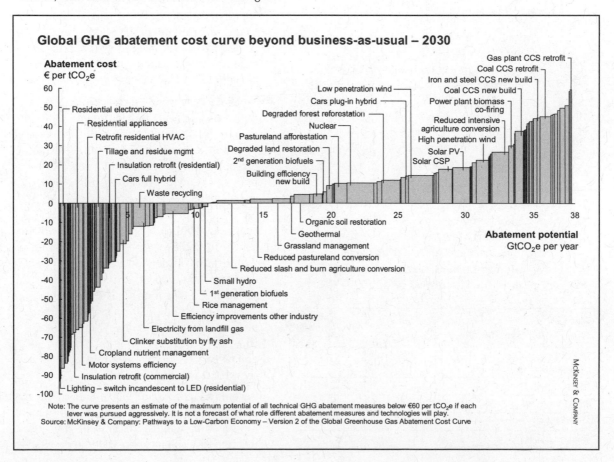

Global GHG abatement cost curve beyond business-as-usual – 2030

Note: The curve presents an estimate of the maximum potential of all technical GHG abatement measures below €60 per tCO₂e if each lever was pursued aggressively. It is not a forecast of what role different abatement measures and technologies will play.
Source: McKinsey & Company: Pathways to a Low-Carbon Economy – Version 2 of the Global Greenhouse Gas Abatement Cost Curve

The Climate Solutions Dividend

No problem of human destiny is beyond human beings. Man's reason and spirit have often solved the seemingly unsolvable — and we believe they can do it again.

— President John F. Kennedy, American University, June 10, 1963.

If we succeed in this great undertaking, the next generation will thank us not only for preventing the looming catastrophe but also for the many benefits that will flow from our success.

The ecological benefits are clear, but no one has calculated the full economic benefits. These are the climate solutions dividends, the enticing rewards for success.

1. We get to avoid the desperate scenarios laid out in *Six Degrees* (see p. 24) and keep human civilization intact, including many species that would otherwise face extinction. What price should we put on the ability of our children to continue the journey of evolution? The avoided costs are known: 5–20% of the US's GDP of $14 trillion is $700 billion to $2.8 trillion a year. In Canada, it is $63 to $254 billion a year. The gains are priceless.

2. We enjoy a managed transition through what would otherwise be the economic trauma of peak oil, avoiding the relentless waves of bankruptcies, evictions, unemployment, poverty and hunger that will be caused by peak oil's sudden arrival, for which — at present — our societies are completely unprepared.

3. We enjoy permanent energy security and the end of dependence on oil from the Middle East and other countries. This means we can bring the troops home, stop irritating the Islamic terrorists, stop being obsessed with security and stop exporting $300 million a day to the oil-rich states of the Middle East. These gains have been costed out by the National Defense Council Foundation. For the US, they come to $825 billion a year. (See p. 38.)

4. We protect the world's forests, which would otherwise face being logged or burned. Under the business-as-usual scenario, the Amazon rainforest starts dying by 2050 and turns into savannah by 2100, because of a combination of drought and fire. Because tackling climate change requires that we preserve the forests' carbon, our children will continue to be enchanted by their magnificence. We will also benefit from the forests' ecosystem services, as a continued contribution to our economy. Globally, their 1200 Gt of stored carbon, priced at $13.60 a tonne, has a $16 trillion value. If we assume a 200-year forest carbon lifecycle, that's $80 billion a year. For the US, with 5% of the world's forest, that's a $4 billion a year contribution. For Canada (10% of the world's forests) it's $8 billion.

5. We enjoy a more secure global food supply by embracing organic farming, which stores more carbon in the soil, reduces farming emissions by up to 30% and increases yields in developing

The Yearly Climate Solutions Dividend	Low range	High range
End of Persian Gulf oil dependence	$825	$825
End smog and air pollution	$59	$690
Peace dividend	$680	$680
Improved efficiency	$338	$338
Total, per year	$1.9 trillion	$2.5 trillion

countries. (See #42.) We also remove the threat to crops from smog, air-pollution and the increasing heat waves, which would reduce the yields.

6. We eliminate smog and pollution, which cause asthma, lung disease, cancer, disability and premature death. For Los Angeles alone, this has been assessed at a $10 billion annual cost. Air pollution and smog also cause crop losses and damage buildings and forests. For Ontario, Canada, the costs have been estimated at $8 billion a year, rising to $250 billion by 2030 as rising temperatures cause more smog.[1] For the US, the full economic savings are up to $690 billion a year.[2] In California, smog and air pollution cause 9,300 deaths, 16,000 hospital visits, 600,000 asthma attacks and five million lost work days every year.

7. We enjoy cheaper driving, more cycling routes, more public transport, more high-speed rail and more friendly, walkable communities, which build neighborhood strength. These benefits have not been costed out.

8. Thanks to our investments in efficiency, we enjoy lower heating and power bills. The American Council for an Energy Efficient Economy has calculated that a 15% increase in efficiency will produce annual savings worth $169 billion a year. By increasing efficiency by 30%, this could rise to $338 billion.

9. We get to end most warfare. This may sound unbelievable, but most modern conflicts are fought over scarce energy supplies. When nations become self-sufficient in renewable energy, we can eliminate 80% of our military expenditures. The US military budget is about $1 trillion a year,[3] of which $138 billion has been included in #3 above. This leaves $850 billion a year, which, trimmed by 80%, produces a peace dividend of $680 billion a year.

10. Finally, we enjoy our first proper experience of working together as a world, and we restore hope to our children. What more can we ask?

Members of the T'Sou-ke First Nation celebrate the installation of its 75 kW solar PV system, the largest in British Columbia. tsoukenation.com

Getting Serious

The challenge is either to build an economy that is sustainable, or to stay with our unsustainable economy until it declines. It is not a goal that can be compromised. One way or another, the choice will be made by our generation, but it will affect life on Earth for all generations to come.

— Lester Brown, Earth Policy Institute[5]

There is no mistaking the warnings, and it is those best positioned to know — the climate scientists — who are issuing the most stark and urgent warnings.

If the changes summarized so starkly in *Six Degrees* by Mark Lynas were being threatened by invasion by a hostile power, we would immediately unite in our determination to stop them. To repeat the warning: once we pass 2°C beyond Earth's pre-industrial temperature, there may be no getting back, as 2°C leads to 3°C, then onwards to 6°C and total extinction; even 2°C is far too dangerous.

Most people live in democracies, so we cannot plan in Technicolor as if we had absolute powers, but the parallels with World War II are very pertinent, for both Britain and the US are democracies, and every major war measure at the time was put to a democratic vote.

In Britain the auto industry was told to switch its entire production to guns, tanks and airplanes, and gas was rationed so that no one could drive more than 200 miles a month. In the US the auto industry switched from cars to tanks in just six months, supported by a government guarantee of continued profits.

In 2006 Sir Nicholas Stern said that we need to invest 1% of our GDP to protect up to 20% of our GDP from being swallowed by multiple climate disasters later. Two years later he warned that new data showed that the cost of averting disaster had doubled to 2% of GDP (see box), because climate change was happening much faster than had previously been thought, so emissions would have to be reduced more sharply.[1] He also said, "If we delay by just 10 or 20 years, the costs will be much higher, and the risks much greater." In 2009 he urged the world to invest $400 billion in green investments to tackle climate change as part of the global fiscal stimulus packages.[2]

During World War II, Britain's government asked people to help pay for the war by buying savings bonds; and used advertising and posters to appeal to patriotic instincts. During a fund-raising drive in 1941, the 15,000 residents of Marple, Cheshire, raised £65,000 toward the cost of a Minesweeper. In 1943 they raised a further £128,360 toward the cost of a Lancaster Bomber, four Spitfires, and four Hurricanes. By the end of 1945, Britain's war savings bonds had raised £1,754 million for the war effort. From a population of 48 million people that was £36 per person, or $1,000 in today's money, representing 3% of Britain's GDP.[4]

This book contains an abundance of solutions, enabling everyone from families and farmers to high financiers to tackle the crisis with the seriousness it deserves.

But above all, we need the world's governments to get engaged. Out of the book's many actions within the section Solutions for Governments,

	GDP $ billion[3]	2% of GDP $ billion	2% per person $
Canada	$1,564	$31	$804
United Kingdom	$2,787	$56	$748
United States	$14,330	$287	$960
World	$78,360	$1,567	$210

Open Day at the Triodos Caton Moor wind farm, UK. If we value their future, we must act now, and urgently.

these are the ten that governments must embrace immediately:

1. Adopt a widespread strategy to win public engagement.
2. Eliminate all fossil-fuel subsidies and phase out all coal-fired power plants by 2020.
3. Plan to achieve 100% renewable electricity by 2020, using feed-in tariffs to speed development.
4. Plan to make all buildings 33% more efficient by 2030.
5. Plan to make the whole country 100% oil free by 2030.
6. Introduce a price on carbon, whether through a carbon tax, cap and trade or cap and dividend, rising to $150 a tonne.
7. Establish incentives to help all farms become organic by 2030.
8. Finance initiatives to protect the world's rainforests.
9. Finance initiatives to reduce black carbon throughout the world.
10. Take urgent steps to reduce methane emissions.

In getting serious, however, we must never forget that tackling climate change is not just about removing a problem so that life can continue as usual. It is about our civilization achieving a historic breakthrough from a world based on fossil fuels to one based on renewable energy, sustainable forestry, sustainable farming and a new respect for our planet's ecology.

Such a breakthrough will bring as much hope, excitement and innovation as the Renaissance did in the 15th century and the Industrial Revolution in the 18th and 19th centuries.

We are on the verge of one of those incredible periods in history when humanity remakes itself, casting off its old decaying forms to emerge with new vision, new hope and a new spirit of optimism. As Carl Sagan says in Solution #101, at the end of this book: "Don't sit this one out. Do something. You are by accident of fate alive at an absolutely critical moment in the history of our planet."

PART 2

The Solutions

A new politics can spark the clean-energy revolution that will serve as a foundation for a new era of human prosperity, protect the world's forests, stabilize the climate, and preserve the diversity of life on the planet.

— Eban Goodstein, author of *Fighting for Love in the Century of Extinction*

1

Calculate Your Carbon Footprint

You may never know what results from your action. But, if you do nothing, there will be no results.

— Mahatma Gandhi

With the help of these ten solutions, you and your family can make a steady transition to a carbon-neutral life that will actually save you money.

The solutions have been designed so that you can use them as the basis for an eight-week course with a group of friends and neighbors. (See #10.) Week One starts with calculating your carbon footprint and setting your goal to reduce it.

Your *ecological footprint* is the total imprint you make on the Earth through the resources you consume and the carbon emissions you produce, which require forests and oceans to absorb them. It is measured as an area of land — a rich family

- Be the Change Earth Alliance: bethechangeearthalliance.org
- Climate Challenge 8-week course: theclimatechallenge.ca
- CO_2 emissions from an average home: rmi.org/sitepages/pid209.php
- Comparative list of carbon calculators: earthfuture.com/calculators
- Global Footprint Network: footprintnetwork.org
- Michael Bluejay calculator: michaelbluejay.com/electricity/carboncalculator.html
- The Atkinson Diet: theatkinsondiet.com
- The Good Life: thegoodlife.wwf.ca
- Together: together.com
- Zero Footprint Calculator: calc.zerofootprint.net/calculator

might be using 25 hectares; a poor family in India as little as a quarter hectare. Across the Earth as a whole, the average human uses 2.7 hectares, but Earth's total biological capacity only allows for 1.8 hectares, which shows how much we are over-consuming our environment — and other species have to eat, too. The average American uses 9.4 hectares; Canadian 7.6 hectares.

Your *carbon footprint* is more limited. It measures your personal and household emissions that cause global warming from five sources:

- The oil you burn when traveling in a car, bus, boat or airplane
- The coal, oil or gas you use to heat your home
- The electricity you use, some of which probably comes from burning coal or gas
- The waste you produce, because rotting garbage produces methane
- The goods you consume, especially meat and dairy

Including the food we eat, the average American produces 20 tonnes of CO_2 a year; a wealthy family with two children could easily produce 200 tonnes. The average Canadian produces 18 tonnes a year, the average European 12.5 tonnes. Most people in Africa produce less than half a tonne a year. The challenge is to get this down to zero. There are many carbon calculators that you can use to work out your carbon footprint, based on how much fuel you use, how many flights you take, etc.

Your first task in Week One is to hold a family meeting, so that you are all on the same page.

Low Carbon Living

Ann and Gord Baird live with their two children and two parents in a 2150-square-foot passive-solar cob-earth home in Victoria, BC, which they built themselves (see photo). A 2 kW solar PV system provides power, and 60 solar tubes provide hot water and heat for their home. They also harvest rainwater, use composting toilets and practice gray water recycling. They are mostly vegetarian and grow much of their own food, with chickens for their eggs. They work from home and use a minivan to ferry their kids to school. They vacation by paddling in local waters. Total: 17.1 tonnes (2.85 tonnes each)

ANN AND GORD BAIRD

The Baird's solar-cob house in Victoria, BC. eco-sense.ca

Start a carbon reduction notebook where you can track your emissions and the changes you make as you work through these solutions. Other alternatives are to join a "Be the Change" group or to use David Gershon's Low Carbon Diet.

Next, use a carbon calculator to work out your household's footprint. I recommend using one from Zero Footprint or Michael Bluejay. (See box.)

Now set a goal to reduce your carbon footprint. If you reduce it by 10% a year, you will achieve a 100% reduction, bringing it down to zero. By then, electric vehicles and other carbon-reducing technologies will be more widely available and affordable.

What will it cost you? In his wonderful book *The Carbon Busters Home Energy Handbook,* Godo Stoyke calculates that, the investments to reduce your carbon footprint by 73% will save you $17,000 over five years. If you include changing your diet, which will cost you nothing, your carbon reduction will increase to over 80%.

Good Reading

- *How to Live a Low Carbon Life* by Chris Goodall: earthscan.co.uk
- *The Carbon Busters Home Energy Handbook* by Godo Stoyke: carbonbusters.org
- *Low Carbon Diet: A 30-Day Program to Lose 5000 Pounds* by David Gershon: empowermentinstitute.net

Actions

- Calculate your CO_2 emissions
- Hold a family meeting
- Set your reduction goal
- Create a carbon reduction notebook

2

Change the Way You Eat

We have to do what we have to do. Miracles happen. The life force of this planet is very strong. Dandelions poke through sidewalks. We don't know enough to give up. We only know enough to know that we have to try to change the course of human events.

— Elizabeth May, leader of Canada's Green Party

There are many inconvenient truths related to global warming, one of which is that livestock animals raised for meat and dairy (chiefly cattle and pigs) are responsible for 14–18% of the climate problem.[1] (See p. 62.)

Each kilogram of conventional beef produces 96 kg of CO_2e; the average American eats 44 kg of beef a year, producing 4.2 tonnes of CO_2e a year. A single quarter-pound hamburger produces 9 kg

of CO_2e — the equivalent of burning a gallon of gas or driving 20–25 miles in a car. Raising animals for food generates more greenhouse gases than all the transportation in the world.[2]

Our overall food consumption may cause up to 33% of global warming, because of:

- The impact of beef, pork and dairy. (Beef has the greatest impact.) (See pp.62, 288.)

- The distance that food ravels before it reaches our plates. Most fruit and vegetables travel 2,500 to 4,000 kilometers from farm to store. A 225-gram package of imported organic spinach flown across the Atlantic produces 3 kg of CO_2.

- The way it is grown. Organic food produces 26% fewer emissions. If you eat locally grown organic vegan food, you could reduce your food's carbon footprint to 0.35 tonnes of CO_2e a year.

- The packaging and waste associated with food.

How big is your food's carbon footprint? Find out at eatlowcarbon.org.

Eat less meat

A meat eater on a bicycle has a larger carbon footprint than a vegan driving an efficient car.[3] Don't just substitute for meat with eggs or tofu. Buy a good vegetarian cookbook, such as *The Moosewood Cookbook*, and visit some vegetarian websites. Vegetarian cooks have developed mouth-watering recipes, and it is rare to find an unhappy vegetarian. It is much better for your health — studies show that vegetarians have less

The author turning compost at his home in Victoria, British Columbia

CAROLYN HERRIOT

Actions

- Calculate your food CO_2 emissions
- Visit a local farmers market, and buy local organic produce
- Read a vegetarian or vegan cookbook
- Eat for one day a week without meat
- Plan to plant a food garden

- 30 Days to a Greener, Healthy Diet: thedailygreen.com/healthy-eating-plans
- *A Year on the Garden Path — 52-Week Organic Gardening Guide,* by Carolyn Herriot: earthfuture.com/gardenpath/Book.htm
- Cool Foods Campaign: coolfoodscampaign.org
- In a Vegetarian Kitchen: vegkitchen.com
- Love Food Hate Waste: lovefoodhatewaste.com
- Low Carbon Diet Calculator: eatlowcarbon.org
- Meat and the Environment: goveg.com/environment.asp
- Meat-Free Monday: supportmfm.org
- Meatless Monday: meatlessmonday.com
- Take a Bite Out of Climate Change: takeabite.cc
- *The Hundred Mile Diet:* 100milediet.org
- VegCooking: vegcooking.com
- Vegsource: vegsource.com

cancer and live longer.[4] It may feel like a leap, but it's a very sound investment. Organic, grass-fed beef produces 40% fewer emissions, as it does not require rainforest destruction to grow soybeans to feed the cows and does not produce N_2O emissions. It is also much healthier for you.

Eat less dairy

The same arguments apply to a vegan diet that cuts out *all* animal products, including milk, cheese and ice cream. Buy a good cookbook, such as *How it All Vegan,* and start experimenting by eating a vegan diet one day a week.

Eat organic food

Organic farming requires 63% less energy than conventional farming, produces no nitrous oxide emissions, and stores 15–28% more carbon in the soil.[5] Until recently, everyone ate organically by default. Since organic food contains more antioxidants and salvestrols, it is also better for your health. If the whole world went organic, yields in the developed world would not fall, and yields in the developing world would double, so there is no reason to worry that we could not feed the world. (See p. 63.)

Grow your own food

If you've got a garden, you can grow your own food — your ancestors did it all the time. The smallest space can house a tub of tomatoes, and you'll discover what they should taste like. Take a "grow your own food" book out of the library, and start sowing seeds. As long as you nourish the soil with good compost and leaf mulch, water regularly in summer, and encourage wildlife to take care of the pests, your plants will reward you with mouth-watering food. To reduce the energy needed to cook your food, try using a solar cooker, haybox or a pressure cooker.

Buy locally grown food

The more you buy from local farmers markets, the smaller will be your food's carbon footprint — and your food will also be fresher and healthier. In a comparison between two Christmas meals, the one with imported food produced 100 times more emissions than the locally grown meal — 11.89 kg versus 0.12 kg.[6] The less food you waste, too, the better: 27% of food in the US is thrown away as waste. Each American throws away 163 pounds of food a year from uneaten portions and food wasted in preparation.[7]

3

Wake Up to Green Electricity

By spending just $30 a year on wind power, an average Colorado family can cut its household CO_2 production by about 10%.

— Colorado Wind Power

Most people take their electricity for granted — you flick the switch, and on it comes. If you follow the wires, however, they often lead to a coal or gas-fired power plant. 85% of the world's electricity comes from burning fossil fuels, producing 8 billion tonnes of CO_2 a year, representing 21% of the global CO_2 emissions and 13% of the cause of global warming.[1]

The solutions are to use less power by conserving and changing to energy-efficient appliances; to switch to green power; and if you can afford it, to install solar PV on your roof. For our purposes, we are assuming that saved power displaces a mixture of coal and natural gas, saving 600 grams (1.34 lbs) of CO_2 per kWh, at 10 cents/kWh.

The chart lists some of the ways you can reduce your emissions. For larger appliance upgrades look for the Energy Star label, and do your homework first. *The Consumer Guide to Home Energy Savings* is a great investment, and their Online Guide is an excellent resource. With every kilogram of CO_2 you save, you help protect our planet from global warming.

Buy Green Power

In many regions, utilities will sell you green power from wind and other sources for a small premium. If you use 10,000 kWh a year, this will cost you $2.50 to $25 a month, and save up to 6 tonnes of CO_2 a year — 10 tonnes if your power is 100% coal-fired. This is one of the most effective ways to reduce your carbon footprint.

Energy-Saving Measures	CO_2 saved per year	Cost	$ saved per year
Switch off five 60 W lights that don't need to be on.	270 kg	$0	$27
Replace 20 light bulbs with efficient ones (each bulb saves 50 kg of CO_2 per year").	1000 kg	$100	$100
Don't use the dryer. Use an outdoor clothes line or indoor rack.	280 kg	$0	$28
Install a low-flow showerhead.	230 kg	$15	$23
Replace a large inefficient fridge with a smaller efficient one.	400 kg	$750	$40
Use your dishwasher only when full, no drying cycle.	45 kg	$0	$4
Upgrade to an efficient front-loading clothes washer.	450 kg	$614	$45
Wash clothes in cold water, not warm or hot.	250 kg	$0	$25
Ditch the waterbed.	480 kg	$0	$48
Put your consumer electronics on power bars, and switch off when not in use to reduce standby power.	240 kg	$40	$24
Turn off your computer and printer when not in use.	420 kg	$0	$42
Totals	4065 kg	$1519	$406

Numbers will vary according to the local price of power and whether you use electricity or gas, etc.

Install Solar PV

Solar PV makes tremendous sense, if you can afford it. In 2009 it cost just under $5 a watt ($8–10 installed), so a 2kW system cost $16–$20,000. There is every hope that prices continue to fall: we dream of the day when solar costs 50 cents per watt and a 2 kW system costs $2,000. To find out about local solar incentives, call a solar dealer. A 1 kW solar system will produce 1000 to 1500 kWh a year, depending on how much sunshine you get, and reduce your carbon footprint by 600 to 900 kg a year.

Clever Technologies

A $40 Kill-A-Watt meter will tell you how much each appliance is using. A motion detector will switch off your lights when there's no one in the room. A solar hot water heater will reduce your hot water bill by 40–60%. A low-flow showerhead will reduce your use of hot water.

- Consumer Guide to Home Energy Savings: aceee.org/consumerguide
- Earth Aid Kit: earthaidkit.com
- Energy Star Appliances: energystar.gov
- Fridge-Freezer Energy Rating Database: kouba-cavallo.com/refmods.htm
- Green Guides: bchydro.com/guides_tips
- Green Made Simple: greenmadesimple.com
- Green Power Network: apps3.eere.energy.gov/greenpower
- Home Energy Saver: hes.lbl.gov
- Install a motion sensor light switch: tinyurl.com/8k4x53
- Install a Low-Flow Showerhead: tinyurl.com/7lkhzl
- Insulate your Water Heater: tinyurl.com/93qogq
- Kill-A-Watt video: tinyurl.com/99feeb
- Lowest Flow Showerhead: tinyurl.com/4ola36
- Office of Energy Efficiency, Canada: oee.nrcan.gc.ca
- Price of Solar PV: solarbuzz.com/ModulePrices.htm
- Real Goods: realgoods.com
- Rocky Mountain Institute Home Energy Briefs: tinyurl.com/3obh7r
- Solar cost calculator: infinitepower.org/calc_pv.htm
- Smart Power Strip: tinyurl.com/42jm9c
- US renewable energy incentives: dsireusa.org

Actions

- Walk around the house. List every appliance, and count the lights you use regularly
- Replace as many bulbs as you can afford to
- Buy or borrow a Kill-A-Watt meter, to see where you are using power
- Make a list of the appliances you'd like to upgrade
- Research the cost of new appliances
- Call a green power company and ask to make the switch
- Call a solar PV and a solar hot water company to get quotes

The Watson Family 5 kW Solar House in Lexington, MA.

GRAY WATSON, 256.COM/SOLAR

4

Keep It Warm, Keep It Cool

Investing in a home on your street
could be more profitable than
investing in Wall Street.

— Home Energy Saver

We used to live in caves or the open air. For many a century, we huddled around a wood fire to keep warm in winter. Next came coal, with its filthy air pollution, and then central heating, with oil or gas furnaces. We also began to realize how important it was to insulate our homes to keep out the cold.

Many of our homes are still very leaky and inefficient. A modern German *Passivhaus* using super-insulation, super-windows and sub-soil heat exchange uses only 9.2 kWh per square meter per year for heat, 95% less than the German average.

In winter when it's –10°C (14°F) outside, it needs no heating inside, and in summer when it's 35°C (95°F) outside, it is only 26°C (79°F) indoors. More than 6,000 passive houses have been built in Europe, so this is not an experimental design.

The average North American single-family home produces almost 12 tonnes of CO_2 a year, 55% of which comes from heating and cooling.[1] Here, we look at ways to keep the heat in — and out. In Solution #5, we look at ways to eliminate your use of fossil fuels altogether.

Keeping Warm	CO_2 saved per year	$ saved per year
Data from Cool Citizens. Numbers will vary according to fuel use and local prices		
Turn down water heater thermostat from 60°C to 49°C (140°F to 120°F)	97 kg	$12
Wash clothes in cold water	148 kg	$18
Wrap the water heater in an insulating jacket; insulate hot water pipes	143 kg	$18
Install water-saving showerheads and faucet aerators	218 kg	$27
Install programmable thermostats in your main rooms, set to 20°C (68°F) daytime, 13–14° C (55–58°F) night. Wear a sweater to keep warm	161 kg	$29
Seal large air-leaks. Caulk and weather-strip/draft-proof around windows, doors, baseboards, attic hatches and power switch plates	862 kg	$104
Add extra insulation in the attic (R-30)	763 kg	$91
Add extra insulation in the basement	338 kg	$45
Add extra insulation in the walls	216 kg	$26
Upgrade to high performance windows	971 lbs	$51
Add thin plastic film to the windows in winter	188 kg	$22
Seal and insulate your heating ducts, and get your furnace tuned	544 kg	$59
Keeping Cool		
Raise your cooling thermostat to 26°C (78°F)	154 kg	$19
Tune up your air-conditioning unit	59 kg	$7
Add an attic radiant barrier	57 kg	$7
Paint your roof white	214 kg	$26

All loaded up and ready to do a home energy assessment in Victoria, Canada.

Financial Planning

Your whole family needs to be involved, so work together to create a plan. There are often grants and incentives available for energy upgrades, so call city hall, a local environmental organization or a local energy company to find out more. According to the Home Energy Saver, a package of ten energy-saving investments will cost you $3,960, save you $597 a year and give you pay-back in 6.6 years, at a 16% rate of return. If you live in a condo or apartment, an energy efficiency upgrade will reduce your carbon footprint by around 30% and pay for itself in 3–5 years.

Keeping Cool

The better your insulation, the cooler your home. The more greenery you plant to shade south-facing walls, the better. Gainesville, Florida, has a strict tree ordinance and twice the tree-cover of nearby Ocala, so Gainesville residents spend $126 less per year on electricity, saving 1.4 tonnes of CO_2 a year per household.[2] When it comes to air conditioning, the Ice-Bear 30 hybrid provides efficient cooling, reducing energy demand by 95%.

- ASE Energy Efficient Financing: ase.org/consumer/finance.htm
- *Cool Citizens: Everyday Solutions to Climate Change,* Rocky Mountain Institute, 2002, rmi.org/images/PDFs/ Climate/C02-12_CoolCitizensBrief.pdf.
- Energy Efficient Financing Info Center: nationalguild.com/residential/hers.html
- Green Home Guide: greenhomeguide.org
- Home Energy Saver: hes.lbl.gov
- Ice Bear: ice-energy.com
- Installing a Programmable Thermostat (video): tinyurl.com/a4v7qq
- Profitability of Energy Efficiency Upgrades: hes.lbl.gov/hes/profitable.html
- Residential Energy Services Network: natresnet.org
- Rocky Mountain Institute Home Energy Briefs: tinyurl.com/3obh7r

Actions

- Use your utility bills and a carbon calculator to work out your CO_2 emissions from heating and cooling
- Ask city hall about retrofits, grants and incentives
- Consider your options from the chart, and create a plan to do as many as you can over the next six months
- Call an energy company for a quote on things you cannot do yourself

Return on Investment

Fluorescent lamps and fixtures	41%
Duct sealing	41%
Energy Star clothes washer	37%
Energy Star thermostat	30%
Energy Star refrigerator	27%
10-point energy efficiency upgrade	16%
Dow Jones industrials, 2000–2007	16%
30-year bond	4.2%

Source: Home Energy Saver

5

Heat Your Home Without Carbon

> Our planet has a rising fever. If the crib catches fire you don't say: 'Hmmm, how fast is that crib going to burn? Has it ever burned before? Is my baby flame retardant?'
>
> — Al Gore

In Solution #4, we looked at ways to reduce your carbon footprint by making your home more efficient. Here, we look at ways to keep warm with no fossil fuels at all.

Solar Hot Water

If you have an area of south facing roof or available land, a solar hot water system can reduce your annual carbon footprint by up to 678kg.[1] There are two types of technology, the older flat plates and the newer evacuated tubes, which are being used by 40 million households in China, with prices starting at $190. They will likely cost $5,000 to $8,000 in North America and reduce your hot water bill by 33% to 75%. There may be local grants or incentives available, and your payback will depend on how much hot water you use. If you have a heated pool, you should definitely invest in a solar hot water system.

Heat Pumps

In a temperate climate where the winters are mild (warmer than –10 C°), an air-source heat pump that extracts heat from the air and circulates it through your home by ducts or under floor pipes is a really good investment, getting a payback in 2–7 years. If your house does not have a duct system to distribute the heat, you will need a mini-split system, which will cost 30% more.

A ground-source heat pump (also known as geo-exchange or geothermal) works in all climates, using pipes filled with a fluid that freezes at a very low temperature installed under your garden or driveway, where the temperature is a constant 7–14°C (45–58°F). For each unit of energy used by the pump, you'll get 3–4 units back. If you live on rock, the system uses deep boreholes, with a slower payback from 8 to 30 years. In summer the system can work in reverse, cooling your home by

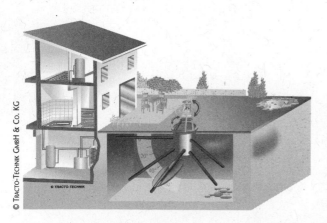

© Tracto-Technik GmbH & Co. KG

Tracto-Technik's innovative technology allows for the installation of a groundsource heat system in a single day. See tracto-technik.de

- Heat Pumps: tinyurl.com/2y5dxe
- Passive Solar Design:
  eere.energy.gov/de/passive_solar_design.html
- Pellet Stoves: naturalheat.ca
- Rocky Mountain Institute HQ:
  rmi.org/sitepages/pid229.php
- Solar Wall: solarwall.com
- Topolino wood-burning Stove: constructionresources.com
- Wood Heat: woodheat.org
- Wood Pellet Association of Canada: pellet.org
- Zero Energy Design: ZeroEnergyDesign.com

pumping unwanted heat into the ground below. If you live next to water, a water-source heat pump can heat your home in the same way with loops that sit in the water. In Zurich and Vancouver, new homes use heat from the sewers, linked to a district heating system.

Solar Heat

The best way to benefit from the sun is to build a super-efficient passive-designed solar house. In Colorado, the 4,000-square-foot Rocky Mountain Institute's headquarters is passive solar designed and super-insulated, with advanced glazing and a large built-in greenhouse roof. Even though it is at 11,000 feet, it has no additional heat-source, and the eco-components only cost an additional $1.50 per square foot, winning their payback in just ten months.

For new builds, it is possible to design a zero-net energy house, using these technologies plus ground-source heating. Some people have designed their own solar heating systems, using solar-heat air or water stored underground and circulated under the floor. There is no standard way of doing this yet, so you need to be creative. It is also possible to install a black solar wall on the south side of any building.

Wood and Biofuels

For the home-owner, a super-efficient woodstove that burns firewood or a pellet stove that burns pellets made from compressed sawdust are carbon-neutral options as long as the wood or pellets have been harvested sustainably, because the timber

Actions

- Call a solar hot water company and ask for a quote, to see what it would cost
- Consider your heating options, and if any seem possible, follow up with phone calls

absorbs the CO_2 before it is released in the burning. Wood smoke is a local pollutant that can be an irritant to nearby residents, however, so a wood stove is really only a rural option, and even then, you need to be mindful of your neighbors. It is important to use a super-efficient stove and to burn only dry, sustainably harvested wood. Another option is to ask for a 20% biodiesel mix in regular heating oil, which Columbia Fuels offers on Vancouver Island, reducing the CO_2 emissions by 16%.

If none of this seems easy, you are right. Retrofitting our homes to make them carbon neutral is a big challenge that needs generous financial incentives, carbon taxes and tax breaks to persuade people to make the necessary investments. This is one reason why we need to become engaged at a political level to push for this kind of change. (See Solution #10.)

6

Change the Way You Travel

There is no science on how we are going to adapt to 4 degrees warming. It is actually pretty alarming.

— Professor Neil Alger, Tyndall Centre for Climate Change Research

A hundred years ago, everyone traveled by foot, horse or bicycle. Today, thanks to the magic of fossil fuels, there are 700 million vehicles on the planet. Together, they are responsible for 2.8 billion tonnes of CO_2 a year, producing 7.6% of the world's CO_2 emissions.

Every time your car burns a liter of gas, it releases 2.34 kg of CO_2 into the atmosphere. Every time it burns a gallon, it releases 19.56 lbs.

If you are an average motorist, driving 20,000 kilometers (13,500 miles) a year in a car that consumes 9 liters per 100 km (26 mpg), you'll burn 2,222 liters (587 gallons), adding 5.2 tonnes of CO_2 to your carbon footprint. Can you reduce this and still get where you need to go?

Martin Golder

Victoria architect Martin Golder sits on his summer chopper. Originally a kid's bike, it sports a 750w wheel motor running on 48 volts, and is capable of some very exciting speeds.

Walk or Cycle

Our ancestors walked around the world: our legs are designed by evolution for walking. For every mile that you walk or cycle instead of driving, you'll eliminate that much CO_2, while becoming generally fitter and happier. For shopping and deliveries, there are backpacks and bicycle trailers. If you cycle to work each day instead of driving a 20 mile round trip, you'll reduce your carbon footprint by 1.7 tonnes a year.

Ride an Electric Bike or Scooter

An electric bikes uses almost no power, so your running costs will be close to zero, and they can zip along at 35 km/hr (22 mph), with a range of up to 50 km. An electric bike kit costs $400 to $1400. If you use green electricity, you'll have no emissions at all. The Vectrix electric motorbike can travel at up to 100 km/hr (62 mph) with a range of 55–90 km (35–55) miles, for only pennies per kilometer.

Take the Bus

In some cities, the bus service is superb, with electronic timetables that tell you when the next bus is coming. In others, it's an embarrassment, and you'll need a lot of citizen pressure to make it effective and reliable. (See #25.) For longer distances, inter-city coaches can be great, especially when they have laptop plug-ins and serve cappuccinos — and when nextbus.com tells you when the bus is coming.

Take the Train

This is a very comfortable way to reduce your emissions, and a great alternative to flying up to 800 kilometers (500 miles.)

- Carpool savings: carpool.ca/calculator.asp
- Carpooling (Canada): carpool.ca
- Carsharing: carsharing.net
- Electric bikes: electric-bikes.com
- Liftshare UK: liftsharesolutions.com
- Nextbus: nextbus.com
- Public Transit Calculators: publictransportation.org
- Ridesharing: erideshare.com
- SPUD: spud.ca
- Teleconferencing: tinyurl.com/y3q9vq
- Top Ten Reasons to Buy an Electric Bike: squidoo.com/electricbicycle

Ridesharing

If you share a ride with one other person, you'll cut your emissions in half. Share with four people, you'll reduce them by 75%. For one-off rideshares or regular commuting, search on erideshare.com or Craigslist, with the name of your city. In Britain, Liftshare has become hugely popular, with more than 300,000 members in 2009.

Carsharing

In 2007, 200,000 people in North America were members of car share groups — 300,000 in Europe, where carsharing started. Philadelphia's PhillyCarShare, has 35,000 members, 10,000 of whom have given up their cars. Carsharing gives you the choice to use a car when you need it, while also using other ways to get around.

Teleconferencing

If you fly economy class from Los Angeles to Washington DC, a round-trip of 4600 miles, you'll produce 1.8 tonnes of CO_2e. (See #9) If you organize a teleconference, you will avoid the emissions as well as the time and hassle. If six people teleconference instead of flying, they'll save $2,400 in return tickets, which could be put toward the cost of a top-of-the-line teleconferencing room.

Shopping without the Car

Carless people do it all the time. Your options include joining a carshare group; renting or buying a bicycle trailer; shopping locally so that you can walk home with a wheeled cart or a backpack; and joining a home delivery service such as Small Potatoes Urban Delivery (SPUD) in Western Canada.

Actions

- Start a travel diary, and see how many trips you could do without a car
- Visit a bike shop and research the choices
- Research electric bikes on the Internet
- Research local ridesharing and carsharing possibilities

CO_2 emissions from travel[1]	Grams per passenger km
Walk/Bike	0
Electric bike	6
Electric scooter	12
Rideshare, EV	23
Bus, full	23
Electric motorbike	25
Inter-city coach[2]	29
Train, full	40
Rideshare 4 people	55
Bus, average	76
Train, average	92
Car, 9.5 litres/100km (25 mpg) driving alone	220

7

Drive a Greener Car

> For the sake of our security, our economy and our planet, we must have the courage and commitment to change.
>
> — President Barack Obama

So which is the best car to drive? The only truly green car is an electric vehicle (EV) powered by renewable energy. We need a flood of EVs to meet the pent-up demand. Having said that, which are the least harmful? In Europe, the way to compare carbon emissions is grams per kilometer (g/km), which will soon become the global standard. The chart shows some of the best cars, not all of which are available yet in North America.

In Europe the goal for all new cars is 120 g/km by 2015. California's goal is 127 g/km by 2016. In 2007 more than 200 European cars produced less than 120 g/km, so they are making good progress.

Biodiesel Cars

Biodiesel is controversial, because the biodiesel you can buy at the pump (B-5, B-20) comes from crops or crop by-products that create CO_2 and nitrous oxide emissions during farming. Some Indonesian biodiesel produces 28 times more CO_2 than gasoline, because rainforest is cleared to grow the crop. Even straight veggie oil from restaurant wastes is not fully green. The CO_2 is carbon neutral because the oil-producing plants absorbed CO_2 while growing, but while most biodiesel emissions are far lower than those of a regular car, the nitrogen oxide emissions, which contribute to smog, are higher.

Electric Cars

The Plug-In Hybrid EV (PHEV) is generating excitement because it runs on both fuel and electricity, eliminating battery constraints. If the electricity is green, it has very low carbon emissions, with typical North American electricity it produces 114 g/km, compared to the Prius at 104 g/km.

When an EV runs on green power, its lifetime carbon emissions are only from its manufacturing, at 6 tonnes of CO_2. When it runs on coal-fired electricity, they are almost as much as a regular car, at 23 tonnes. The best of all options is a DIY EV conversion.

Is it Worth Buying a New Car?

For a Ford Taurus, assuming 22 mpg and a 14-year vehicle lifetime, 10% of its lifetime energy use

How Much CO2?[1]	g/km
EV green power	0
Biodiesel SVO B-100	0
Plug-in Prius, green power[2]	56
Smart fortwo	88
EV grid power[3]	94
Ford Fiesta Econetic	98
VW Polo BlueMotion	99
Toyota iQ 2	99
Toyota Prius	104
Mini One Cooper	104
Toyota Aygo	108
Peugot 107	109
Honda Civic	109
Ford Fiesta	110
Fiat 500	110
Plug-in Prius, grid power	114
Average sedan	165
Hummer H3	346

Actions

- Calculate your car's CO_2 emissions, using a carbon calculator
- Consider your options. Is it time to buy a greener vehicle?
- Practice climate friendly eco-driving
- Buy a Scan Gauge, for real-time MPG data
- Buy 4 LED tire pressure lights (search Amazon)

- 15 Green Cars: grist.org/feature/2007/08/13/cars
- CalCars (PHEVs): calcars.org
- Car Fuel Data (UK): vcacarfueldata.org.uk
- DIY Electric Car: diyelectriccar.com
- EcoDriving: ecodrivingusa.com
- EV World: evworld.com
- Fuel Economy Guide: fueleconomy.gov
- Greener Cars: greenercars.org
- Green Car Congress: greencarcongress.com
- Green Vehicle Guide: epa.gov/autoemissions
- Hypermiling: hypermiling.com
- New Car CO_2 emissions (UK): dft.gov.uk/ActOnCO2
- New vehicles guide (Canada): oee.nrcan.gc.ca/transportation/personal
- Scan Gauge: tinyurl.com/47tkee
- Solar Vehicles: builditsolar.com/Projects/Vehicles/vhehicles.htm
- The Green Car Guide (UK): carpages.co.uk/co2

occurs during manufacturing, and 90% during use.[4] Renault estimates that the CO_2 from manufacturing their cars comes to 6 tonnes — the equivalent of driving 40,000 kilometers (25,000 miles).[5]

EcoDriving Questionnaire

Ecodriving can reduce your emissions by up to 30%. Score 0–10 for each question:

1. Do you cut your car's weight by removing roof racks, trailer hitches and assorted junk? ___
2. Do you check your tire pressure regularly? With too little air, you'll use more fuel. ___
3. Do you avoid driving distances less than three kilometers (two miles?) A cold engine produces 40% more emissions. ___
4. Do you avoid revving up like a Formula 1 car when you start the engine? ___
5. Do you drive straight off and not waste energy waiting for the engine to warm up? ___
6. Do you accelerate gently rather than pushing your foot to the floor? ___
7. Do you try to optimize your speed at 55 mph / 90 kph? For every 5 mph you travel faster, you burn 5–10% more fuel. ___
8. Do you avoid idling by switching off if you have to wait more than 10 seconds? ___
9. Do you coast up to a red traffic light, to minimize braking? ___
10. Do you keep a safe driving distance from other cars, to minimize braking? ___

Your Eco-Driving Score: _____

Dr. Mary-Wynne Ashford enjoying her Smart car. Mary-Wynne is author of Enough Blood Shed: 101 Solutions to Violence, Terror and War *(New Society Publishers).*

8

Take a Climate - Friendly Vacation

> Nobody made a greater mistake than he who did nothing because he thought he could only do a little.
> — Edmund Burke

Hands up who has never flown to visit a loved one. And what about that oh-so-tempting holiday in Paris or Costa Rica?

It's a dilemma that we all face, once we learn how flying contributes to global warming. On a one-way flight from San Francisco to New York, a distance of some 4,200 km (2,600 miles), your share of the aircraft's kerosene will produce 426 kg of CO_2.

This is not the whole story, however. When a plane flies at altitude, its exhaust produces methane, water vapor, ozone and soot, all of which trap heat. It also produces contrails, which create high-level clouds, trapping more heat. When you include these, the overall impact (or radiative forcing) is up to 2.7 times greater than the CO_2 alone — so that one-way flight really produces 1.15 tonnes of CO_2e, or 2.3 tonnes for the return flight.[1] Short flights use more fuel per passenger, because an aircraft needs proportionally more fuel to rise to altitude than to cruise at altitude.

What are the solutions? For shorter distances, use a bus or train. Don't fly unless you really have to. A friend in Canada with a granny in Holland found that talking to her every week using Skype and a webcam, so that her granny could see her face, removed the pressure to fly so often.

When you do fly, be sure to buy offsets to neutralize your emissions. (See #10). This requires the use of a good carbon calculator, and most, alas, are *not* good when it comes to flying. Most count only CO_2, not the full radiative forcing, and even the plain CO_2 numbers can vary by as much as 100%. The two best calculators are at chooseclimate.org/flying and atmosfair.de.

Because a first class seat takes up more space than an economy seat, its share of an aircraft's emissions is 1.5 times more, which is reflected

Solar Sailor's 100-passenger charter vessel on Sydney harbour, Australia

CO₂e from Flying Short — under 1000 km Medium — up to 5,000 km Long — more than 5,000 km	Grams of CO₂e per passenger-kilometer
Short, economy class	400
Short, business class	600
Medium, economy class	250
Medium, business class	375
Long, economy class	270
Long, business class	405

only in the chooseclimate.org calculator. Toronto to Ottawa return (708 km) will produce 283 kg of CO_2e in economy, or 424 kg in first class. The same journey by train will produce 65 kg of CO_2, or 34 kg if the train is full. For the longer-term solutions for flying, see Solution #50.

Climate-Friendly Holidays

Cycling, hiking, camping or lazing in the sun by a local beach or lake all have a zero carbon footprint: this kind of holidaying will almost certainly reduce your carbon footprint by more than if you had stayed at home.

If you like boating, the most climate-friendly ways to travel on water are sailing boats, kayaks, canoes, rowboats and boats with an electric motor powered by solar or other green energy. The Canadian Electric Boat Company produces the Fantail 217, a very graceful inland-waters boat that can run for eight hours before it needs a recharge. For motorboats, the two-stroke engines are terrible water polluters and really should be phased out: one liter of spilt oil can contaminate up to two million liters of water.

Other holiday habits have a larger carbon footprint. An hour of jet-skiing will burn about 7.5 liters (2 gallons) of gas and produce 17.55 kg of CO_2. A week on a cruise ship will burn 454 liters (120 gallons) of fuel, producing 1.4 tonnes of CO_2.[2]

If you like to go RVing, a large vehicle may burn 31 liters per 100 km (7–8 mpg), so every 1,000 kilometers that you travel will produce 725 kg of CO_2. A smaller RV might produce half of this. The less you travel and the more you laze, the smaller your footprint will be.

- Atmosfair Flight Calculator: atmosfair.de
- Biodiesel/SVO Discussion Forum: biodiesel.infopop.cc
- Boat Carbon Footprint: boatcarbonfootprint.com
- Choose Climate Flight Calculator: chooseclimate.org/flying
- Electric boats, Canada: electricboats.ca
- Electric Boats, UK: electricboats.co.uk
- Electric Boat Association (UK): electric-boat-association.org.uk
- Straight vegetable oil: journeytoforever.org/biodiesel_vehicle.html

Actions

- List all the flights you took last year, and calculate their carbon footprint
- Plan your close-to-zero carbon vacation

CO_2 per hour of travel	Kg of CO_2
Boat — 10 HP 4-stroke[3]	5.9
Boat — 200 HP 4-stroke	118
Boat — 1000 HP 4-stroke	590
Car — 20 mpg/11.4 liters per 100 km (90 kph/55 mph)	11
Car — 50 mpg/4.7 liters per 100 km	4

If you power your RV — or any diesel vehicle — with biodiesel in the form of straight veggie oil (SVO) from waste restaurant fats, your carbon footprint will be effectively zero. Keith Addison and Midori Hiraga from Japan are engaged in an overland "Journey to Forever" educational tour from East Asia to southern Africa, linking schools and communities along the way and sharing their knowledge about a practical sustainable innovations — while driving two 30-year-old Land Rovers powered by biodiesel SVO.

9

Change Your Consumer Habits

Besides the noble art of getting
things done, there is the noble
art of leaving things undone.
The wisdom of life consists in the
elimination of non-essentials.

— Lin Yutang

What a fabulous world. The shops are stuffed with everything a person could ever want. But where does it come from? From Earth's forests, fields and oceans. And where does it go when it's no longer wanted? If not recycled, to Earth's landfills and incinerators. Earth has only 1.8 hectares of available productive biological capacity for every human, but our average footprint is 2.7 hectares, and Earth's other species have to live as well.

We worry about financial debt, but not ecological debt. Our lifestyles have overdrawn our account with the Earth by 28% — and our debt increases every year. Each American, on average, is responsible for 11 tonnes of CO_2 a year from the manufacture and delivery of the things we consume.[1] Creating and shipping junk mail in the USA produces more CO_2 than 9 million cars, and 44% goes straight to the landfill, unopened.[2]

- Carbon Rally Challenge: carbonrally.com
- Change Everything: changeeverything.ca
- Conscious Consumer: newdream.org/cc
- Ecological Footprint Quiz: myfootprint.org
- Ecological Footprint (UK): ecologicalfootprint.com
- Freecycle Network: freecycle.org
- New Road Map Foundation: financialintegrity.org
- North West Earth Institute: nwei.org
- Red Dot Campaign: reddotcampaign.ca
- Simple Living America: simplelivingamerica.blogspot.com
- Simple Living Network: simpleliving.net
- Simplicity Scorecard: simplelivingamerica.org/survey
- Stop Junk Mail: 41pounds.org
- Sustainable Travel International: sustainabletravelinternational.org
- *The Story of Stuff* (20-minute video): storyofstuff.com
- *YES! Magazine, A Journal of Positive Futures:* yesmagazine.org

The Carbon Impact of Small Things

Item	CO_2e
100 Google searches	20 grams[3]
Plastic shopping bag	45 grams[4]
Small packet of chips	75 grams[5]
Liter plastic bottle	80 grams[6]
Paper shopping bag	135 grams[7]
Liter of imported water	250 grams[8]
Daily paper	326 grams[9]
Large carton orange juice	1.7 kg[10]
1 kg of beef	16 kg[11]
One year's junk mail	42 kg[12]
52 bags of garbage a year	338 kg[13]

Five Ways To Reduce your Carbon Footprint

- Use cloth shopping bags instead of plastic and paper: save 30 kg a year.[14]
- Cancel all your junk mail: save 42 kg a year.[15]
- Recycle five bottles a week: save 117 kg a year.[16]
- Recycle one newspaper a day: save 41 kg a year.[17]
- Buy 100% post-consumer recycled paper, instead of virgin paper: save 1.48 kg per kg of paper.[18]

... must we create so much garbage?

Live More Simply

There's a quiet revolution happening called voluntary simplicity. By reorganizing their priorities and spending less on stuff, people are reclaiming their lives, discovering nature, art, families, local communities and time for meaningful activity. One way to explore the movement is to join or form a Northwest Earth Institute study group.

Are You Winning the War against Stuff?

Score yourself on each question from 0–10.

1. Do you take time each day to pause, breathe in and be grateful? ___
2. Do you know how big your ecological footprint is? ___
3. Do you avoid buying bottled water? ___
4. Do you always use cloth shopping bags? ___
5. Do you recycle everything you can? ___
6. Do you Freecycle or give away stuff you want to get rid off? ___
7. Do you compost food and garden wastes? ___
8. Have you eliminated your junk mail? ___
9. Do you sell or give something away to balance each new purchase? ___
10. Have you visited your local landfill? ___

Total: ___

Invest in a Climate-Friendly World

Which would you rather invest in: Tobacco? Oil? Or companies working to develop solutions to global warming? 1,000 mutual funds specialize in

- CERES, Investors and Environmentalists: ceres.org
- Climate Friendly Banking (Canada): climatefriendlybanking.com
- Community Investing: socialinvest.org/projects/communityinvesting.cfm
- Green Money Journal: greenmoneyjournal.com
- Social Funds: socialfunds.com
- Social Investment Forum: socialinvest.org
- Social Investment Organization, Canada: socialinvestment.ca
- Socially Responsible Mutual Fund Performance Chart: socialinvest.org/resources/mfpc
- Sustainable Stocks: sustainablebusiness.com/Stocks

Actions

- Take the Ecological Footprint Quiz
- Take the Carbon Rally Challenge
- Fill in Simplicity Scorecard
- Ask your financial adviser about moving your investments into green funds

socially responsible investment. You can invest in a green mutual fund or choose from the Top 20 Sustainable Stocks listed on the Sustainable Business website.

Ask your investment adviser to green up your investments. If he or she tells you they won't make money, it's time to change your advisor, for the sector is exploding, with $117 billion invested in renewable energy in 2007 — a 20-fold increase since 1995.

You can also invest in community funds, which bring capital and financial services to low-income areas. In 2007 the Self-Help Credit Union had a foreclosure rate of only 0.62%, compared to a market average of 8% for sub-prime mortgage lenders.

If your savings are with a pension fund, get together with your colleagues and ask your employer, college or labor union to divest itself of businesses that contribute to global warming and switch to socially responsible stocks.

10

Become a Climate Champion

> When people ask me what they
> should do, I reply "Get informed, get
> outraged, and then get political."
> — Joseph Romm

Finally! If you have worked your way through these Personal Solutions, you deserve a warm handshake of appreciation. You should have no problem in reducing your carbon footprint by 10% — or more. So now two questions arise. First, what else can I do? And second, can I offset my remaining emissions in a reliable manner?

The answer to the first question opens a door to a whole new landscape, with several paths that you could take. One leads to a living room, where you invite a group of neighbors to organize an eight-week Climate Challenge Circle, based on these Solutions. (See #11.)

Another leads to a meeting in your school (#12), college (#13), place of worship (#14) or workplace (#32), where you discuss setting up a Climate

Action Team, following the ideas laid out in the pages that follow. Both of these paths will bring you new friends, deepen your learning and lead you into new areas of understanding and change.

A third path leads to an existing group, either a local independent group or one connected to a larger network such as the Sierra Club, Step It Up or Focus The Nation (see #20), which will introduce you to the amazing network of groups all over the world that are working to make a difference.

As you proceed along these paths, your creativity will come alive and you will hopefully get a sense that it is possible to change the world. One day you may hear someone refer to you as a champion, and you'll think "Me?" And the answer will be "Yes — you."

The Solar Electric Light Fund helps villagers in the developing world install their own solar systems, replacing the use of kerosene, for which it also sells carbon offsets. self.org

SOLAR ELECTRIC LIGHT FUND

Actions

- Choose a carbon offset organization, and buy offsets to neutralize your emissions
- List five reasons why you should become a Climate Champion, and five reasons why not. Decide which list carries the most weight
- Congratulate yourself on having come this far, and invite some friends over to tell them what you've been doing

Good Carbon Offset Organizations

- AgCert/Driving Green: agcert.com
- Atmosfair: atmosfair.de
- Carbon Offset Companies: tufts.edu/tie/carbonoffsets/carboncompanies.htm
- Climate Care: climatecare.org
- Climate Trust: climatetrust.org
- Las Gaviotas: gaviotasoffsets.org
- Native Energy: nativeenergy.com
- Offsetters: offsetters.ca
- *Purchasing Carbon Offsets — a Guide for Canadians*: davidsuzuki.org/Publications/offset_vendors.asp
- Solar Electric Light Fund: self.org
- Voluntary Carbon Offsets: tufts.edu/tie/carbonoffsets/index.htm
- *A Consumer's Guide to Retail Carbon Offset Providers*: cleanair-coolplanet.org
- The Gold Standard: cdmgoldstandard.org

Offset Your Personal Emissions

Before we move on, there is an important piece of business to take care of. Even if you are doing your best, you will still have a carbon footprint. Can you "offset" it and become carbon neutral?

As author of this book, I work to reduce my emissions, but I also fly to conferences where I speak about the solutions to global warming, hopefully inspiring people to act. Each January, I tally up my emissions and pay $15 a tonne to the Solar Electric Light Fund (SELF), based in Washington DC.

SELF is a non-profit organization that brings solar power to villages in Bhutan, Nigeria and elsewhere. When a family installs a 50-watt solar system they stop using kerosene to light their lamps. This ends the air pollution and risk of fire associated with kerosene and stops the use of a fossil fuel. SELF calculates how much CO_2 is saved by not burning kerosene, and it makes the saved emissions available as offsets.

SELF's offsets meet two critical requirements: (a) The carbon reduction *would not have happened otherwise*. If SELF had not made it possible for the villagers to install solar power, they would still be burning kerosene. (b) The reduction is real, and not something that might happen in the future.

Climate Care, based in Oxford, UK, is another good offset seller. They support projects in the developing world such as substituting treadle pumps for diesel pumps, and ovens that burn briquettes made from farm wastes instead of natural gas. Other good carbon offset organizations support energy efficiency in the developing world and low-income communities and protect tropical forests that would otherwise have been cut.

Tree-planting offsets sell the hope that the trees will absorb your emissions when they are big enough — provided they don't die from drought or are cut down for firewood. Offsets based on the sale of renewable energy certificates (RECs) are also questionable, because it is hard to understand how the income gathered is used to generate new green power that would not have happened otherwise.

This is the question that must be asked of every carbon offset project: "Would this project be happening anyway, without my offset?" If the answer is yes, you have a piece of paper, but not a valid offset. The best carbon offset programs are verified by a third party and approved by "The Gold Standard" so that you know they are for real. For an in-depth report, see the *Consumer's Guide*.

11

Solutions for friends, family and neighbors

Start locally and build. Start small and grow. Start in your house, then move to your school, your book club, your gym, your church, your temple, your city.
— Laurie David, Founder of the Stop Global Warming Virtual March

You understand what's at stake, and the urgency with which we must reduce our carbon footprint — but how do you communicate this to others to get them involved?

The easiest way to start is with your friends and neighbors. This solution offers four possible approaches. Whichever you choose, these principles will help:

1. Be well informed. Do your homework in advance, so that you are confident in your understanding and able to respond to people's questions.

2. Be positive. Don't come over all heavy and gloom laden. What matters is not whether you are optimistic or pessimistic but whether you are determined or defeated.

3. Be inclusive. You will obtain lots of support if you find out what else is happening and meet the people involved.

4. Be quick. Don't spend too long on planning — you can learn as you go along.

5. Be irreverent. Remember the Fifth Law of Sustainability: if it's not fun, it's not sustainable.

Host a Small Climate Party

Put together a short presentation on global warming and why you are concerned, and invite friends and neighbors to join you for a potluck dinner and discussion in your living room.[1] Don't invite people you know to be skeptical or cynical — stick with those you feel good about. After introductions, give your presentation. PowerPoint is good if you've got the skills; if you've not got a projector, people can cluster around your computer. You could either make your own; or adapt one made for this purpose at theclimatechallenge.ca. After you have eaten, open it up for discussion, and invite people to think ways in which they could reduce their personal carbon footprints.

Form a Climate Challenge Circle

Take the material in Solutions #1–#10 and use it as the basis for an eight-week group that meets in someone's home. Being a group will enable you to support each other, share local research and report on your progress as you reduce your carbon footprints — a bit like Weight Watchers. If you go to theclimatechallenge.ca, you will find a week-by-week guide to the eight-week course designed especially for this purpose, which has been tested by the Cowichan Carbon Busters group in British Columbia, Canada.

As an alternative, you could host:

- A four-week Low Carbon Diet course organized by the Empowerment Institute: empowermentinstitute.net/lcd
- A discussion course developed by the Northwest Earth Institute titled Global: Warming: Changing CO2urse, nwei.org
- A DIY National Teach-In: nationalteachin.org/diyteachin.php
- A Be the Change Discussion Circle: bethechangeearthalliance.org/circles/start

Start a Climate Action Group

Local groups are forming all over the place, from the Bovey Climate Action in Bovey Tracey, Devon, England, to the Bass Coast Climate Action Group

Best Reading

- *An Inconvenient Truth* (book and DVD), by Al Gore: climatecrisis.net
- *Fight Global Warming Now: The Handbook for Taking Action in Your Community* by Bill McKibben: billmckibben.com
- *Heat — How to Stop the Planet from Burning* by George Monbiot: monbiot.com
- *Our Choice*, by Al Gore
- *Six Degrees — Our Future on a Hotter Planet* by Mark Lynas: marklynas.org
- *Ten Technologies to Save the Planet,* by Chris Goodall
- *The Solution is You! An Activist's Guide* by Laurie David: lauriedavid.com
- *The Weather Makers: How we are changing the climate and what it means for life on Earth* by Tim Flannery: theweathermakers.com
- *Now or Never: Why we Need to Act Now to Achieve a Sustainable Future* by Tim Flannery:

Best Climate and Energy News

- Climate Progress: climateprogress.org
- BBC: news.bbc.co.uk
- CalCars: CalCars.org
- EV World: EVWorld.com
- Planet Ark: planetark.org
- Real Climate: realclimate.org
- Renewable Energy Weekly: renewableenergyaccess.com/assets/newsletter
- The Daily Grist: grist.org
- Guardian: guardian.co.uk/environment
- The Independent: independent.co.uk
- Worldwatch Institute: worldwatch.org

Best Videos

- *Airsick:* thestar.com/fpLarge/video/294982
- *An Inconvenient Truth:* climatecrisis.net
- *Burn Up:* imdb.com/title/tt1105836/
- *Climate Change (AAAS):* tinyurl.com/23btk2
- *e2 PBS Film series:* e2-series.com
- *Environmental videos:* ecofootage.com/green-online-video
- *Escape from Suburbia: Beyond the American Dream:* escapefromsuburbia.com
- *Everything's Cool:* everythingscool.org
- *Fuel:* thefuelfilm.com
- *Homegrown:* homegrown-film.com
- *Kilowatt Ours:* KilowattOurs.org
- *Marching for Action on Climate Change: Five Days Across Vermont* with Bill McKibben and Friends: jancannonfilms.com/climatechange.htm
- *Revolution Green:* revolutiongreen.com
- *The 11th Hour:* 11thhouraction.com
- *The Age of Stupid:* ageofstupid.net
- *The Great Warming:* thegreatwarming.com
- *Too Hot Not to Handle:* hbo.com/docs/programs/toohot
- *Wake Up, Freak Out:* wakeupfreakout.org
- *Who Killed the Electric Car?* whokilledtheelectriccar.com

in Australia and the Boston Climate Action Network in Massachusetts. In most of these groups, the organizers are ordinary people who jump in and learn as they go along. (See #16.)

Organize an Event

It could be a rally, a walk, a festival, a bike-ride, a polar-bear plunge, a petition-drive, a teach-in, a mural, a fast, a concert, a day of prayer, a film festival, a 100-Mile Feast or improvised street art with 30 people in polar bear costumes demanding a climate-friendly earth by 2030. It doesn't have to be big — small can be very effective if it wins the attention of the media and politicians. For a great account of the 1,400 "Step It Up" actions that happened across the US on April 14, 2007, and practical advice about how to organize an event, read Bill McKibben's great book, *Fight Global Warming Now.*

12

Solutions for Schools

Greening school design provides an extraordinarily cost-effective way to enhance student learning, reduce health and operational costs and, ultimately, increase school quality and competitiveness.

— Gregory Kats, Greening America's Schools

In February 2009, hundreds of schools across the US took part in the National Teach-In on Global Warming Solutions, participating in the national webcast and following up with local discussions about how they could contribute.

School Carbon Calculators

- earthteam.net/GWCampaign/calculate.html
- epa.gov/climatechange/wycd/school.html
- dott07.com/flash/dott_1024.htm

Green Schools

- Build Green Schools: buildgreenschools.org
- EnergySmart Schools: 1.eere.energy.gov/buildings/energysmartschools
- Green Schools Alliance: greenschoolsalliance.org
- Green Schools Checklist: epa.state.il.us/p2/green-schools/green-schools-checklist.pdf
- Green Schools Program: ase.org/section/program/greenschl
- International Walk to School: iwalktoschool.org
- Kids for Saving Earth: kidsforsavingearth.org
- Solar Schools: solarschools.com
- Students Leading the Way — Energy Saving Success: tinyurl.com/2bbxml
- The Edible Schoolyard: edibleschoolyard.org
- The Green Squad: nrdc.org/greensquad
- Wind Energy for Schools: windpoweringamerica.gov/schools_projects.asp

At Akron Westfield Community School, Iowa, students helped to install their school's 600-kW wind turbine. Many schools have installed solar systems, integrating the data into their science and business studies. In a video contest in 2008, McTavish Elementary School, near Victoria, BC, was voted the greenest school in North America for reducing its waste by 80% by composting, paper recycling and soft-plastics recycling.

In Britain all seven classes at the St. Francis of Assisi Academy in Liverpool have a garden, where teenagers grow food and plants, integrating it into their math and geography classes.

All over Europe and North America, schools are taking the initiative to stop buses and cars from idling, to walk and cycle to school, to eat local organic food and grow their own food, to increase recycling, to stop using toxic cleaners and pesticides, and to bring global warming into the curriculum. Schools are also working to rebuild the broken connection with nature, with students spending time in nearby forests, wetlands, rivers and farms.

Getting Started

Week 1: Organize a lunchtime meeting with your friends. Go around the circle and gather everyone's ideas as to what you could do as a school. Download copies of this Solution from theclimatechallenge.ca, and ask each student to explore one of the listed websites.

Week 2: Brainstorm for ideas. Focus on those that are achievable within three months, and choose the one that is most doable and has the most support. Give your group a name, decide on

Students from Oak Bay High School, Victoria, Canada, with the solar PV array they fundraised to install on their school's physics lab.

GARRETT BRISDON

your long-term goals, and create a page for your group on Facebook or your favorite networking site.

Week 3: Meet with a teacher you know is sympathetic, and ask for his or her support. You will need everyone's support, including the support staff, teachers, principal, school board and parents.

Week 4: Get to work on the project you have chosen. At the same time, start discussions with staff about how the whole school could become carbon neutral and how long it would take.

If you are a member of staff, organize a meeting with other staff members and discuss how you could make your school more green, using the resources listed below. There is a free *Climate Challenge Teacher's Guide* at TheClimateChallenge.ca that uses this book as the basis for class projects.

Build Green Schools

A national US survey of 30 green schools demonstrated that they cost less than 2% more to build than conventional schools, but provide financial benefits that are 20 times as large, saving money by increasing earnings, retaining teachers, reducing colds, flu and asthma, and using less energy and water. A Washington State study showed that high performance lighting caused a 15% fall in absenteeism and a 5% increase in student test scores.[1] "For the average conventional school, building green would save enough money to pay for an additional full-time teacher."[2]

Green Curriculum
- *Climate Challenge Teachers Guide:* earthfuture.com/challenge
- Climate Curriculum: worldwildlife.org/climate/curriculum/item5944.html
- Climate Change Education Portal: climatechangeeducation.org
- Energy Kid's Page: eia.doe.gov/kids
- EPA Teaching Center: epa.gov/teachers
- Focus the Nation: focusthenation.org
- Green Learning: greenlearning.ca
- Green Teacher Magazine: greenteacher.com
- *How We Know What We Know About Our Changing Climate: Scientists and Kids Explore Global Warming,* by Lynne Cherry and Gary Braasch, Dawn Publications, 2008
- Lesson Plans from California Green Schools: ase.org/content/article/detail/2053
- Roofus' Solar & Efficient Home: 1.eere.energy.gov/kids/roofus
- Sustainable School (UK): suschool.org.uk
- Teaching About Climate Change: greenteacher.com/tacc.html
- The Climate Challenge Game, by Guy Dauncey: tinyurl.com/2h42nj
- Wind with Miller: windpower.org/en/kids
- *The Down-to-Earth Guide to Global Warming* by Laurie David and Cambria Gordon: scholastic.com/downtoearth

13

Solutions for Higher Education

C'mon kids! Wake up and smell the CO_2! Take over your administration building, occupy your university president's office, or storm in on the next meeting of your college's board of trustees until they agree to make your school carbon neutral.

— Thomas Friedman

Leith Sharp managed Harvard's Office for Sustainability from 2001–2008, during which time she was able to achieve some deep organizational change and many practical results, building her staff from one to 24 plus 30 part-time students. Harvard's many successes include 50 LEED projects, staff green teams, a sustainability pledge, a $12 million green campus loan fund — even a Green Skillet Competition between the kitchens. Her experience makes her one of the best sources of advice. Among other things, Leith has put her finger on the need to overcome financial barriers and disincentives, and end the disconnect between capital and operating budgets.

Buildings and Energy: At State University of New York, Buffalo, students and staff saved $10,000 in a single day as part of an ongoing campaign that has saved more than $100 million. In Vancouver, UBC's $30 million EcoTrek energy retrofit was financed by annual savings of $2.6 million. Carleton College, in Northfield, Minnesota, built a 1.65 MW wind turbine that meets 40% of the college's power needs. Napa Valley College, California, installed a 1.2 MW solar array. Many institutions are buying 100% green energy. At Harvard, a green team is now required on all new building projects.

Leith Sharp

Transport: When Cornell University, in Ithaca, NY, raised its parking fees, favoring carpooling and providing free public transit to anyone without a parking pass, it reduced its car commuter miles by 10 million miles a year, saving $36 million.[1] Many colleges have adopted UPASS, where students pay for transit in their student fees. Stanford added 0.2 million square meters of new buildings in the 1990s without any increase in peak period auto trips, thanks to integrated transport planning.

Procurement: The US's 4000 universities spend $190 billion a year on goods and services — more than the GDP of all but 20 nations. Rutgers University, NJ, has developed comprehensive green contract specifications and established a Green Purchasing Institute. At Camosun College, Victoria, Canada, students voted to pay a 15-cent monthly levy on their student fees so that the whole college would use 100% recycled paper.

Waste: At Dartmouth College, New Hampshire, students who buy a $20 kit with reusable mugs and napkins save more than $1 a day. Penn State University, Pennsylvania, is collecting its residential organic wastes. All ten campuses at the University of California have pledged to achieve zero waste by 2020.

Food: The University of California in Berkeley has a certified organic kitchen and organic salad bar. Evergreen State College, WA, has a half-hectare organic farm. Students at Swarthmore College, Pennsylvania, have created an organic food garden in front of their dorm.

Lifestyle: At Harvard, the peer-to-peer Graduate Green Living program encourages graduate and

- *Green Campuses: The Road from Little Victories to Systemic Transformation,* by Leith Sharp: greencampus.harvard.edu/about/documents/green_universities.pdf
- *Greening the Ivory Tower,* by Sarah Hammond Creighton, MIT, 1998
- *Planet U: Sustaining the World, Reinventing the University,* by Michael M'Gonigle and Justine Stark, New Society, 2006
- *Harvard Green Campus Presentation* by Leith Sharp: tinyurl.com/7mdpt8
- *Transportation and Sustainable Campus Communities,* by Will Toor and Spenser Havlick, Island Press, 2004

- Campus Climate Challenge: climatechallenge.org
- Campus Ecology: nwf.org/campusecology
- Campus Zero Waste: grrn.org/campus
- Cornell Sustainable Campus: sustainablecampus.cornell.edu
- Focus The Nation: focusthenation.org
- Greening of the Campus: bsu.edu/greening
- Harvard Office for Sustainability: greencampus.harvard.edu
- Harvard Sustainability Pledge: greencampus.harvard.edu/pledge
- Higher Education Associations Sustainability Consortium: heasc.net
- International Sustainable Campus Network: international-sustainable-campus-network.org
- New Energy for Campuses: apolloalliance.org/downloads/resources_new_energy.pdf
- Oberlin's Office of Sustainability: oberlin.edu/sustainability
- Presidents' Climate Commitment: presidentsclimatecommitment.org
- Princeton Green Rating Honor Roll: princetonreview.com/green-honor-roll.aspx
- Second Nature: secondnature.org
- Student Environmental Action Coalition: seac.org
- Tufts Gets Green: sustainability.tufts.edu
- UBC EcoTrek: ecotrek.ubc.ca
- UBC Sustainability Office: sustain.ubc.ca
- University Leaders for a Sustainable Future: ulsf.org

undergraduate students to increase their recycling and reduce their use of water and energy. In 2006 more than 7,000 students, staff and faculty took the Sustainability Pledge to reduce their environmental impact.

Teaching: In 2006 St. Olaf College, Minnesota, made sustainability the guiding theme for the whole academic year. Green Mountain College, Vermont, uses the environment as a central theme in its core curriculum. All architecture and engineering school assignments should require carbon zero designs.

To achieve good results, you need (a) a strong commitment by the vice chancellor and senior management; (b) full-time staff dedicated to environmental management; (c) an inspiring and compelling goal; (d) a written environmental policy that is publicly available; (e) the active engagement of students and staff; and (f) a complete review of the institution's environmental impacts and a comprehensive strategy to address them.

The Presidents' Climate Commitment, signed by more than 600 presidents and chancellors, is detailed and specific. If your president has not signed yet, apply pressure until he or she does.

If a coalition of green colleges was to announce that starting in three years, they would only accept students who had taken a basic Environment 101 course, high schools would be obliged to start teaching it. Now, that would be leverage.

14

Solutions for Faith Communities

God has charged us to be loving and responsible stewards of creation — but global warming threatens all creation.

— Rev. Milton Jordan,
United Methodist Church, Texas

What would Jesus Drive? This is the question that many evangelical Christians are putting to each other as they discuss global warming. In 2006 more than 100 mainstream evangelical leaders signed the Evangelical Climate Initiative, declaring that they had seen and heard enough to be convinced that real perils threaten the Earth.

Around the world, religious groups and congregations are organizing around climate change. There are prayer groups, study groups and Earth Sabbaths. Methodists are buying green energy. Synagogues are celebrating Lo-Watt Shabbats. Churches are moving their investments to socially responsible companies.

Across the US, Interfaith Light and Power (ILP) is inspiring congregations to express their thankfulness for creation in practical terms by changing light bulbs, buying green energy, carpooling and making their sacred spaces more energy efficient. In Michigan, Interfaith members reduced their carbon footprints by more than 13,500 tonnes in 2005 and 2006, saving their congregations more than $2 million over the life of the products they invested in. By pooling their numbers, ILP members

21 Leading Religious Voices Speak Out on Our Sacred Duty to Protect the Environment

LOVE GOD HEAL EARTH

The Rev. Canon Sally G. Bingham

have negotiated discounts through shopipl.org. In 2006, Interfaith groups in 24 states organized for 4,000 congregations to see Al Gore's movie *An Inconvenient Truth*.

How Do We Start?

The first step is to meet together and identify opportunities where your church, synagogue or temple could make a difference. There are four major directions to explore — prayer, learning, practical action and global action. For each, there are many resources you can draw on. Download a copy of this Solution from theclimatechallenge.ca, and make a copy for everyone. Share out the websites as homework and ask everyone to bring ideas to the next meeting.

Jesus was an active person. Praying about clean water and air is fine. But taking action to make sure that the air and water are clean, that's where we put our faith into action.

— Rev. Sally Bingham,
Grace Cathedral, San Francisco

When God created Adam, he showed him all the trees of the Garden of Eden and said to him: See my works, how lovely they are, how fine they are. All I have created, I created for you. Take care not to corrupt and destroy my universe, for if you destroy it, no one will come after you to put it right.

— Ecclesiastes Rabbah 7

Greater indeed than the creation of man is the creation of the Heavens and the Earth.

— Koran, Verses 40:57

As a bee gathering nectar does not harm or disturb the colour and fragrance of the flower, so do the wise move through the world.
— Dhammapada: Flowers, verse 49

- 350 — The Sound of Hope and Spirit: 350.org/en/sound-hope-and-spirit
- Alliance of Religions and Conservation: arcworld.org
- Caring for God's Creation: nccbuscc.org/sdwp/ejp
- Christian Ecology Link: christian-ecology.org.uk
- Coalition on the Environment and Jewish Life: coejl.org
- Cool Congregations: coolcongregations.com
- Cooling Creation: coolingcreation.org
- Earth Ministry: earthministry.org
- Earth Renewal: earthrenewal.org
- Earth Sangha: earthsangha.org
- EcoCongregation: ecocongregation.org
- Evangelical Climate Initiative: christiansandclimate.org
- Evangelical Environment Network: creationcare.org
- Faith and the Common Good: faith-commongood.net
- Faith in Public Life: faithinpubliclife.org
- Faithful America: faithfulamerica.org
- Forum on Religion and Ecology: environment.harvard.edu/religion
- GreenFaith: greenfaith.org
- Greening Congregations: earthministry.org/Congregations/handbook.htm
- Greening Spirit: greeningspirit.ca
- Hanukkah Guide: gipl.org/pdf/Study_Guides/OneForEachNight.pdf
- Indigenous Environmental Network: ienearth.org
- Interfaith Center on Corporate Responsibility: iccr.org

- National Council of Churches EcoJustice: nccecojustice.org
- National Religious Partnership for the Environment: nrpe.org
- Network Alliance of Congregations Caring for the Earth: nacce.org
- Network of Spiritual Progressives: spiritualprogressives.org
- Prayer Guide for Global Warming: pub.christiansandclimate.org/pub/PrayerGuide.pdf
- Re-Energize — Time for a Carbon Sabbath: re-energize.org
- Religion, Science, and the Environment: rsesymposia.org
- Religious Witness for the Earth: religiouswitness.org
- *Serve God, Save the Planet:* servegodsavetheplanet.org
- Shop Interfaith Power and Light: shopipl.org
- Southern Baptist Environment and Climate Initiative: baptistcreationcare.org
- Target Earth: targetearth.org
- The Great Story: thegreatstory.org
- The Green Bible: greenletterbible.com
- The Green Rule: faith-commongood.net/rule
- The Omega Climate Change Course: omegaclimate.org.uk
- The Regeneration Project: theregenerationproject.org
- Thomas Berry and the Earth Community: earth-community.org
- Unitarian Ministry for Earth: uuministryforearth.org
- Unitarians Ministry for the Earth: uuministryforearth.org
- United Church of Canada: united-church.ca/ecology
- Web of Creation: webofcreation.org
- What Would Jesus Drive? whatwouldjesusdrive.org

Green Inspiration

- *Love God, Heal Earth: 21 Leading Religious Voices Speak Out on Our Sacred Duty to Protect the Environment,* by Rev. Canon Sally G. Bingham, St. Lynn's Press, 2009
- *Saving God's Green Earth: Rediscovering the Church's Responsibility to Environmental Stewardship,* by Tri Robinson, World Wide Video and Books, 2006
- *The Great Work — Our Way into the Future,* by Thomas Berry, Three Rivers, 2000
- *The Greening of Faith: God, the Environment, and the Good Life,* by John E. Carroll, New Hampshire, 1997

15

Solutions for Artists, Musicians, Comedians and Athletes

Everything we know and love is at risk if we continue to ignore the warnings.

— Laurie David

In the 1970s physicist and ecologist Barry Commoner formulated four laws of ecology: (1) everything is connected to everything else; (2) everything must go somewhere; (3) nature knows best; and (4) there's no such thing as a free lunch. These are all solid and true. But life is more than eating, mating, and reading frightening books about global warming. Life also needs delight and the occasional burst of ecstasy. Hence the Fifth Law of Sustainability: *If it's not fun, it's not sustainable.*

We need songs to inspire us, musicals to stir our hearts, symphonies for our souls and anthems that arouse our deepest feelings. We are not just fighting global warming; we are building a new culture rooted in sustainability.

We need good drama to make us think and question our values. We need good comedy, such as Robert Newman's *History of Oil*. And we need dance. In Belfast, Maine, a Step It Up group organized a square dance evening that they used to spread the word: "80% by 2050." The Canadian Youth Climate Coalition has a song-and-dance routine that you can teach to as many as 1,000 people:

- 2 degrees of Fear and Desire: headlinestheatre.com/past_work/2Degrees08
- Aerial Art: spectralq.com
- Artists Project Earth: apeuk.org
- Athletes Play It Cool: davidsuzuki.org/Climate_Change/play_it_cool.asp
- Christmas carols: ourclimate.ca/main/resources/AllIWant.pdf
- Earthsongs: planetpatriot.net/earth_songs.html
- Environmental responsibility plays: dreamridertheatre.com
- *History of Oil:* tinyurl.com/9kou8d
- Green Thing: www.dothegreenthing.com
- Little Earth Charter for Kids: littleanimation4kids.com
- LiveEarth: liveearth.org
- Pete Seeger, My Rainbow Race: tinyurl.com/dbspko
- Rosie Emery: interconnected.ca
- Songs for Environmental Education: geocities.com/RainForest/Vines/2400
- *The Boycott:* theboycottplay.com
- The Carbon Tax Song: earthfuture.com/carbontaxsong
- Ubuntu Choirs Network: ubuntuchoirs.net
- World Naked Bike Ride: worldnakedbikeride.org

Ooh, it's hot in here! There's too much carbon in the atmosphere!
(Shake your body, use movements to show heat and sweat)
I said Ooh, it's hot in here! There's too much carbon in the atmosphere!
(Repeat)
Take action, take action, to get some satisfaction
(Two hands up, fingers pointing, circle in one direction)
Take action, take action, to get some satisfaction
(Ditto, circle in the other direction)
Cut carbon now!
(Hands raised high for final cheer)

We also need beauty. Photos of starving polar bears and rising sea levels are important, but we

Canadian Youth Climate Coalition performing "Ooh, it's hot in here!" on the steps of the Canadian Parliament in Ottawa.

also need art that celebrates solar power, sustainable communities and protected rainforests, inviting us to dream of a new future. Humans are motivated by syntropic dreams of life and wholeness, not entropic nightmares of death and collapse.

When we organize public actions, the more creative they are, the better. At Key West, Florida, the Step It Up group organized an underwater rally at the coral reefs with divers video-linked to millions on TV. In Cambridge, MA, a group organized a "Winds for Change" rally with people holding 1,000 kite tails. Elsewhere people have created aerial art using an outline taped on the ground, a bullhorn, a tall building, tree or crane, a camera and lots of volunteers. In the Comox Valley, BC, an activist made a ten-foot-long plastic bag using vapor barrier and duct tape, which he attached to the exhaust pipe of his car. In 20 minutes the bag was full of CO_2, giving a visual demonstration of how much CO_2 one car creates.

Video is valuable. YouTube allows anyone to make a short film that can be seen by thousands. Key in "global warming," "solar energy" or "green finger" and ask yourself if you could do likewise.

Carbon Neutral Concerts and Sports

Many artists and musicians travel a lot, so it is good to neutralize those carbon footprints. The Rolling Stones, Coldplay, Pink Floyd, Atomic Kitten and the Dave Matthews Band have all offset the emissions associated with their concerts and albums — often only by tree-planting, alas — as did the global LiveEarth concerts in 2007.

In the world of professional sports, Justin Rose became the first professional golfer to offset his emissions as well as those of his wife, coach, caddy and manager. So have 500 NHL hockey players and the members of Canada's alpine and cross-country ski teams. In 2007, the Kodak Gallery Pro Cycling Team became carbon-neutral, purchasing wind-energy credits to offset the emissions of the team and staff.

In 2006 the FIFA World Cup became carbon neutral by helping rural villagers in a region of India hit by the tsunami to obtain biogas digesters that generate cooking gas from cow manure, instead of using kerosene and firewood. The NFL SuperBowl and Honolulu's ProBowl have embraced carbon neutrality — and it has become mandatory for the Olympic Games to be carbon neutral.

There is so much more that athletes, artists, musicians and comedians could do — this game is just beginning.

16

Start a Climate Action Group

> Never doubt that a small group of thoughtful, concerned citizens can change the world. Indeed, it is the only thing that ever has.
>
> — Margaret Mead[1]

At some point in her life, American anthropologist Margaret Mead spoke the much-quoted words "Never doubt...."

This is how the anti-slavery and civil rights movements started — and this is how we are starting the climate movement. Climate action groups are forming all across North America and Europe, from the smallest village to the largest city, and soon across the world. There will be groups that lobby their politicians, groups that work to reduce their carbon footprints, groups that organize protests outside coal-fired power plants.

These groups go by many names — climate action groups, sustainable energy groups, post-carbon groups, transition towns. Climate action groups focus on solutions to climate change, while post-carbon groups focus on solutions for a world beyond oil. Either way, the solutions are the same.

In November 2005 the parish council of the small village of Ashton Hayes, Cheshire, England (population 900), voted to try to become the first carbon neutral community in England. In January 2006 they launched their Going Carbon Neutral Project with 400 residents participating, with plans for wind and solar power, biofuels, home energy reduction, composting and tree-planting. In their first year they achieved a 20% community-wide reduction in emissions by cutting back on home energy use, car travel and flying; increasing their recycling; and setting up a video conferencing room, inspiring other communities to do the same — which shows what can you can do if you put your mind to it.

Getting Started

What is the best way to start a local group? There is no single formula — it will differ for each group of people. When I invited a group of friends to form the BC Sustainable Energy Association in December 2003, we met weekly in my home for several months over beer and pizza while we developed our plans. We agreed on our goals, chose our name, established a non-profit society, invited well-respected experts to become honorary directors, created a content-rich website, invited members and then launched projects and local Chapters. Did we know what we were doing? No, we learned as we went along — how to work with the media, how to work with government, how to manage our budgets, how to apply for funds — and we are still learning.

There are four pieces of advice that will serve you well, whatever kind of group you start:

Ashton Hayes
Please drive carefully
through the Village

Aiming to become England's
first carbon neutral village
www.goingcarbonneutral.co.uk

GARRY CHARNOCK

1. Be positive. People are inspired by hope, not fear. If you are excited by the solutions, others will share your excitement.

2. Be personal. If you nourish and support each other, people will look forward to your meetings and be pleased to work with you.

3. Be polite. We need everyone to be involved, so don't create unnecessary enemies, not even the manager of the local coal mine.

4. Be good partners. If you support similar groups in town, they will support you.

Being part of an effective group can be a truly memorable experience. Many dedicated groups lose members through burnout, so be sure to embrace the Fifth Law of Sustainability: if it's not fun, it's not sustainable.

Some Ideas:

- Create a website with information on how to reduce your carbon footprint.
- Hold monthly meetings with speakers and videos.
- Create a mobile display and take it to schools, colleges and shopping malls.
- Launch a carbon reduction challenge, involving families, businesses and schools.
- Run a carbon-neutral study circle for businesses.
- Meet with your local council, and encourage them to adopt the Solutions for Cities.
- Train your members as speakers, learning from the Green House Network. (See #20.)
- Meet with your local politicians, and encourage them to support the best legislation.

- Work with your community to create an Energy Descent Action Plan or a Climate Action Plan to guide your path into the future.
- Participate in initiatives such as Step It Up, Earth Hour and International Days of Climate Action. (See #20.)

- Ashton Hayes: goingcarbonneutral.co.uk
- BC Sustainable Energy Association: bcsea.org
- Carbon Coalition (NH): carboncoalition.org
- Carbon Rationing Action Groups: carbonrationing.org.uk
- Carbon Reduction Groups: cred-uk.org
- Chesapeake Climate Action Network: chesapeakeclimate.org
- Chew Magna Go Zero: gozero.org.uk
- Climate Camp (UK): climatecamp.org.uk
- Climate Lists: climate-l.org
- Climate Precinct Captains: local.1sky.org
- Green House Network: greenhousenet.org
- Massachusetts Climate Action Network: massclimateaction.net
- Midwest Renewable Energy Association: the-mrea.org
- North Carolina Climate Action Network: nc-can.net
- Northeast Sustainable Energy Association: nesea.org
- One Sky Florida: 1skyfl.org
- Rising Tide (Australia): risingtide.org.au
- Seattle Climate Action Now: seattlecan.org
- Transition Towns New Zealand: transitiontowns.org.nz
- Transition United States: transitionus.ning.com
- Zero Emission Network (Australia): zeroemissionnetwork.org

17

Take the Initiative

> Every person is the right person to act. Every moment is the right moment to begin.
>
> — Jonathan Schell,
> author of *The Fate of the Earth*
> and other titles

There are so many ways in which we can make a difference. Whether we know it or not, each of us is blessed with a rich collection of talents and strengths. As soon as we step out of the shadows of self doubt, each of us can make a difference in the world.

It doesn't need to be as big as the initiative taken by 15-year-old Malkolm Boothroyd, who worked hard to do three school years in two and then persuaded his parents to accompany him on a 10,000-mile bike ride from the Yukon to Florida, looking for new birds, working to protect them and encouraging others to watch birds without using fossil fuels. (birdyear.com)

Nor does it need to be as world-changing as the initiative taken by Felix Kramer, a marketing strategist from Silicon Valley, CA, who founded the non-profit CalCars in 2002 and worked with a team to convert a Toyota Prius into a Plug-In Hybrid that did more than 100 mpg, stimulating a sea change in the world's motor industry. (calcars.org)

It might be as simple as inviting friends over to watch a video or calling a friend and making an appointment to meet your city councilor to discuss what more might be done to tackle global warming locally. It might be turning your birthday party into a fundraiser for a climate action group. Out of small initiatives larger ones grow. Every oak tree starts its life as a seemingly insignificant acorn.

From One Man's Despair to a National Movement

For Bill McKibben, author of several great books about global issues, it started with a feeling of despair in the summer of 2006 and the question "What more can I do?" He thought he would walk across his home state of Vermont to the capital, Burlington, and get himself arrested for protesting Washington's lack of action. What happened was that he called some friends, who offered to join him — and his idea turned into 300 people walking through Vermont's winding country roads for five days, with singing and potluck dinners and great conversations along the way, ending up with a rally in Burlington where all of the major state candidates for federal office showed up and publicly pledged their support for legislation calling for cars that get 40 mpg, 20% renewable energy by 2020 and an 80% reduction in carbon emissions by 2050.

From tiny acorns... In January 2007 Bill met with six college students at Middlebury College, where he is scholar in residence, and together they dreamed up Step It Up 2007, hoping to inspire perhaps 100 demonstrations across the country demanding that Congress cut carbon emissions by 80% by 2050. They sent out emails,

Ted Glick is one amazing man who has participated in and led hundreds of actions. He has been arrested 15 times for nonviolent civil disobedience, including four times on climate issues. In 2007 he fasted for 107 days, 25 days on plain water and the rest liquids only, as part of a campaign for strong climate legislation. "If there is one overriding thing that I've learned over these years, it's this: together, we really can move mountains. We need to do so now. Let's make it so."

— Ted Glick (tedglick.com)

and out of nowhere the idea went viral. On April 14, 2007 there were 1400 demonstrations across the US in all 50 states, and Bill's small acorn became one of the biggest days of grassroots environmental protest since the first Earth Day in 1970, covered extensively in the national media and in thousands of local stories across the country. (See stepitup07.org.) Bill shares everything they learned in *Fight Global Warming Now: The Handbook to Taking Action in Your Community*, so that the oak tree can continue to grow until it is able to protect the entire world. The impulse from Step It Up has been taken up by 1Sky and Powershift (see #19). In 2008, Bill launched 350.org, to galvanize support for the reduction of atmospheric emissions to 350 ppm (See 350.org.)

If you are a musician, write a song or organize a concert. If you are an artist, create a poster celebrating the world beyond fossil fuels. If you are retired, invite your friends to form a study circle and see how much trouble you can stir up together, instead of assuming your days of action are over.

Whenever you take an initiative, you step onto new ground. You do not need to know how to achieve your dream. All you need is a clear picture in your mind of your final goal and the confidence to put one foot in front of another. You will learn as you go along.

This is how it has always been. In his book *The Scottish Himalaya Expedition* (1951), pioneering mountain climber William Murray wrote:

Until one is committed, there is hesitancy, the chance to draw back, always ineffectiveness.

Concerning all acts of initiative and creation, there is one elementary truth the ignorance of which kills countless ideas and splendid plans: that the moment one definitely commits oneself, providence moves too. A whole stream of events issues from the decision, raising in one's favor all manner of unforeseen incidents, meetings and material assistance, which no man could have dreamt would have come his way. I learned a deep respect for one of Goethe's couplets: 'Whatever you can do or dream you can, begin it. Boldness has genius, power and magic in it!'

Bill McKibben speaking at Rochester Institute of Technology, November 2008.

18

Become Political

If we stand for change, we can spark a popular movement with power, influence, magic and genius.

— Van Jones

We cannot stop global warming by voluntary actions alone. We must also have strong policies and legislation from our governments.

This means we have to learn how to elect green leaders, propose good legislation, and be good advocates for the legislation, mobilizing public support. Since almost all politicians belong to a political party, it is useful to join one and make your voice heard there.

Elect Green Leaders

One of the most powerful things you can do might seem the most scary — to run for public office. If you have a powerful green vision for the future, and you like working with others to get things done, you have all the qualifications that are needed. Every city mayor started out as an ordinary citizen who had the courage to put his or her name in the hat.

Whenever there is an election, ask yourself, "Who are the greenest candidates, and who has the best chance of being elected?" Then contact their campaign office and offer to help.

In the US, the League of Conservation Voters has a rating system that tells you which candidates for Congress or Senate have the best green voting record. At the city level, groups often prepare questionnaires that they put to candidates, publicizing their responses. Inviting other organizations to become cosponsors will increase your credibility.

Propose New Legislation

If you want to craft a new piece of legislation you will need to put together a three- or four-page draft paper arguing why it is needed, what it will achieve and what it might cost. You will then need to approach like-minded organizations and invite them to support the legislation. If you study the advocacy resources listed, you will get a good understanding of how the legislative process works.

Become a Strong Advocate for Your Cause

There are five dimensions to good advocacy:

1. Create a dedicated webpage with information on your proposal, and an "Act Now" button where people can send instant faxes and emails to politicians.

2. Arrange personal meetings with politicians, their staff and the bureaucrats who are responsible for getting things done. These

A RESULTS letter-writing circle at work.

BLAISE SALMON

relationships are all-important because they build credibility and respect.

3. Meet with your newspaper's editorial board to engage them in the topic, and write an op-ed piece to stimulate public debate. A good piece of visual theatre will attract the media's attention, as they are always on the lookout for pictures.

4. Ask members of local environmental groups to write to their politicians, send letters to the editor, and call into local radio shows.

5. Organize public meetings, events and rallies to build momentum and public support for your proposal.

Learn from RESULTS

RESULTS is one of the world's most effective advocacy societies. Its members work to create the public and political will to end world hunger and the worst aspects of poverty, and they do it by meeting in someone's home each month to study a particular issue and write a letter to their political leaders, followed by a request for a meeting. They also learn how to write letters to the editor, and the op-ed articles that appear opposite the editorial page. In 2004 for every $1 that a supporter donated, RESULTS Canada leveraged $72 in increased government funding to fight diseases such as TB and malaria. In *Reclaiming Our Democracy: Healing the Break Between People and Government*, RESULTS founder Sam Daley-Harris tells the inspiring story of how RESULTS empowers ordinary citizens to have a clear and lasting impact on government policies and funding decisions.

- Advocacy On-Line: web.advocacyonline.net
- Congressional Directories: capitoladvantage.com/capub
- Contact your Canadian MP: webinfo.parl.gc.ca
- Contact your US Representative: house.gov
- Contact your US Senator: senate.gov
- Contact your State Governor: nga.org
- Developing Relationships with Reporters: spinproject.org/downloads/Reporters.pdf
- *Going Public: An Organizer's Guide to Citizen Action,* by Michael Gecan, Anchor, 2004
- Governments on the WWW: gksoft.com/govt
- *How to Save the World in Your Spare Time,* by Elizabeth May, Key Porter Books, 2006
- League of Conservation Voters: lcv.org
- Media Advocacy Manual: apha.org/about/news/mediaadvocacy.htm
- Reclaiming Our Democracy: results.org/website/article.asp?id=433
- RESULTS Activists Toolkit: results.org/website/article.asp?id=1745
- The Art of Advocacy — A Handbook for Non-Profit Societies: cleonet.ca/resources/355
- *The One-Hour Activist: The 15 Most Powerful Actions You Can Take to Fight for the Issues and Candidates You Care About,* by Christopher Kush, Jossey-Bass, 2004
- The Spin Project: spinproject.org
- Washington Conservation Voters: wcvoters.org

The RESULTS Activists Toolkit is also full of useful advice.

19

Build a Nation-Wide Movement

We are now faced with the fact, my friends, that tomorrow is today. We are confronted with the fierce urgency of now.
— Rev. Martin Luther King

What does it take to turn a scattering of local actions into a movement that can change a nation's history?

It has happened before. The end of slavery ... the birth of labor unions ... votes for women ...

- Affinity Groups: actupny.org/documents/CDdocuments/Affinity.html
- Alliance for Sustainable Jobs and the Environment: asje.org
- Apollo Alliance: apolloalliance.org
- Canadian Youth Climate Coalition: ourclimate.ca
- Climate Crisis Coalition: climatecrisiscoalition.org
- Climate Movement Links (US): stepitup2007.org/links
- Clinton Global Initiative: clintonglobalinitiative.org
- Demonstrators Manual: actupny.org/documents/demomanual/Demomanual.html
- Focus the Nation: focusthenation.org
- Freedom from Oil Campaign: freedomfromoil.org
- Green for All: greenforall.org
- Green House Network: greenhousenet.org
- National Polar Bear Plunge: keepwintercold.org
- National Teach-In: nationalteachin.org
- Natural Resources Defence Council: nrdc.org
- Step it Up 2007: stepitup2007.org
- Stop Climate Chaos (UK): stopclimatechaos.org
- The Alliance for Climate Protection (Al Gore's initiative): climateprotect.org
- The Climate Project: theclimateproject.org
- US Climate Emergency Council: climateemergency.org
- We Can Solve It: wecansolveit.org

civil rights ... the end of the Vietnam War. All of these victories happened because a few determined people holding a scattering of candles gradually grew in numbers until they created a powerful blaze of light.

Global warming is the same — it needs a nation-wide movement. It is also different, however. As well as legislation, new technologies, a change in the way we live, and a change in our spiritual relationship to the planet — it needs the Renaissance, the Reformation, the Industrial Revolution, and the Environmental Movement, all rolled into one.

We need a nation-wide wave of rallies, teach-ins, pray-ins, pilgrimages, polar bear plunges, petitions, and lobbying. We need mountain-top rallies and underwater rallies — both of which happened with 1400 other actions in Step it Up in April 2007. (See #17)

Success in building the movement also requires architects who can design zero-energy buildings, and investors who can put their money into large-scale technologies such as solar thermal in the world's deserts and hot rocks geothermal energy.

Every movement needs three things: a well-educated public that understands what the issues are; the ability to mobilize people across the nation at critical moments; and a clear picture of "success".

A Well-Educated Public

Global warming is far more complicated than "Votes for Women" or "Stop the War". Most people are on a steep learning curve while facing confusing statements by deniers and nay-sayers.

Chesapeake Climate Action Network fast at the White House, 2005.

In 2008 and 2009, the National Teach-In at Global Warming Solutions organized three national days of engagement, involving over 2500 educational institutions and a well over million Americans, and they continue to organize similar events.

This kind of focused personal learning is sorely needed to get people fully engaged. The science and the solutions of climate can both seem complicated, until they are unraveled and presented in a clear way by knowledgeable people.

Architecture 2030 has done a similar job of educating future architects and engineers; Interfaith groups have educated religious congregations; and Al Gore's The Climate Project has trained over 2,600 people who are giving presentations across North America and around the world, reaching over five million people by the end of 2009.[1]

The Ability to Mobilize

Once people understand how grave the situation is, and what the solutions are, they want to act. Acting to reduce their personal carbon footprints is great, but we also need to organize locally, and to participate in nation-wide events that stir the heart and send a resounding message to the nation's capital.

No-one is in charge of building such a movement. Each nation-wide event has a person or group that dreams it up, and if they catch the right moment they are joined by a host of others who want to join in. Bill McKibben's book *Fight Global Warming: A Handbook for Taking Action in Your Community* tells the story of Step it Up 2007 and passes on their best advice, from building partnerships to making the best use of the Internet. "Think like a fellowship", he says. Use music. Have fun. Make it collaborative.

A Clear Picture of "Success"

Every movement needs a goal that can be clearly visualized and held in the mind, ideally in four words or less. This is difficult for global warming. We can't stop it, and even if we reduce our carbon footprint dramatically, it may not be enough. We must focus on the solutions, and redefine success at each stage of the journey, such as "80% by 2050", and "20% Green Energy by 2020", as 1Sky is calling for, and the very simple "350". This is not a one-shot movement: it must be a strong, steady, persistent movement that will — over time — transform our world.

20

Build A Global Movement

Coal is the single greatest threat to civilization and all life on our planet.
— James Hansen (NASA)

I believe we have reached the stage where it is time for civil disobedience to prevent the construction of new coal plants.
— Al Gore, 2008

In June 2009, the streets of Teheran, Iran, were bursting with millions of protesters, demanding a re-run of their election. Imagine the streets of cities all over the world being similarly filled with people demanding action on climate change.

Nine — Step Organizing Plan

(adapted from 350.org)

1. Choose your action and location — a walk, a rally with speakers and music, a hike, a potluck, a community discussion, a service project.
2. Invite friends, neighbors and local organizations to assist. Maybe a local church, mosque, synagogue, labor union, business, sports team, university or arts cooperative would like to get involved?
3. Work out the details — timing, directions, transport, bathrooms, sound system, permits, sponsorships.
4. Invite local, regional and national leaders. You never know who might become a climate champion if you don't try to engage them.
5. Spread the word — write editorials, get on local radio, ask organizations to include the details in their newsletters. Put up posters all over town.
6. Make a Banner — make it bright, large and legible.
7. Tell the Media — what print, radio, television and blogs do you want to cover your event?
8. Have a fun and meaningful day, knowing that you're part of a giant global movement to stop climate change. Use a sign-up sheet to stay in touch with people.
9. Select your best photo and upload it to the 350 website to deliver the strongest possible message to the media and the world's decision-makers.

We're getting there, but oh, so slowly. Each December, there's a Global Day of Action on Climate Change to coincide with the big annual United Nations Climate Conference. In 2005, there were events in 20 countries. By 2009 events were being planned in 96 countries.

In 2007, in Australia, protestors succeeded in disrupting coal exports from Newcastle Harbour, and 115,000 people walked against warming in 60 towns and cities. In Auckland, New Zealand, a Climate Rescue Bus organized an open air exhibition with stalls, food, drinks, music, a biodiesel powered bouncy-castle, and a samba band that led 600 people to form a human banner spelling 'Climate SOS'.

In China, 1000 students from universities in twenty provinces participated in a variety of actions, including launching a kite to promote wind power in front of the Temple of Heaven in Beijing, skating and cycling to a wind farm, and forming the sign CO_2 to show the need for action to reduce our emissions.

In Lebanon, in spite of an extremely tense political situation, 2000 people took part in the biggest environmental march in the country's history. In Poland, volunteers stuck 8000 stickers in cities saying, "Emergency exit — will be flooded" and "You can't escape from the Earth. Stop climate change."

These actions are critically important, for without a clear demand from the public, politicians are being swayed by corporate lobbyists and head-in-the-sanders who oppose real action. In the run-up to the UN Copenhagen Conference in December 2009, the head of the UN climate change branch, Yvo de Boer, told non-governmental organizations,

we can solve the climate crisis.

"If you could get your members out on the street before Copenhagen, that would be incredibly valuable."

On October 24, 2009, the world's largest climate protest was inspired by 350.org, an international campaign dedicated to building a movement to unite the world around solutions to the climate crisis, and reduce atmospheric CO_2 to the safer level of 350 parts per million. At the time of writing, over 1,427 actions were being planned in 110 countries.

As well as large, movement-building actions, we need focused actions to close down the world's coal-fired power plants, stop the export of coal from countries such as Canada and Australia, and stop the really dirty industries such as Canada's tar-sands. Actions like these need knowledge of how to organize a non-violent action, how to build consensus, and how to keep the hotheads away, who are sometimes agent provocateurs, paid by industry or the police.

With so much at stake, we are entering astonishing times. As the global climate movement builds, we will need the most creative non-violent mobilizing the world has ever seen. For inspiration take five minutes and listen to Pete Seeger, as he sings "One blue sky above us..." (See opposite)

- 350: 350.org
- Arab Climate Campaign: indyact.org/environmental.php
- Australia's Climate Action Hub: climatemovement.org.au
- Austria's Climate Movement: sos-klima.at
- Avaaz: avaaz.org
- Beyond Talk: beyondtalk.net
- Campaign Against Climate Change UK: campaigncc.org
- Civil Disobedience: en.wikipedia.org/wiki/Civil_disobedience
- Climate Action Network Canada: climateactionnetwork.ca
- Climate Action Network Europe: climnet.org
- Climate Alliance Germany: die-klima-allianz.de
- Earth Hour: earthhour.org
- Fossil Fools Day: fossilfoolsdayofaction.org
- Friends of the Earth International: foei.org
- Global Climate Campaign: globalclimatecampaign.org
- Greenpeace: greenpeace.org
- India's Climate Movement: whatswiththeclimate.org
- International Climate Action Network: climatenetwork.org
- Network for Climate Action UK: networkforclimateaction.org.uk
- One Blue Sky Above Us — Pete Seeger: tinyurl.com/98a7wu
- *One Sky*: 1sky.org
- Rainforest Action Network: ran.org
- Rising Tide Australia: risingtide.org.au
- The Ruckus Society: ruckus.org
- US Climate Action Network: usclimatenetwork.org
- WWF: panda.org
- Youth Climate Movement: itsgettinghotinhere.org

21

Accept the Challenge

We as mayors have the opportunity to push the envelope and get people thinking, even when it is not politically popular. Cities hold the key.
— Greg Nickels, Mayor of Seattle

Our world is a multitude of communities, and if each works to reduce its carbon footprint we will make good progress. It doesn't matter whether you are a village like Ashton Hayes, England (see #16) or a city the size of Shanghai.

It starts with making the commitment and setting your goal. Berlin's goal is a 50% city-wide reduction by 2010. Växjö, Sweden, has set the same goal, raising it to 70% by 2025. The Kalmar Region of Sweden in which Växjö sits is aiming at a 100% reduction by 2030, and had achieved 60% by 2008. *This is the goal we should all pursue: 100% by 2030.*

London's goal is a 60% reduction by 2025. In the US, cities that sign the Mayors' Climate Protection Agreement commit to 7% by 2012. In Britain, cities that sign the Nottingham

Seattle Mayor Greg Nickels, founder of the Mayors' Climate Protection Agreement, takes delivery of a PHEV Prius, converted by Hymotion.

Declaration are chasing 12% by 2010. All goals relate to the 1990 level.

1. Involve the Public

The task is huge but achievable, provided you reach out to your whole community, including households, businesses, schools, colleges, clubs and societies. Your staff may write reports, but if you proceed without community involvement, your citizens will never get engaged.

Form a **climate action task force**, co-chaired by the mayor and another community leader, and meet weekly for 12 weeks following an organized program. (See box.) Appoint members on the basis of their practical expertise in the key focus areas; smaller communities might invite neighboring communities to join them in a shared process.

Public involvement does not stop when the task force has completed its work. Your community will need an ongoing **climate action council**, supported by 12 **climate action teams** to develop plans for public engagement, corporate emissions, transport, cycling, walking, transit, buildings, energy, zero waste, food, greenery and climate protection.

2. Appoint Staff

This is an essential piece of the puzzle: you need dedicated staff who can give it full-time attention. Many staff feel swamped with their existing workload, and this is not something that can be done on the side of the desk.

In San Francisco (population 750,000) the department of the environment had 66 staff in 2007, including eight who worked on energy and

12-Week Climate Action Task Force

1: Overview
2: Engaging the Community
3: Corporate Emissions
4: Green Mobility
5: Walking, Cycling, Transit
6: Green Buildings
7: Green Energy
8: Sewage and Zero Waste
9: Local Food
10: Trees and Green Space
11: Protection and Adaptation
12: Framework, Goals and Strategy

- Boulder's Carbon Tax: newrules.org/environment/ climateboulder.html
- C40 Climate Leadership Group: c40cities.org
- Cities for Climate Protection: iclei.org/co2
- Cities Go Green: citiesgogreen.com
- City Best Practices: usmayors.org/uscm/ best_practices/EnergySummitBP06.pdf
- Climate Protection Manual: climatemanual.org/Cities
- Community Climate Solutions: cleanair-coolplanet.org /for_communities/toolkit_home.php
- Cool Counties: kingcounty.gov/exec/coolcounties
- EcoBUDGET: ecobudget.com
- ICLEI USA: coolmayors.com
- Local Governments for Sustainability: iclei.org
- Low Carbon Innovation Network: carbon-innovation.com
- Quick Action Guide (Canada): tinyurl.com/2dx78d
- SustainLane Best Practices: sustainlane.us
- SustainLane City Rankings: sustainlane.com/us-city-rankings
- US Mayors Climate Action Handbook: seattle.gov/ climate/docs/ClimateActionHandbook.pdf
- US Mayors Climate Protection Center: usmayors.org/climateprotection
- Useful Knowledge for Greener Cities: sallan.org
- Visible Strategies see-it: visiblestrategies.com

climate and six on clean air and transportation — that's one sustainability staff person per 11,000 residents. In Portland (population 568,000) the office of sustainable development had 41 staff who were responsible for solid waste and recycling, global warming, energy conservation, renewable energy and biofuels, green building, sustainable food, sustainable city government and citizen advisory committees (1 per 14,000 residents). A separate department covered transportation. By 2007 Portland and the neighboring Multnomah County had reduced their carbon footprint to 1% above the 1990 level (chasing 10% below 1990), but on a per capita basis their emissions had fallen by 12.5% since 1993, compared to a US-wide increase of 13%.

3. Allocate a Budget

San Francisco's department of the environment has an annual budget of $20 million ($26 per resident). On this basis, a community of 5,000 should budget $130,000 for its work on sustainability and climate change. Don't be shy to tell people why their taxes need to rise: $26 per person is only 50 cents a week for a world our children and grandchildren can live in. Boulder, Colorado, uses a local carbon tax that generates $1 million a year

to finance its Climate Action Plan. Local governments in Europe are increasingly using ecoBUDGET organize their sustainability work.

4. Develop a Plan

There are many resources that can help you develop a Climate Action Plan, to be integrated into your city's development and transportation plans. (See #30.) The four keys to success are senior management support, motivated staff, solid funding and support from community organizations. Don't get bogged down in disconnected documents and spreadsheets — use an integrative program such as Visible Strategies to pull it all together.

22

Engage the Community

> Reversing global warming will take a World War II level of mobilization. It is the work of tens of millions, not hundreds of thousands.
>
> — Van Jones

It is very encouraging when someone makes the commitment to a more sustainable lifestyle — so how can we encourage every family, school and business to do the same?

Ashton Hayes (population 900) is doing it by volunteer involvement at the village level. (See #16.) So is the Saxon village of Chew Magna, Somerset, England, with just 1,200 people. Their Go Zero initiative has four focus groups and twelve projects, including a sustainable education centre, a waste and energy audit, a local food guide, a car-sharing club and Go Zero Junior Clubs for children. It is easier to organize a small community — but how do we mobilize people in a city the size of Chicago or Vancouver?

In World War II, when people in Britain had to learn how to build bomb shelters and cook with meager wartime rations, the Women's Voluntary Service (WVS) provided a solution. Launched in 1938, it had 165,000 members by the time war started in September 1939. By 1943 more than a million women had joined, caring for children, helping those who had been bombed out of their houses, and volunteering for myriad tasks. Out of Britain's population of 48 million, that's one woman in 24. In a city of a 100,000 people, this level of engagement would produce 2,000 volunteers.

We need volunteers who can run climate action circles (see #11) to reach every household. In a city of 100,000 people, 200 volunteers could run 2,000 circles. Spread over five years the right incentives, it is an achievable task.

Live Green Toronto is investing $25 million in grants for neighborhood green projects. Vancouver has established One Day Vancouver, an interactive website where people share carbon reduction tips, and sign up for daily advice by email.

In New Hampshire, the New Hampshire Carbon Challenge helps residents reduce their carbon footprints by 20% with trained presenters, a carbon calculator, a Climate Action Kit and an Employee Carbon Challenge. In Burlington, Vermont, the Alliance for Climate Protection has issued a 10% Challenge in support of the city's goal, using a carbon calculator, a pledge drive and resources for residents, students and businesses. In England, Manchester is My Planet has accepted tens of thousands of pledges from residents.

Britain's Carbon Reduction (CRed) initiatives work in a similar manner, gathering pledges,

Climate Heroes at the Oyster River Carbon Challenge, New Hampshire

New Hampshire Carbon Challenge™

encouraging local groups and supporting them with personalized web pages where you can track your reductions. CRed has more than 100 partner organizations, including cities, businesses and schools. In 2006 they gathered 25,000 pledges.

The business community also needs to become engaged. In Cambridge, MA, businesses are joining the Climate Leader Program, receiving technical assistance from the city and a Climate Leader logo when they meet their goals. In Fort Collins, Colorado, the ClimateWise program offers similar support for local businesses.

Calculators and pledges work well for those who are committed, but community-wide reduction needs an extensive network of trained volunteers, similar to Britain's WVS. The public is willing to change, but they need help to understand their kilowatt hours and insulation techniques.

Get Started

- Set up **Team #1: Climate Action Community Engagement**, choosing people with carbon reduction and social marketing experience.
- Invite them to study the best models and recommend the best solutions.
- Approach the leaders of community groups, environmental groups, churches, benevolent societies and clubs, and seek their support.
- Develop a project plan, and seek your council's approval.
- Ask local community leaders to help you recruit volunteers to run the program.

- 10 Tips for Sustainability Communications: futerra.co.uk/downloads/10-Rules.pdf
- Ashton Hayes: goingcarbonneutral.co.uk
- Berkeley Climate Action: berkeleyclimateaction.org
- Chew Magna Go Zero: gozero.org.uk
- Citizen Participation and Community Engagement: tinyurl.com/3y4qsd
- ClimateWise, Fort Collins: fcgov.com/climatewise
- CRed: cred-uk.org
- LED City: ledcity.org
- Live Green Toronto: toronto.ca/changeisintheair
- Logicity: logicity.co.uk
- Manchester is My Planet: manchesterismyplanet.com
- New Hampshire Carbon Challenge: carbonchallenge.sr.unh.edu
- "New rules, new game: Communications tactics for climate change": futerra.co.uk/downloads/NewRules:NewGame.pdf
- One Day Vancouver: onedayvancouver.ca
- Seattle Climate Action Now: seattlecan.org
- Seattle Climate Partnership: seattle.gov/climate/partnership.htm
- Seattle Neighborhood Climate Protection Fund: seattle.gov/neighborhoods/nmf/climate.htm
- Toronto: toronto.ca/environment

Build Partnerships

The wider your circle of partnerships, the better will be your results. In pursuit of its Local Action Plan on Global Warming, Portland is working with green builders, cyclists, recyclers, solar activists, Business Green Teams, The Natural Step, local food activists and others. Seattle's Climate Partnership is engaging with some of the city's largest employers, and its Neighborhood Climate Protection Matching Fund is supporting local initiatives. Saanich, in Victoria, Canada, has established a Carbon Neutral Fund to support local initiatives.

Create a directory of all the groups in your community that are working on climate change, and reach out to them. Build a living network, and show that their work is appreciated.

23

Reduce Your Corporate Emissions

Our experience shows that making buildings energy efficient requires more thinking, not more money.

— Ronald J. Balon, Senior Energy Engineer, Montgomery County

To provide good leadership, a city must be a role model in reducing its own carbon footprint. This is about walking the talk.

By switching to LED streetlights that use 85% less energy, Chicago is saving $2.5 million and 23,000 tonnes of CO_2 every year. In Essex, England, street lights are turned off from midnight to 5 a.m. San Francisco is running all its garbage trucks on biodiesel. Calgary is powering its light rail transit system with wind energy.

There are some great resources that can help communities wrap their minds around what's needed. If it feels overwhelming, it's only because this is all so new. The C40 Cities Climate Leadership Group has well-organized best-practices sheets. Cities for Climate Protection is assisting 1,000 communities around the world to inventory their emissions, set reduction targets and develop Action Plans. The New Rules Project

has detailed text for policies such as carbon taxes and climate neutral bonding.

In 2003 Vancouver set out to reduce its corporate emissions to 20% below 1990 levels by 2010. By 2006 it had got them to 5% below, largely by doing a comprehensive energy retrofit of its facilities and capturing its landfill gas. In 2007 the manager of sustainability estimated that they would reach 18% below by 2010 but would need further initiatives to reach 20%.

Other communities have done equally well. Woking, England, achieved a 77% reduction in its corporate emissions by 2004. Seattle achieved a 60% reduction by 2007. In Australia, Melbourne's new city hall uses 87% less energy and 72% less water.

The key to progress is to get everyone on board. As soon as your community has made the commitment to reduce its emissions, form **Team #2: Municipal Climate Action**. Involve leaders from each department and develop a Corporate Climate Action Plan with a series of projects. The full support of your mayor, council and chief executive is essential.

The International Center for Local Environmental Initiatives (ICLEI) works with municipalities around the world to help them reduce their emissions. By joining the US Mayors Cool Cities Program, or Partners for Climate Protection in Canada, your community can benefit from ICLEI's support and sign onto its Five Mileposts: (1) establish a baseline; (2) set a target; (3) develop a local action plan; (4) implement the local action plan; and (5) measure results.

Wherever you use energy there are opportunities to become more efficient and switch to green

Yea, Berkeley! See BerkeleyClimateAction.org

power. In 2001 San Francisco voters approved a $100-million bond initiative that paid for solar panels, energy efficiency and wind turbines for public facilities, to be repaid entirely from energy savings at no cost to taxpayers.

Wherever there is a fire station, library, courthouse, school, office or city hall, it can be retrofitted for great energy efficiency. In 2005 Chicago completed an energy retrofit in every library; in 2006 it did a lighting retrofit in every fire station. Every community should appoint a local energy conservation officer. Every building should have an energy conservation leader who knows how the systems work. Whenever a new building is planned, build it to LEED Gold standard, as Scottsdale, Arizona mandated in 2005.

Wherever vehicle fuel is used, there are opportunities to save and to switch to biofuel and electric vehicles. In Graz, Austria, the entire city bus fleet uses local biodiesel made from wastes. In Saco, Maine, the building inspectors use a ZENN electric vehicle. In many communities in England, city staff receive a per mile cycling allowance. Wherever there are employees, they can be persuaded to commute to work in an eco-friendly manner.

Many cities are also taking steps to green their procurement policies, so that their purchasing dollars support non-toxic, recycled, energy efficient goods and services.

And finally, don't be shy of telling your community what you are doing. Create a clear, easy link on your municipal website, and tell the media each time you undertake a new initiative. The public needs — and wants — to see your leadership.

- American Council for an Energy Efficient Economy: aceee.org
- Buyer's Guide on Green Purchasing: tinyurl.com/856qr8
- C40 Cities: c40cities.org
- Cities for Climate Protection: iclei.org/co2
- Clean Air and Climate Protection Software: icleiusa.org/cacp
- Climate Neutral Bonding Resolution: newrules.org/de/climateneutral.html
- Energy Performance Contracts: tinyurl.com/7ky9lm
- Energy Star for Government: energystar.gov/government
- Environmentally Preferable Purchasing Guide: rethinkrecycling.com/government/eppg
- ICLEI Canada: iclei.org/canada
- King County's Environmental Purchasing Program: metrokc.gov/procure/green
- *Lessons from the Pioneers: Tackling Global Warming at the Local Level:* newrules.org/de/pioneers.pdf
- Local Governments for Sustainability USA: icleiusa.org
- National Association of Energy Service Companies: naesco.org
- Partners for Climate Protection, Canada: sustainablecommunities.fcm.ca/partners-for-climate-protection
- Portland's Climate Action Plan: tinyurl.com/333oxa
- Portland's Trip Reduction Incentive Program: tinyurl.com/yp4b8q
- Public Employees for Environmental Responsibility: peer.org
- Richmond's Environment Purchasing Guide: richmond.ca/services/Sustainable/environment/policies/purchasing.htm
- Seattle's Climate Action Plan: seattle.gov/climate
- Sustainable Procurement Europe: iclei-europe.org

24

Encourage Green Mobility

Most comprehensive and objective analysis tends to rank mobility management strategies among the most cost-effective emission reduction options.

— Todd Litman

Transportation is responsible for 25% to 50% of a city's carbon footprint, and it keeps growing as people move to the suburbs and then hit the road on the weekend in search of the glorious parks and mountains shown in SUV adverts.

We need to design a very different future, where suburbs and neighborhoods become pedestrian communities, cycling is safe and easy, rail and transit link our communities, and cars and trucks are mostly electric.

Establish **Team #3: Climate Action for Green Mobility**. Develop an integrated transportation plan with a clear order of priorities: walking, cycling, transit and LRT, ride-sharing and, last of all, single-occupancy vehicles, as Vancouver has done. Set clear goals. Create a community development plan that encourages compact, transit-oriented development instead of sprawl. Transportation-demand management and mobility management are the most cost-effective ways to reduce emissions, while also meeting many other goals. Any initiative that makes it easier to drive tends to encourage more driving.

18 Ways to Reduce Transport's Local Carbon Footprint	
1. Anti-idling policies	Common in many North American communities. Idle-Free Zone: tinyurl.com/2adnqy
2. Car-free planning	In the Vauban neighbourhood of Freiburg, Germany, 70% of residents live without a car, and most streets are child friendly. vauban.de
3. Carsharing	In Philadelphia, 30,000 people share 450 cars through PhillyCarShare. Cities can support carsharing with start-up funding, preferential parking and direct participation.
4. Commuter trip reduction	Washington State's Commuter Trip Reduction Law requires employers with 100+ employees to reduce their peak-period trips. Portland's SmartTrips Program pays $30 a month to staff who walk, cycle, take the bus or carpool on a regular basis.
5. Compact development	Compact, transit-oriented development can be encouraged with development and zoning incentives, fast-track approvals and smart-growth scorecards.
6. Congestion pricing	London's congestion tax reduced traffic by 20% and CO_2 emissions by 16%, as people switched to transit and cycling. In Stockholm, reduced CO_2 emissions by 14%.
7. Electric vehicle incentives	Vancouver has upgraded building codes to require EV charging posts. Many cities have approved the use of Neighborhood Electric Vehicles. Stuttgart is making light electric scooter-bikes available throughout the city. San Francisco is working with Better Place to turn the city into the electric vehicle capital of the US. Berlin and Paris are equipping their cities with community electric vehicles. Portland is challenging, to win the EV contest.
8. Green car incentives	Stockholm has a shared procurement policy with businesses and car manufacturers to increase the use of hybrids and clean fuel vehicles to 5% of all vehicles. Cities are supporting green cars with preferential parking, HOV lanes, purchasing pools and direct purchases. Irvine, CA, provides a pool of electric vehicles for participating employers. Mexico City is replacing 10,000 older taxis with fuel-efficient models.

PhillyCarShare's members are using 500,000 fewer gallons of gas a year and driving 53% fewer miles, thanks to carsharing. phillycarshare.org

- Association for Commuter Transportation: actweb.org
- Canadian Telework Association: ivc.ca/cta
- Car-Free Cities: carfree.com
- *Car-Sharing:* carsharing.net
- City Repair Project: cityrepair.org
- Less Traffic: lesstraffic.com
- Municipal Actions for Efficient Transportation: vtpi.org/tdm/tdm204.htm
- Online TDM Encyclopedia: vtpi.org/tdm
- Project Get Ready (Electric Vehicles): rojectgetready.com
- Smart Communities Network: smartcommunities.ncat.org
- "Smart Transportation Emission Reductions": vtpi.org/ster.pdf
- Surface Transportation Policy Partnership: transact.org
- *Transport Revolutions 2025:* transportrevolutions.info
- Trolley-Lorry: tbus.org.uk/trolleylorry.htm
- Victoria Transport Policy Institute: vtpi.org
- World Car-Free Day: worldcarfree.net/wcfd

9. Green freight	Amsterdam is launching a freight-carrying streetcar plan with special tracks, transferring to electric vans for the final leg of the journey. citycargo.nl
10. HOV lanes	Widely used in many cities to encourage carpooling, ridesharing and transit.
11. Light rail transit	Popular wherever built. Saint Paul and Minneapolis are about to build inter-city LRT, encouraging transit villages.
12. Location-efficient mortgages (LEMs)	LEMs offer mortgage incentives to people who buy homes in compact walkable communities. locationefficiency.com
13. Optimizing traffic signals	Portland used special software to optimize the traffic lights at 135 intersections, saving 15,000 tonnes of CO_2 a year.
14. Parking cash-out	In California, employers who provide subsidized parking are required to offer employees the option to choose cash instead. Surveys show that carpooling increases by 64%, transit ridership by 50%, walking and cycling by 33%, and commuter travel miles fall by 12%. arb.ca.gov/planning/tsaq/cashout/cashout.htm
15. Parking solutions	Smart parking and pricing solutions can help reduce emissions while also meeting many other goals. vtpi.org/tdm/tdm72.htm
16. Ridesharing	In Britain, *liftshare* has more than 200,000 members. Many cities are developing ride-share websites as well as carpooling programs. liftshare.com.uk
17. Smart-trip information	Portland's SmartTrips and Ottawa's TravelWise programs help people make car-free decisions. Seattle's One Less Car Challenge provides incentives for people to leave a car at home for a month. seattle.gov/waytogo/onelesscar.htm
18. Teleworking	Flexible work programs and teleworking can be adopted anywhere. See teletrips.com

25

Encourage More Walking, Cycling and Transit

> The Earth cannot wait 60 years. I want a future for my children and my children's children. The clock is ticking.
>
> — Richard Branson

People have become accustomed to walking being unpleasant, cycling dangerous and riding the bus inconvenient. We need to create a very different future.

Encourage More Walking

People *like* to walk, and wherever there is a safe, visually attractive environment, they will do so. Copenhagen, Denmark, has been making itself people friendly since 1962, when it pedestrianized its main street. By increasing its car-free space six-fold, it has created four times more public life. With the Danish architect Jan Gehl's help, London plans to become a world-class walking city by 2015. North American cities have a long way to go, but wherever cities embrace the new urbanism, with attractive streetscapes and a conscious effort to encourage walking, success follows.

Invite citizens to form **Team #4: The Walkable City Team**. Ask them to study Copenhagen's example, do a local walking audit, and report back with a list of recommended improvements. Create a ten-year plan to achieve as many as possible. In Massachusetts, Boston Walks has made a huge contribution by acting as a leader for pedestrian design.

Encourage More Cycling

Hire a cycling planner, form **Team #5: Cycling Action Team**, and develop a ten-year plan to increase cycling's share to 20% of all trips.

Copenhagen is also the world leader in cycling, with 1.2 million kilometers cycled every day. In 2007, 36% of all commuter trips were by bike, chasing a goal of 50% by 2015. If Copenhagen's citizens cycled an extra 10%, the health care system would save $11 million a year, productivity

Copenhagen's Ten — Step Program to become a Walkable City

1. Convert streets into pedestrian thoroughfares.
2. Reduce traffic and parking gradually.
3. Turn parking lots into public squares.
4. Keep scale dense and low.
5. Honor the human scale.
6. Populate the core.
7. Encourage student living.
8. Adapt the cityscape to changing seasons.
9. Promote cycling as a major mode of transportation.
10. Make bicycles available.
    See metropolismag.com/html/content_0802/ped

TriMet's MAX Light Rail system in Portland, OR. Trains run every 5-15 minutes between 5am and midnight.

would benefit by $30 million and there would be 57,000 fewer annual days of absence and 46,000 fewer years of prolonged, severe illness.[1]

Davis, CA, (population 64,000) is one of North America's cycling leaders, with two paid cycling staff and 17% of commuter trips being by bike. The entire university campus is closed to vehicle traffic, and the city has 100 miles of bike lanes, trails and routes. In 2007 Paris introduced 26,000 VELIB public bicycles for hire in 1400 bike racks, to an overwhelming public response.

Encourage More Transit

Form **Team #6: Community Transit Solutions**, to develop a ten-year plan to increase transit's share to 20% of all trips.

Redesign your service so that buses are frequent and low-cost, bus shelters are comfortable and electronic timetables tell you when the next bus is coming. When Boulder, CO, redesigned its bus service, introducing small colorful buses that run regularly on popular routes, ridership increased by 500%. A Neighborhood EcoPass costs less than $150 a year, and if enough neighbors sign on, the cost is covered in city taxes, making the service free. When the Belgian town of Hasselt made its bus service free in 1997, costing 1% of the municipal budget, ridership increased twelve-fold, supported by an integrated transport policy that includes free access to bicycles, scooters and wheelchairs.[2] Student U-Pass systems, which provide free transit paid for in student fees, are essential. Community-wide U-Passes are best.

Light rail transit and bus rapid transit (BRT) are key components of future transportation planning.

Walking
- America Walks: americawalks.org
- International Walk to School: iwalktoschool.org
- Is Your Neighborhood Walkable? walkscore.com
- Livable Streets Initiative: livablestreets.com
- Walkability Checklist: walkableamerica.org/checklist-walkability.pdf
- Walking Audits: walkable.org/assets/downloads/walking_audits.pdf

Cycling
- Bike Plan Source: bikeplan.com
- Copenhagen Cycle Chic: copenhagencyclechic.com
- Copenhagen, City of Cyclists: tinyurl.com/8exwsu
- Cycling and Walking Encouragement: vtpi.org/tdm/tdm3.htm
- National Bicycle Greenway: nationalbicyclegreenway.com
- National Center for Bicycling and Walking: bikewalk.org
- Safe Routes to School: saferoutestoschool.ca

Transit
- Boulder's EcoPass: bouldercounty.org/ecopass
- Bus Rapid Transit Central: busrapidtransit.net
- Bus Rapid Transit Policy Center: gobrt.org
- Light Rail Now: lightrailnow.org
- National Alliance of Public Transport Advocates: napta.net
- Packing Pavement: swt.org/share/bguard.html
- Trams in France: trams-in-france.net
- TramTrain City: lightrail.nl/TramTrain

Bogota's Transmilenio BRT carries 1.4 million passengers a day, using dedicated lanes, large-capacity buses and elevated bus stations, reducing greenhouse gas emissions by 40%.[3] We also need electric trolley buses, street cars and luxury commuter coaches whose passengers can use their laptops while riding in peace. Portland's free downtown streetcar is a part of what makes Portland great.

26

Become a City of Green Buildings

The ten strongest green building markets in the US are in cities that have established public policies that promote green building.

— C40 Cities

A huge green building movement is underway. The world's first carbon neutral eco-cities are being built in Dongtan, China and Masdar, Abu Dhabi. In Germany 6,000 *Passivhaus* buildings use only 5% of the energy used by a typical German house. The benefits are enormous, not only for the world's climate but also financially and for cleaner air, improved health and increased happiness.

Adopt a Visionary Framework

Britain has laid down its Carbon Challenge: all new buildings and ten new towns must be zero carbon by 2016. In Austin, Texas, all new homes must be Zero Energy Capable by 2015 — 65% more efficient than the current code and have protected roof space for solar PV and hot water.

In England it was a long-term green commitment by the London Borough of Sutton that attracted the pioneering Beddington Zero Energy

TOM CHANCE BIOREGIONAL DEVELOPMENT GROUP

The Beddington Zero Energy Development in Sutton, England.

Development (BedZED). In Victoria, Canada, it was the city's commitment to the "triple bottom line" for land-use decisions that attracted the equally pioneering Dockside Green development. In Port Coquitlam, Canada, and Aspen, CO, all permit applicants must complete a sustainability scorecard, with points needed for approval.

Architecture 2030 is calling for all new buildings and retrofits to be carbon neutral by 2030, a challenge that has been endorsed by the American Institute of Architecture, the US Conference of Mayors and a growing list of cities. These commitments must be backed by engagement with the local building community, as Austin Energy's Green Building Program has done since 1991.

Create Green Building Incentives

In Seattle, developers can add extra density or height to a LEED-certified building. In San Francisco, all larger proposed buildings must demonstrate the highest level of green performance in America before they can receive a building permit. Other communities using fee rebates, tax incentives and grants. It is essential to identify the regulatory barriers to green building so that they can be eliminated.

Adopt Green Building Codes

In Boston and Los Angeles, all projects over 50,000 sq ft must be LEED certified. In England, the London Borough of Merton requires the use of renewable energy for 10% of the heat whenever ten or more buildings are constructed. The Merton Rule is being adopted by hundreds of communities, with Kirklees, Yorkshire, adopting a

- Architecture 2030: architecture2030.com
- Austin Energy Green Building: tinyurl.com/2v7r9a
- BedZED: tinyurl.com/5zjr7
- Berkeley's RECO: tinyurl.com/8cffq4
- Boston Green Building: bostongreenbuilding.org
- Building Technologies Program: www1.eere.energy.gov/buildings
- Cambridge Energy Alliance: cambridgeenergyalliance.org
- Dockside Green: docksidegreen.ca
- Energy Efficiency Building Retrofit Program: clintonfoundation.org
- Energy Savings Plan: saveenergynow.ca
- European Center for Renewable Energy (Gussing): eee-info.net/cms
- LED City: ledcity.org
- Playbook for Green Buildings and Neighborhoods: greenplaybook.org
- Port Coquitlam's Sustainability Checklist: tinyurl.com/2clt29
- San Francisco's RECO: tinyurl.com/2p36zl
- The Carbon Challenge (UK): englishpartnerships.co.uk/carbonchallenge.htm

35% rule. In Fingal, Ireland, all new buildings must use less than 50 kWh/m^2 a year for space and water heating, 30% of which must come from renewable energy. In Freiburg, Germany, where a 1992 Low Energy Housing Construction code limits houses to 65 kWh/m^2, new buildings use 80% less energy than average. White roofs are also important. Hashem Akbari, a physicist at the Lawrence Berkeley National Laboratory, has calculated that making a 1,000-square-foot roof reflective rather than absorbent color would offset ten tonnes of carbon dioxide emissions.[1]

What about existing buildings? There are three successful solutions — ESCOs, RECOs and Utility Programs.

Energy Services Companies (ESCOs)

An ESCO is a business or non-profit society that specializes in building energy retrofits, financing the work by the energy saved. Toronto's Better Buildings Partnership works with 40 ESCOs. In Cambridge, MA, a $100 million ESCO partnership is visiting 23,000 buildings and offering free energy audits, with retrofits financed through the savings. Portland's Multifamily Home Energy Solutions and the Berlin Energy Agency in Germany operate in a similar manner.

Residential Energy Conserving Ordinances

In Berkeley and San Francisco, since 1981, every building has been required to have an energy upgrade whenever it is sold, transferred or renovated. In Berkeley, by 2006, 12,000 residences had been upgraded (30% of the building stock), resulting in a 25–50% energy saving. Berkeley also has a Commercial Energy Conservation Ordinance.

Utility Programs

If your community owns its utility, much can be achieved. In Austin, TX, Austin Energy provides free home-energy improvements to customers with low to moderate incomes and gives rebates for energy investments to 48,000 apartments. Seattle City Light has numerous programs that assist with energy upgrades, including the Neighborhood Power Project in targeted neighborhoods. In Colorado, Fort Collins Utilities provides Zero-Interest Loans for Conservation Help (ZILCH).

How to Proceed?

Form **Team #7: Green Building Solutions,** including local green builders, citizens and municipal staff. Ask them to research the best practices and make recommendations to council. See also Solutions #53 and #67.

27

Go for Green Energy

People here feel less vulnerable because they know their energy's coming from renewable sources and not imports. This should be the top priority of anyone who goes into politics, anywhere in the world.

— Peter Vadasz, Mayor of Güssing

The small forest-based town of Güssing, in eastern Austria (population 7,800), reduced its community-wide carbon footprint by 93% between 1995 and 2007 (not counting flying). They did it by generating heat, electricity and vehicle fuel from sawdust, maize, cooking oil and solar energy, creating more than 1,000 jobs in 50 new businesses and turning the town into a magnet for climate-based ecotourism.[1] The Güssing Energy Network includes more than 30 energy projects, and the town now exports energy instead of importing it.

In southern Sweden, the city of Växjö (population 78,000) achieved a 30% reduction in its carbon footprint between 1993 and 2006 in pursuit of the goal of being a fossil-fuel-free city. Now 90% of its

- C40 Cities: c40cities.org
- City of Växjö: vaxjo.se
- Drake Landing: dlsc.ca
- Energie-Cites Association: energie-cites.org
- Güssing European Centre for Renewable Energy: eee-info.net
- Güssing: en.wikipedia.org/wiki/Güssing
- Local Energy: localenergy.org
- Solar America Cities: solaramericacities.org
- Solar Energy Incentive Program: sfsolarsubsidy.com
- San Francisco Solar Map: sf.solarmap.org
- Solar City: solarcity.org
- Solar Neighbourhoods (Toronto): solarneighbourhoods.ca
- Sunnyside Neighborhood Energy: sunnysideneighborhoodenergy.wikispaces.com

heat comes from renewable energy, mainly biomass, and more than 50% of its energy comes from a mix of biomass, geothermal and solar. If you know where to look, communities are pioneering a new energy revolution all over the world.

In India, in the Mumbai suburb of Thane, 16,000 households had installed solar hot water systems by 2007, following a decision to make it mandatory on all new buildings, supported by a 10% discount on property taxes. In Spain, every new building must install a solar hot water system, following the lead established by Barcelona. In Rizhao, China (population 3 million), 99% of buildings in the city center have solar hot water, following the decision to make it mandatory.

The Hague, Amsterdam and Toronto are using cold seawater or lake water for air conditioning and cooling, reducing electricity use by up to 90%. In Lund, Sweden, geothermal power plants provide 40% of the heat for the city's district heating system; similar plants operate in Ferrara, Italy, and Prenzlau, Germany. In Richmond, BC, Canada, waste heat from the Olympic ice rink heats local buildings.

San Francisco has a 645 kW solar system on the roof of its Convention Center and is planning a 38 MW tidal energy turbine under the Golden Gate Bridge. In Freiburg, Germany's solar city, solar systems on homes, schools, public facilities, the railway station and the soccer stadium cover 37,000 square meters, in pursuit of the city's goal to obtain 10% of its energy from renewables by 2010. New York's potential for solar PV has been estimated at up to 15,000 MW, capable of meeting 20% of the city's power needs.[2]

Güssing, Austria.

The biomass gasification powerplant at Güssing, Austria.

Gainesville, Florida, has adopted Europe's feed-in tariff to accelerate the take-up of solar energy and other renewables. (See #69.) Los Angeles is planning to do likewise as part of its plan to build or purchase 1,300 MW of solar energy by 2020. Berkeley has created Berkeley FIRST (Financing Initiative for Renewable and Solar Technology), enabling a property owner to borrow from the city's Sustainable Energy Financing District to install a solar PV system, with repayment through property taxes. San Francisco's Solar Energy Incentive Program is one of the best in the world. In Portland, a plan is being developed to retrofit a school with a solar and geothermal district energy system that would also heat and cool the surrounding 38 blocks, covering 500 homes in the Sunnyside neighborhood.

In Okotoks, Alberta, homes in the Drake Landing subdivision are 90% heated with solar hot water gathered in summer on community garages, stored underground, and pumped back in winter. In Copenhagen, 97% of the city's heat comes from a citywide district heating system that distributes waste heat from electricity generation, biofueled combined heat-and-power plants and waste incineration, using a 1,300 km network of pipes.

This raises a challenging point, because burning garbage produces cancer-causing dioxins and furans that everyone downwind of the plant is exposed to. It also undermines the push to reduce, re-use, recycle and pursue the more sustainable goal of Zero Waste. (See #28.)

In England, the city of Woking supports distributed energy through decentralized small-scale power generation projects close to homes and businesses, using gas-burning combined heat-and-power plants and solar PV roofs. Combined with energy efficiency measures, this lowered Woking's energy consumption by almost 50% between 1991 and 2004, and cut CO_2 emissions across the community by 17% (70% in council-owned buildings).

Are there any factors that the communities have in common? In Sweden, local governments have a legal obligation to promote the efficient use of energy, helped by the Swedish Energy Agency. The key to the stories, however, is that they tap into the ingenuity of local people, who dig in and become inspired. In Barcelona, it was Josep Puig, a Green Party city councilor. In Güssing, it was Mayor Peter Vadasz and a local engineer, Rheinhard Koch. In Woking, it was engineer Allan Jones, director of the city's Energy and Environmental Services Company.

This is the secret. Appoint knowledgeable local people to **Team #8: Green Energy Solutions,** and when they come up with ideas, help them find partners in the university, government and business.

28

Worship Your Wastes

Sewage contains ten times the
energy needed to treat it.

— Dr. David Bagley[5]

For millions of years our human footprint was low, and what little waste we produced dissolved easily into compost. When we moved into cities, we started the lazy habit of dumping our wastes outside the city walls.

Today, we are drowning in garbage. Each North American household produces more than a tonne of trash a year. New York City spends $1 million a day hauling garbage to landfills. Our rivers of sewage are an equally big problem. To live sustainably, we must create a world in which we consume far less, all goods are 100% recyclable, and all wastes are reclaimed as a useful resource.

Zero Your Garbage

First, set your goal. In New Zealand, 51 of 71 local councils have adopted Zero Waste policies, with most aiming at Zero Waste by 2015. Opotiki District Council (population 9,600) cut its waste by 85% between 1998 and 2002, using curbside recycling, organic waste recycling and widespread public education. In Berkeley, CA, the city has abandoned the term "solid waste" altogether and has a Zero Waste Commission.

Establish **Team #9: Zero Waste Solutions**, and develop a strategy to reach your goal. By 2009 San Francisco had achieved a 72% recycling rate, en route to the goals of 75% by 2010 and zero waste by 2020. They use a three-bin system — black for regular garbage; blue for all paper, bottles and cans, which go to a state-of the art recycling facility; and green for food and yard waste, which is composted. They use financial incentives, so the more a business recycles, the lower its garbage bill: the Fetzner winery has reduced its waste by 95%.

The city has banned the use of plastic bags and styrofoam take-out food containers and restaurants must now use biodegradable, compostable or recyclable containers. They have squads of friendly recycling missionaries who inspect people's garbage and teach the gospel of recycling to any backsliders, and they have staffed the city's recycling department with social activists, rather than engineers, which may be their real secret. Zero-waste strategies also generate innovation and new jobs.

Copenhagen, which has achieved a 97% reduction in its solid waste, also uses street-level Waste Caretakers who educate residents and businesses on a block-by-block basis. They achieve their high number, however, by incinerating 39% of their waste.

The Swedish company Envac has developed a system of vacuum tubes that transport garbage underground.

Don't Burn Your Garbage

It is tempting to believe the pitch when a company promises to convert garbage into "green energy," but incineration, pyrolysis and plasma gasification produce 0.35 to 0.8 million tonnes of greenhouse gases for every tonne of garbage converted.[1] Far from being "pollution free," they produce cancer-causing dioxins and furans; they undermine the drive to reduce and recycle; and they are very expensive. There are some fascinating new technologies that don't incinerate, such as Global Renewables' mechanical biological treatment, which Sydney, Australia, is using to process 11% of its waste.

Utilize Your Landfill Gas

Landfills produce methane, which traps 100 times more heat than CO_2 over its 10-year life in the atmosphere. So capture it and turn it into useful green heat and power. One Toronto city landfill generates 24 MW of power, enough for 18,000 homes. Toronto reduced its garbage-related emissions by 92% by capturing its landfill gas. It's a no-brainer, which should be required by law.

Extract Energy from Your Sewage

Sewage is a river of resource gold. In Oslo, Norway, heat from the city's sewers is piped through a hill and warms 9,000 apartments using heat pump technology, reducing heating costs by 50%; this is also being done in Zurich, Switzerland, and in Vancouver's Olympic Village.

In Stockholm, Sweden, 200 cars run on biogas from the city's sewage works.[2] In Kristianstad, Sweden, 22 buses do likewise, for $1.21 a gallon.

- Eco-Cycle: ecocycle.org
- Global Alliance for Incinerator Alternatives: no-burn.org
- Global Renewables: globalrenewables.eu
- Grassroots Recycling Network: grrn.org
- Inspiration from Sweden, by Stephen Salter: georgiastrait.org/?q=node/359
- Landfill Gas as Green Energy: energyjustice.net/lfg
- Landfill Gas in Canada: tinyurl.com/3exykl
- Landfill Gas Industry Alliance (Canada): lfgindustry.ca
- Landfill Methane Outreach Project: epa.gov/lmop
- "Resource Recovery from Sewage," by Stephen Salter: georgiastrait.org/?q=node/567
- San Francisco Zero Waste: sfenvironment.org/our_programs/overview.html?ssi=3
- Sewer Heat: rabtherm.com
- Stop Trashing the Climate: stoptrashingtheclimate.org
- Zero Waste America: zerowasteamerica.org
- Zero Waste around the World: ecocycle.org/zero/world.cfm
- Zero Waste California: zerowaste.ca.gov
- Zero Waste International Alliance: zwia.org
- Zero Waste New Zealand Trust: zerowaste.co.nz

In Linkoping, Sweden, buses, cars and taxis are powered by biogas from anaerobic organic waste digestion.[3] In Gothenburg 1,000 cars run on biogas, and a wood waste gasification plant planned for 2010 will provide enough for 75,000 cars.[4] In the city of Victoria, Canada, engineer Stephen Salter, inspired by a visit to Sweden, calculated that smart sewage treatment could power 200 buses with biodiesel and 5,000 cars with biogas, reducing the city's carbon footprint by 33,000 tonnes of CO_2e a year.

29

*When we plant trees, we plant the
seeds of peace and seeds of hope.*
— Wangari Maathai

Grow More Food, Plant More Trees

When you sit down to a meal in North
America, the average mouthful has traveled
1,500 miles before it reaches you, dumping carbon
all the way. Unless is has been grown organically,
it has also dumped nitrous oxide emissions. And
if it's beef, its journey has produced clouds of
methane emissions. It's not a pretty picture.

In most communities, these emissions will not
show up in the local carbon reckoning. They happen on someone else's tab, and in consequence,
most Climate Action Plans turn a blind eye, saying "not our problem." Globally, this is disastrous.
We must take responsibility for our full carbon
footprint and not hide behind the convenience of
"somewhere else."

Grow More Food

Climate action plans must include targets to grow
more local organic food, like Berkeley's Climate
Action Plan, which calls for most food consumed

in the city to be produced within a few hundred
miles. In nearby Oakland, the Food Policy
Council's goal is that 30% of the city's food is produced in or near the city. Chicago's Green Food
Resolution supports local organic gardens, farmers'
markets, community supported agriculture, community gardens, and other venues that provide
healthful plant-based food.

In Hanoi, 80% of the vegetables are grown in
or near the city. In Kolkata, India, 3,500 hectares
of wastewater ponds are used to raise fish. Paris
has many community gardens on its outskirts.
Seattle's 65 P-Patch Community Gardens provide
space for 2,000 urban gardeners, and the city's
Comprehensive Plan calls for a community garden
for every 2,000 households within urban villages.
In Britain, councils are required to provide 15
allotment gardens for every 1,000 households.

Shanghai manages 300,000 hectares of farmland that recycle the city's night-soil, producing
50% of their pork and poultry, 60% of their veg-
gies and 90% of their milk and dairy. In Havana,
city dwellers produce 50% of their vegetables.[1] In
Japan, the city of Iida has an entire street that students planted with apple trees in 1947, which is
now a pedestrian street.

Most cities have vacant lots. Chicago has
70,000, Philadelphia 40,000 — and when you
improve a down-at-heel neighborhood with trees,
parks and community gardens, property prices rise
by 30% as hope returns.[2]

Amsterdam's Food Strategy sets targets for
locally grown organic food to be available in all
hospitals, care institutions, schools and municipal
cafeterias. New schools must have kitchen ameni-

PAUL SYMINGTON

P-Patch Community Garden in Highpoint, West Seattle.

ties, and every primary school must have a working school garden.

Create **Team #10: Local Food Solutions**, and invite people to get involved. Set a goal to produce more local food and a target to create new community gardens. Encourage local farmers' markets and pocket markets, and work to protect all local farms and community gardens. Write planning policies to support all this.

Grow and Protect More Trees

Trees store carbon — the average city tree absorbs 20–40 lbs of CO_2 a year. Trees also produce shade, reducing the energy needed for air-conditioning and its associated carbon footprint. In Gainesville, Florida, which has a strict tree protection ordinance, residents spend $126 less a year on electricity than they do in nearby Ocala, which has a weak one, and only half as many trees.[3]

The bad news is that we are losing ground. From 1988–2003 the number of trees in many US cities fell by 30%, while the cold dead space occupied by concrete and asphalt increased by 20%.

Sacramento's residents save $20 million a year in air conditioning costs thanks to 500,000 trees planted by the Sacramento Municipal Utility District and the Sacramento Tree Foundation, which also store 240,000 tons of CO_2 a year. The energy saved by planting trees costs 1 cent/kWh, compared to 2.5 cents for saved energy and 10 cents for new energy. A 1990 study found that if American cities enjoyed an additional 100 million mature trees, energy costs would be cut by $2 billion — or maybe $4 billion today.[4] Chicago has planted 500,000 trees since 1989.

- Alliance for Community Trees: actrees.org
- American Forests: americanforests.org
- Billion Trees Campaign: unep.org/billiontreecampaign
- Canopy, Palo Alto: canopy.org
- Chicago's Green Food Resolution: tinyurl.com/lny5jg
- City Farmer News: cityfarmer.info
- Foodprint USA: foodprintusa.org
- Friends of the Urban Forest, San Francisco: fuf.net
- Green Roofs for Healthy Cities: greenroofs.org
- Greening of Detroit: greeningofdetroit.com
- Michael Pollan: michaelpollan.com
- National Arbor Day Foundation: arborday.org
- Pocket Markets: foodroots.ca/pocket_market.htm
- Policy Guide on Regional and Community Food Planning: planning.org/policy/guides/adopted/food.htm
- P-Patch: seattle.gov/neighborhoods/ppatch
- Seattle Tilth: seattletilth.org
- *Shading Our Cities: A Resource Book for Urban & Community Forests*, by Gary Moll, Island Press, 1989
- Share Trees Benefits Estimator: usage.smud.org/treebenefit/calculate.aspx
- Tree Canada Foundation: treecanada.ca
- Treefolks, Austin: treefolks.org
- Treelink: treelink.org

Include tree-planting targets in your Climate Action Plan. Invite citizens to form **Team #11: Trees and Green Spaces**. Set Green Roof targets — Chicago has planted 200,000 square meters of rooftop gardens. Establish strong tree protection ordinances and bylaws, and make building permits and development approvals conditional on tree protection. Use American Forests' Urban Ecosystem Analysis and CITYgreen software to help you protect your local green spaces and threatened ecosystems. Make a commitment to bring trees and green space back to your city.

30

Plan Ahead for a Green, Stormy Future

> The ice caps are melting now. They're not going to refreeze next year just because we reduce our emissions. We're going to live in that world. So plan for it.[1]
> — Ron Sims, King County, WA

It needs to be green, and it's going to be stormy — so we must plan for both.

Before we had cars, all human communities were designed for easy walking. Then we discovered fossil fuels, invented the automobile and encouraged our cities to sprawl outward, eating up forests and farmland.

In the USA, green space is being lost to low-density sprawl at 365 acres an hour. The lower the housing density per acre, the more roads are needed, the harder it is to make transit work, and the more CO_2 is released.

The alternative is compact, pedestrian-oriented communities where you live surrounded by green space, walk into town to do your shopping, pause to chat in a neighborhood café, cycle with ease,

and work in the local economy. Vermont is full of this kind of community, and Vermonters love it. Well-planned smart growth also costs less for roads, utilities and schools.

In Portland, thanks to its urban growth boundary and smart growth policies, the city has seen a 20% increase in population since 1990 with only 2,200 extra acres being developed, while Chicago has expanded its land by 40% to accommodate just 4% more people. In Vancouver, which adheres to smart growth policies, downtown residents increased by 62% between 1991 and 2002, while the number of car trips remained constant.

The challenge is to turn around 50 years of planning regulations that discourage compact, smart growth. Help is just a click away, however, through the Smart Growth Network and other resources. *This Is Smart Growth* lays out the vision and principles in a clear, easy manner, while *Getting to Smart Growth: 100 Policies for Implementation* makes it easy to integrate smart growth principles into planning and transport master plans, land-use plans, and official community plans.

- Map your region for its ecological assets. Develop a plan of where you want to be in 50 years, set urban containment boundaries and plan long distance greenways.
- Say no to out-of-town big-box stores, as 330 US communities have done. Support locally owned small businesses instead. Sprawl-Busters offers help and examples of successful legal and zoning decisions. Build a strong local economy.
- Use development codes and financial incentives to encourage reduced carbon footprints,

PEDBIKEIMAGES.ORG/CHARLIE ZEGEER

La Rambla, a 1.2 km long pedestrian street in Barcelona, Spain.

ecological assessment, compact pedestrian communities, neighborhood centers, affordable housing, greenways; community-based economic development and green buildings.

- Set development cost charges high, and reduce them for each aspect of sustainability that a developer meets.

- Integrate smart growth planning into your community's comprehensive and climate action plans.

- Adopt a sustainable development matrix that sets targets and tracks progress on a wide range of climate and sustainability issues, as Santa Monica and Denver have done.

Prepare for the Deluge

Whether we like it or not, climate change is coming. The more we reduce our carbon footprint, the less will be the impact, but it's still coming. We must prepare for sea-level rise, forest fires, hurricanes, tornadoes, floods, salt-water intrusion into drinking water, droughts, health crises from infectious diseases, heat waves, habitat loss and the arrival of invasive species. We must also prepare for the consequences of peak oil, including the inability of people on lower incomes to afford fuel or food.

Form **Team #12: Climate Protection**. In Miami-Dade County, Florida, the 32-member Climate Change Adaptation Task Force met monthly for a year, enabling county and state officials, scientists, engineers, regional planners and other experts to wrap their minds around what was coming, and identify ways to prepare

- *Better, not Bigger: How to Take Control of Urban Growth,* by Eben Fodor, New Society, 1998
- Big Box Toolkit: bigboxtoolkit.com
- *Big-Box Swindle,* by Stacey Mitchell, Beacon Press, 2006
- Congress for the New Urbanism: cnu.org
- *Getting to Smart Growth: 100 Policies for Implementation*: smartgrowth.org/pdf/gettosg.pdf
- Greenprint Denver: greenprintdenver.org
- *Green Urbanism: Learning from European Cities,* by Morra Park, Island Press, 1999
- *Growing Cooler: The Evidence on Urban Development and Climate Change,* by Reid Ewing, 2008 uli.org
- *The Hometown Advantage — Reviving Locally Owned Business*: newrules.org/retail
- New Urban News: newurbannews.com
- Smart Growth America: smartgrowthamerica.org
- Smart Growth Network: smartgrowth.org
- *Solving Sprawl,* by F. Kaid Benfield, Jutka Terris and Nancy Vorsanger, Island Press, 2002
- Sprawl-Busters: sprawl-busters.com
- Sprawlwatch Clearing House: sprawlwatch.org
- Stopping Sprawl: sierraclub.org/sprawl
- *The Small-Mart Revolution,* by Michael Shuman, Berrett-Koehler Publishers, 2006
- Urban Advantage: urban-advantage.com
- Climate Impacts Group: cses.washington.edu/cig
- National Climatic Data Center: ncdc.noaa.gov
- *Preparing for Climate Change: A Guidebook for Local, Regional and State Governments*: iclei.org

for them, including 100-year events that now come every 10 years. King County, WA, Boston, Anchorage and Chicago have formed similar groups. King County prepared a county-wide flood control plan, merging several small flood-control districts into one that serves the whole county, funding it with a property tax increase. It just has to be done.

31

Join the Challenge

> Seventy-four percent of people think businesses are not doing enough to cut their carbon emissions and tackle climate change.
> — 2006 Carbon Trust Survey

What is a business? Is it simply a means to make money, or is it also part of a more noble effort to make a difference in the world?

Our predicament is grim: if we don't steer a rapid path to a climate-friendly world, our world will be hit so badly that most business owners will wish they had paid more attention.

And here's a prediction: soon there will be a green sticker that approved companies will display in their windows to show their customers they are part of the solution. Customers will take note and gradually walk away from businesses that are not playing their part.

It's already starting. Climate Counts, a business supported initiative, is using a sophisticated scorecard to rank businesses for their climate change contribution, as Stuck, Starting or Striding.

In 2008 Burger King and Wendy's scored a fat red "Stuck," with zero points out of 100. Amazon

and CBS also scored absolute zero. Levi Strauss managed just 1, while the best that Apple and eBay could muster was 2. Canon came top with 77, followed by Nike at 73, Unilever 71 and IBM 70.[1]

The solutions in this section are intended for all businesses, from the largest corporations to the tiniest pub or corner store. They are intended to sweep away the confusion and provide a clear path that any owner or manager can follow to shrink your business's carbon footprint.

Ten Solid Reasons

But first — why bother? Here are ten components of a solid business case that you can take to the board. A good carbon reduction program will enable you to:

1. Tackle the global climate crisis.
2. Reduce your costs by making you more energy efficient.
3. Protect your business against rising oil and gas prices, and the future price on carbon.
4. Be part of the exploding business of saving the planet, and the innovation it brings.
5. Make use of incentives that are becoming available and be better situated to avoid new regulations.
6. Give your employees reason to feel good about working for you.
7. Give your customers reason to feel good about dealing with you. In a 2006 survey for the Carbon Trust in Britain, 64% of respondents said they would be more likely to use a business that claimed to have a low-carbon footprint.[2]

YOUR CHOICE. YOUR VOICE.
OUR COMPANY SCORES HELP YOU
VOTE WITH YOUR DOLLARS.

GROUPE DANONE · HITACHI · KELLOGG · COCA-COLA · CLOROX · NEWS CORPORATION · CONAGRA · GENERAL MILLS · L'OREAL · KIMBERLY-CLARK · LIMITED BRANDS · YUM! BRANDS · LEVI STRAUSS · AMAZON.COM · LIZ CLAIBORNE · IBM · KRAFT · JONES APPAREL GROUP · SAB MILLER · STONYFIELD FARM · NESTLE · NOKIA · APPLE · MICROSOFT · EBAY · DELL · TIME WARNER · UNILEVER

8. Reap the free publicity that comes with being a leader.

9. Sell a greener product. 67% of the survey respondents said they would be more likely to buy a product with a low-carbon footprint.

10. Respond positively to your shareholders and investors when they ask what you are doing.

Step Up the Urgency

The carpet company Interface has set the benchmark for what's needed: zero emissions by 2020. Reducing your carbon footprint by 10% a decade will not cut it: that'll take a century. Nor can it be done by buying offsets. They're like training wheels — a valuable intermediate step but evidence that you still need help.

The role of large companies is critical — in Britain, 2% of companies account for 80% of the emissions from industrial processes and the business use of buildings. The contribution from smaller companies is also important, for the message it sends that we are all in this together.

In Britain the Carbon Trust has worked with the British Standards Institution to create a carbon footprinting standard called PAS 2050, which provides a consistent way to measure GHG emissions from products over their whole life cycle, including the sourcing of raw materials, manufacture, transport, use and final recycling or disposal.

Eliminate the Excuses

I don't know where to begin: There are many excellent resources, including case studies. Spend a week doing some reading.

- "Beyond Neutrality — Moving Your Company Towards Climate Leadership": bsr.org
- Business Environmental Leadership Council: pewclimate.org/companies_leading_the_way_belc
- Carbon Disclosure Project: cdproject.net
- Carbon Reduction Label: carbon-label.com
- Carbon Trust: carbontrust.co.uk
- Climate Action Programme: climateactionprogramme.org
- Climate Change Corp: climatechangecorp.com
- Climate Change Intelligence: svantescientific.com
- Climate Counts: climatecounts.org
- Climate Group: theclimategroup.org
- Combat Climate Change: combatclimatechange.org
- Environmental Leader: environmentalleader.com
- EPA Climate Leaders Program: epa.gov/stateply
- "Getting Ahead of the Curve — Corporate Strategies That Address Climate Change": pewclimate.org
- Global Roundtable on Climate Change: grocc.ei.columbia.edu
- Green Biz: greenbiz.com
- State of Green Business: stateofgreenbusiness.com
- Sustainable Business: sustainablebusiness.com
- US Climate Action Partnership: us-cap.org
- WWF Climate Savers: worldwildlife.org/climatesavers

Our CEO is not interested: Build a strong business case, and work to get support from senior managers and departmental leaders.

Our chief accountant won't look at anything less than a two-year ROI: Seek management approval to establish a climate-solutions investment budget that will allow your company to undertake carbon reduction projects that have a three- to seven-year return on investment.

We've not got spare staff capacity: Seek a volunteer who will set up a low-carbon luncheon club where staff can develop a plan together.

32

Four Solutions for Every Business

> We're operating this planet like a
> business in liquidation.
>
> — Al Gore

These four fundamental actions offer a practical path that any business can follow to become a leader in the emerging low-carbon economy.

1. Form a Green Team and Develop a Plan

Staff involvement is crucial. Ask your best brains to sit down together to design a zero-carbon future for your company. In Britain, when BT (formerly British Telecom) started a company-wide initiative, the staff formed 28 carbon-busting clubs within three weeks. Be sure to include the finance department, who might otherwise be foot-draggers, and to get engagement from the highest level of management. Write an environmental policy, and create a coordinated management structure that involves everyone.

What goals should you adopt? Interface, which sells carpet products in 100 countries around the world, has made a commitment to Mission Zero — "to eliminate any negative impacts the company may have on the environment by 2020." Alcan, the aluminum producer, is aiming to help the entire industry become carbon neutral by 2020. DuPont reduced its CO_2e emissions by 65% below 1990 by 2008 — and achieved it by 2008.

2. Reduce Your Footprint

Use a carbon calculator to help you work out your carbon footprint, and then adopt a range of programs that can be effective across a wide front.

- Nike found a way to remove SF_6 (a powerful greenhouse gas) from the process it used to fill the air pockets in its running shoes.[1]

- Google installed a 1.6 MW solar array on its roof and uses on-line forums to encourage employee carpooling.
- Hyperion pays a $5,000-a-year bonus to 200 employees who drive a car that averages 45 mpg or more.
- The owner of the AOK Auto Body Shop, Philadelphia, changed light bulbs, installed motion sensors, put timers on equipment, switched to Energy Star appliances, and trained his staff to turn the lights off, saving $5,577 a year on a $7,832 investment.
- New Society Publishers, who published this book, use 100% post-consumer recycled paper for all their books and offset 100% of their emissions.
- HP has redesigned its print cartridge packaging, saving 17,000 tonnes of CO_2 a year.

When you've done all this, buy reliable carbon offsets to neutralize your remaining emissions.

3. Reduce Your Supply Chain's Footprint

Your supply chain's carbon footprint may be far larger than your own.

- Vodaphone is working with its suppliers to develop more energy-efficient equipment.
- The food and retail company Unilever, whose raw materials' carbon footprint is ten times larger than its own, gives preference to suppliers with lower emissions.
- The healthcare company Novo Nordisk requires its suppliers to complete a self-evaluation questionnaire as part of its Sustainable Supply Chain Management program.

FROM BOB WILLARD, THE BEST SELLING AUTHOR OF
THE NEXT SUSTAINABILITY WAVE & THE SUSTAINABILITY ADVANTAGE

BUSINESS CASE FOR SUSTAINABILITY

DIANE MCINTOSH

The Business Case for Sustainability *presents a compelling case for corporate responsibility, framing sustainability arguments so that they are relevant to executives concerned with minimizing risks while maximizing profits.*

- Nike is working to factor its suppliers' emissions into its scoring system for partner selection.

4. Communicate Transparently

The Carbon Disclosure Project represents institutional investors with combined assets of $57 trillion, who want to know the risks and opportunities that climate change presents. Since 2000, it has become the gold standard for carbon disclosure and is the world's largest repository of corporate greenhouse gas emissions data. Your investors, customers, staff and the public all need to know what you are doing. The Global Reporting Initiative's Sustainability Reporting Guidelines include carbon footprint metrics. Timberland, the shoe company, is working to carry "Green Index" labels on all its shoes by 2010.

In Britain, starting in 2010, all quoted companies will be obliged to disclose their carbon emissions in their annual reports, and their performances will be compared in an annual league

- Canada's Voluntary GHG Registry: ghgregistries.ca
- Climate Smart: climatesmartbusiness.com
- Carbon Trust Solutions: carbontrust.co.uk/solutions
- ClimateBiz — Business Resource: climatebiz.com
- Climate Neutral Business Network: climateneutral.com
- Center for Sustainable Business Practices: lcb.uoregon.edu/csbp
- Curbing Your Climate Impact: seattle.gov/climate/docs/Climate_partnership_resource_guide.PDF
- Energy MAP: sei.ie/energymap
- Energy Star@Work: energystar.gov/work
- Global Reporting Initiative: globalreporting.org
- Green Power Market Development Group: thegreenpowergroup.org
- HSBC Climate Partnership: hsbccommittochange.com
- Low Carbon Innovation Network: carbon-innovation.com
- Resource Smart Businesses: tinyurl.com/62ss33
- Step-by-Step Interactive Guide to Energy Saving Solutions: cool-companies.org/guide
- Start a Green Team: tinyurl.com/mp7djv
- Sustainable Business.com: sustainablebusiness.com
- *The Business Case for Sustainability:* by Bob Willard: sustainabilityadvantage.com
- *The Truth about Green Business,* by Gil Friend: natlogic.com/truth
- Who's Going Carbon Neutral? bsr.org/reports/BSR_Carbon-Neutral-Chart.pdf

Business Carbon Calculators

- Business Travel Carbon Calculator: carbonneutral.com/sbc/buscalc.asp
- Climate Care: climatecare.org/business
- Footprinter (UK): footprinter.com
- Seattle Carbon Calculator: seattle.gov/climate/SCPresources.htm
- Workplace Eco-Audit: cupe.ca/forms/ecoaudit.php
- Zero Footprint: zerofootprint.net

table summarizing the best and worst performers. It is only a matter of time before this approach is adopted all over the world.

33

Solutions for the Financial Sector

You need to really scrub your investment portfolios, because I guarantee you, many of you are going to find them chock-full of subprime carbon assets.

— Al Gore

The world's energy economy is a $6 trillion affair, all of which requires financing. Without investors, new coal mines, oil projects and livestock operations would grind to a halt. Similarly, without investors, solar and wind projects would cease.

If you work in the financial sector, you have more ability to reduce the world's carbon footprint than almost any other business sector. If you are a climate activist, you will achieve more by persuading the banks to stop fossil-fuel lending than by any other tactic.

Stop Financing Global Warming

In 2008, when most banks were still asleep at the switch, CERES (a coalition of investors and environmentalists) analyzed 40 of the world's largest banks for their climate change governance practices. Twenty-eight were addressing their internal carbon footprint and 29 were investing in alternative energy, but none had a policy to stop investing in high-climate-risk projects such as coal-fired power or the Canadian tar sands. This makes their executives like tobacco executives who have stopped smoking but still peddle tobacco around the world.

There is a global call for a halt to all new conventional coal-fired power plants, and the Rainforest Action Network is calling on the banks not to invest in them. Citi had pledged to invest $50 billion over ten years in renewables and other solutions, but this is only 2.5% of its $1.86 trillion assets — in 2006 Citi financed 200 times more dirty energy than clean energy, including coal projects all across the US. The Bank of America says it will reduce the greenhouse gases associated with its lending portfolio by 7%, but in 2006 it invested 100 times more in dirty than in clean energy. The report *Banks, Climate Change and the New Coal Rush* spells out all the details. In 2008, it announced that it would refuse loans to mining companies that destroy mountaintops to get at the coal.

Invest in The Future

Most of the world's electricity and almost all the world's transport has to be retrofitted for a climate-friendly world, making it a $6 trillion investment project that needs to be completed by 2030. This is a huge opportunity. In 2008, the McKinsey Global Institute found that investments in energy productivity could earn a 10%+ rate of return, and achieve half the reduction needed to stop the global temperature rising by more than 2°C.

Institutional investors have formed the Investor Network on Climate Risk. In 2007, their Carbon Disclosure Project, representing 315 institutional investors with $40 trillion at their

- Banktrack: banktrack.org
- *Banks, Climate Change, and the New Coal Rush*: ran.org
- Climate Friendly Banking (Canada): climatefriendlybanking.com
- *Curbing Global Energy Demand — The Energy Productivity Opportunity*: mckinsey.com
- *Corporate Governance and Climate Change — The Banking Sector*: ceres.org
- Ending Destructive Investment: dirtymoney.org
- Environmental Finance: environmental-finance.com
- Investor Network on Climate Risk: incr.com

disposal, wrote to 2,400 corporations asking what they were doing about climate change. There is pressure on the Securities and Exchange Commission to make climate-risk disclosure mandatory, and in 2007 there were 43 shareholder resolutions asking major corporations to set reduction targets. At the 2008 Investor Summit on Climate Risk, nearly 50 US and European institutional investors, managing $1.75 trillion in assets, backed a climate-change action plan that called for more green investments and tougher scrutiny of carbon-intensive investments.

Pension funds matter too. The California Public Employees Retirement System has invested more than $1 billion in clean technology and housing retrofits through its Green Wave program. In Pennsylvania, the Heartland Labor Capital Network has done likewise.

Offer Green Consumer Products

Citizens Bank of Canada offers a $10,000 line of credit at prime for efficiency upgrades. Vancity Credit Union, based in Vancouver, offers a Bright Ideas low-interest line of credit for up to $20,000 for home retrofits and Clean Air Auto Loans for fuel-efficient vehicles. The Development Bank of Japan offers green loans for companies that buy energy-saving products in bulk and lease them to households at low cost, with the lease fees being added to the electricity bills. This kind of product needs to become standard across the industry.

Become a Climate Friendly Bank

HSBC is one of the world's largest finance groups, with $1.86 trillion in holdings. It became carbon

Scott Ridley at the Caton Moor wind farm in Lancashire, UK, owned and operated by Triodos Renewables, a subsidiary of the Triodos Bank. Triodos a member of the Global Alliance for Banking on Values. triodos.co.uk

neutral in 2005 by reducing its energy use, buying green electricity and carbon offsetting, and it has established a long-term carbon management plan to reduce its net internal emissions to zero. It has also launched a carbon finance strategy to invest in non-fossil fuel energy solutions, and created the $100 million HSBC Climate Partnership to respond to the threat of climate change. Even though it ranked highest in the CERES survey, at the time of writing it had not announced a policy to stop investing in the cause of the problem.

34

Solutions for Retail Stores

Any difficulties which the world faces today will be as nothing compared to the full effects which global warming will have on the world-wide economy.

— Prince Charles

We are history's greatest consumers. People who return from less-developed parts of the world are frequently thrown into culture shock by the abundance of our consumerism. The retail sector works hard to encourage this, stoking the furnace of global warming. In an eco-conscious world, retail stores would be centers of education where we learn to balance good housekeeping with good planetary housekeeping.

The best role models are small and local. Small Potatoes Urban Delivery (SPUD) provides home delivery of organic food to customers in western Canada and the US, sourcing more than 50% of its products locally and delivering by bicycle where possible. Each delivery van carrying 80 orders saves 160 car trips, reducing fuel use by 40%. Larger retailers are also making a big effort, however.

SPUD is the largest organic food delivery company in North America, serving more than 19,000 customers.

Set Meaningful Goals

In Britain, Marks & Spencer (M&S) has pledged to reduce its CO_2 emissions by 80% and to send zero waste to the landfill by 2012. Tesco has pledged to reduce the carbon footprint of its stores by 50% by 2020. Wal-Mart has pledged to make its stores 20% more energy efficient, to have them run on 100% renewable energy by 2013, to double the fuel efficiency of its trucks by 2015, and to achieve zero waste by 2025. The challenge is not easy. In 2006 Wal-Mart's carbon footprint increased by 8.6%.[1] Among other things, eliminating use of the refrigerant HFC-134a is essential. (See #97.)

Use Energy Sustainably

Many retailers are building model factories and green stores. Because the margins in food are so low, $100,000 saved by efficient lighting has the same impact as increasing sales by $10 million. Wal-Mart has invested $17 million in a partnership with GE to develop an LED refrigerator lighting system that will reduce its carbon footprint by 28,000 tonnes a year.

Travel Sustainably

Stores can make their trucking operations more fuel efficient — but what about their customers' footprints? Every time an out-of-town box store opens, two local stores close down. Stacey Mitchell from The New Rules Project has calculated that Wal-Mart's yearly share of Americans' travel produces as much carbon as Wal-Mart itself.[2] IKEA —

- *Best Practices in Greening Retail:* tinyurl.com/2s5f5o
- Canada Grocery Stores Green Scorecard: static.corporateknights.ca/CK23.pdf
- Global Forest and Trade Network: gftn.panda.org
- Green Retail (UK): talkingretail.com/news/green-retail-news
- IKEA: ikea-group.ikea.com
- M&S: plana.marksandspencer.com
- Retail Energy and Environment Club (UK): thereec.co.uk
- SPUD: spud.ca
- Wal-Mart: walmartstores.com/Sustainability

perhaps the greenest large retail chain in the world — is encouraging 10% of its customers to travel to its stores by public transport by 2009, helped by a free bus and home delivery service.

Support Local Suppliers

M&S has pledged to reduce its imported food and double its regional food sourcing. Tesco has pledged to give its customers more local products than any other retailer. Still in Britain, Asda (Wal-Mart) has opened 15 local sourcing hubs to deliver local products directly to its stores.

Sell Green Products

Home Depot plans to have 6,000 products in its Eco-Options line by 2009. At Tesco, shoppers who buy organic, Fairtrade and biodegradable items are rewarded with green loyalty card points. In 2008, the British stores Waitrose and Sainsbury's each gave a million efficient light bulbs away for free, while Tesco sold them for 2 cents each, supported by an electric utility. Wal-Mart sold 100 million bulbs by 2008, using lower prices, more shelf space at eye level and end-of-aisle showcase displays. It also plans to double its selection of organic products.

Green up Your Supply Chains

All of IKEA's 1,600 suppliers have to sign onto the IKEA Way, which lays down green rules for emissions, waste and chemical management, working conditions and child labor. Wal-Mart is asking its electronics suppliers to fill in a sustainability scorecard, which it will use to influence purchasing decisions, and challenging its suppliers to make their products with renewable energy. It has also

joined the Global Forest and Trade Network, pledging to phase out all illegal and unwanted wood sources from its supply chain and to increase the proportion of wood products that come from credibly certified sources. The British retail chain Boots runs supplier workshops on packaging and energy efficiency.

Reduce Your Packaging and Waste

M&S and Wal-Mart have both set ambitious goals (see above), and Wal-Mart is using a packaging scorecard with its 2,000 private-label suppliers. In 2007, Sainsbury's gave 6.5 million reusable bags to its customers, reducing plastic bag use by 50%, and is packing its own brand products in compostable packs. In the US, IKEA is aiming to increase its reclaimed store waste from 67% to 90% by 2010.

Educate the World

Tesco is developing carbon footprint labeling for all the goods it sells. M&S is helping its customers reduce their energy use by running a Carbon Challenge with the Women's Institute. Wal-Mart employees are undertaking personal sustainability practices to improve their wellness and the health of the environment. When M&S started to go green, its CEO took his 100 top executives to see *An Inconvenient Truth* to get them motivated. These are small beginnings — but that's how oak trees grow.

35

Solutions for Architects, Builders and Developers

We are proposing buildings that, like trees, are net energy exporters, produce more energy than they consume, accrue and store solar energy, and purify their own waste, water and release it slowly in a purer form.
— Bill McDonough

What a difference a few years makes. In 2000 carbon-neutral housing was just an idea. In 2010 architects all over the world are designing zero-carbon buildings.

Zero-carbon cities are being planned at Dongtan, China and Masdar, Abu Dhabi. In Britain, all new homes must have zero net CO_2 emissions from energy use in the home by 2016, including appliances, following a 25% improvement in efficiency by 2010 and 44% by 2013. By 2011, all new Chinese buildings will need to reduce their energy use by 50%.

This is the new normal. The technologies needed are in widespread use — airtight construction, super-insulation, heat exchange, passive solar orientation, solar PV and hot water, combined heat and power units, and compact, pedestrian, mixed-use neighborhood designs that reduce travel.

At Dockside Green, in Victoria, Canada, where all the buildings in the planned downtown community of 2,500 residents will be LEED Platinum, the savings from building green in an integrated manner has kept costs stable.

As a company, what can you do?

- Join Architecture 2030's Challenge to make all new buildings carbon neutral by 2030.
- Adopt Architecture 2030's 2010 Imperative Curriculum, requiring all design studios to design in a way that dramatically reduces or eliminates the need for fossil fuel and achieves a carbon-neutral design school campus by 2010.
- Require all your architects and engineers to design toward a carbon-neutral goal, using LEED or Green Globe to keep on track.
- Form partnerships with local colleges so that the trades are prepared.
- Form partnerships with your local government, as green builders have done in Austin, Portland and Los Angeles.
- Write energy efficiency and zero-carbon incentives into your contracts, enabling designers to keep a portion of the energy savings as a bonus fee.[1]
- Work with your sales or other departments to ensure that everyone is working to the same goals. Don't allow green considerations to be arbitrarily overruled.
- Take your staff, partners and investors to see An Inconvenient Truth and The 11th Hour.

Laurie and Anibal Guevara-Stone's 1,947 sq. ft. super-efficient, passive solar straw-bale and adobe home in Carbondale, Colorado. They have a 2 kW solar PV system in the back yard and a 2-panel solar thermal heater on the roof.

- Integrate LEED, zero-carbon and the new urbanism into all your projects.
- Don't accept projects where the developer does not want to pioneer.
- When working outside your culture, learn from local people and their traditions.
- Build a sense of community, not just houses. If it's not fun, it's not sustainable.

The ING Bank

In Holland's ING bank building, completed in 1987, water trickles down channels in the stair railings and daylight pours in. With no conventional air-conditioning, it uses 80% less energy than other new buildings in Amsterdam. The energy-saving features cost $700,000, but they save $2.9 million a year and paid for themselves in three months.

Village Homes, Davis, California

- 70 acres, 240 homes, 12 acres of green space
- 12 acres common agricultural land
- 4,000 sq. ft. commercial center
- Natural swales; car access by back lanes
- Passive-solar designs, tree-shaded streets
- Houses clustered around green spaces
- 80% participation in community activities
- Many work locally or cycle to work
- Estimated reduced CO_2 per household: 75%
- *Designing Sustainable Communities: Learning from Village Homes* by Judy and Michael Corbett villagehomesdavis.org

- Architecture 2030: architecture2030.org
- Building America (50% less energy): eere.energy.gov/buildings//building_america
- Canada Green Building Council: cagbc.ca
- Earthship Biotecture: earthship.net
- Energy & Environmental Building Association: eeba.org
- Energy Star Buildings: energystar.gov
- Environmental Building News: buildinggreen.com
- Garbage Warrior (film): garbagewarrior.com
- *Green Development: Integrating Ecology and Real Estate,* by Alex Wilson/RMI, Wiley, 1998
- New Urbanism: newurbanism.org
- Oikos Green Building Source: oikos.com
- Rocky Mountain Institute Built Environment Team: bet.rmi.org
- The Sustainable Condo: sustainablecondo.com
- UK-Sweden Initiative on Sustainable Construction: ukswedensustainability.org
- US Green Building Council: usgbc.org
- Zero Carbon City: britishcouncil.org/zerocarboncity
- Zero Carbon House: zerocarbonhouse.com

Some Green Architects and Projects

- Dockside Green: docksidegreen.ca
- Village Homes: villagehomesdavis.org
- Bill McDonough: mcdonoughpartners.com
- Busby Perkins + Will: busby.ca
- Duany Plater Zyberk: dpz.com
- Oshara Village: osharavillage.com
- Masdar: masdaruae.com
- SOM (Pearl River Tower): som.com
- Dongtan: tinyurl.com/34ufug
- ZedFactory: zedfactory.com

36

Solutions for the Cement Industry

> Instead of being part of the problem, the industry could be part of the solution.
>
> — John Harrison, TecEco

The Romans used cement to build aqueducts that still stand today, but their knowledge was almost lost before being reinvigorated in the 19th century. Today, the global cement industry produces almost 3 billion tonnes of cement a year, and each tonne produces 900 kg of CO_2, totaling 2.75 billion tonnes, or 7.5% of global CO_2 emissions (3.9% of the cause of global warming).[1]

Making cement requires limestone, which comes from the crushed shells of ancient sea-creatures, which store ancient carbon. The limestone is mixed with clay and heated in a kiln to 1,450°C, transforming it into a dry clinker that can be ground and stored. The energy needed to grind the limestone, fire the kilns and transport the cement produces half the CO_2 emissions. The other half arises when the limestone is kilned,

- British Cement Association: cementindustry.co.uk/sustainability.aspx
- Calera: calera.biz
- Carbon Sense Solutions: carbonsensesolutions.com
- Cement Sustainability Initiative: wbcsdcement.org
- EcoSmart Concrete: ecosmartconcrete.com
- Geopolymer Institute: geopolymer.org
- Hemp-Crete Construction: hempcrete.ca
- Hycrete: hycrete.com
- limetechnology: limetechnology.co.uk
- TecEco: tececo.com
- *Towards a Sustainable Cement Industry*: wbcsd.org
- Zeobond: zeobond.com

releasing its ancient carbon. A bag of cement that costs $7 produces 35 kg of CO_2.[2]

The European cement industry has cut its energy use by 30% since 1970,[3] and the Canadian industry has achieved an 11% improvement since 1990,[4] but the gains have been swamped by the tripling in global production since 1970, led by the growth in China, which produces half the world's cement. A 33% improvement is possible in both India and China,[5] but the limiting factor is cost in a competitive market, pointing to the need for new global standards.

Use Supplementary Materials

One way to reduce cement's CO_2 emissions is to add fly ash, an abundant waste from the coal industry, most of which is dumped in landfills. In Vancouver and Seattle, Lafarge is replacing 25–30% of its cement with fly ash, increasing this to 50% if a developer is willing to accommodate its slightly slower setting time, offset by greater final strength. In 2001 the industry used 11.4 million of the 600 million tonnes of fly ash that are globally available.[6] In Canada, EcoSmart trains concrete suppliers, designers and contractors to use supplements without fear of failure. There is also a product called Hycrete, which applies a waterproof membrane to cement, preventing it from water damage and making it possible to recycle larger quantities of cement, reducing the need for new cement.

A second approach is to burn materials such as scrap tires — but this needs constant monitoring to ensure that there are no harmful toxic emissions. Switching to biogas or biomass energy from

Earthship Brighton, in Britain, floors made with TecEco eco-cement. lowcarbon.co.uk

JO MITCHELL

wood wastes is another possibility. In Melbourne, researchers at the Australian Sustainable Industry Research Centre hope to capture kiln CO_2 and feed it microalgae to produce biodiesel to power the kilns, creating a closed loop system.[7]

Design a New Cement

The long-term solution is to change the way cement is made. One approach, developed by the Australian company Zeobond, adds an alkali to the silicates and aluminates found in fly ash and slag waste to create geopolymeric E-Crete, which produces 80% less CO_2 than regular cement.[8]

A second approach, developed by the Australian company TecEco, replaces 50–90% of the cement with magnesium carbonate (magnesite), a naturally occurring rock. Magnesite cement absorbs CO_2 like a tree, and the kiln heat needed is only 650°C. When magnesite is used to make masonry blocks, all the material eventually carbonates, absorbing up to 400 kg of CO_2 per tonne of cement within a year. Because TecEco cement can also contain up to 90% supplementary material, it makes for a dramatic reduction in cement use.[9] The chief drawback is the poor global distribution of magnesite.

Working along similar lines, the Californian company Calera, founded by Brent Constantz and funded by the venture capitalist Vinod Khosla, is developing a technology to siphon waste CO_2 from a gas-fired power plant through seawater from the Pacific Ocean. When the CO_2 combines with the dissolved magnesium and calcium in the seawater, it replicates the way ocean coral is made, turning the CO_2 into carbonic acid and then carbonate. This eliminates the need to heat limestone, and for every tonne of cement made this way, half a tonne of CO_2 can be sequestered into the cement.

A Canadian company, Carbon Sense Solutions, is exploring ways to allow precast concrete to store CO_2, and scientists at MIT are exploring the nanostructure of concrete in the hope that this might lead to breakthroughs. Hemp mixed with lime to create hemp-crete is another CO_2 reducing innovation. The biggest constraint to these alternatives is the natural conservatism of the building industry, as engineers need to be very sure that a building won't crumble in 50 years.

The best facilitating solution is for governments to apply a carbon tax to the cement industry and place the funds in a dedicated cement solutions fund, to be used for research, field-testing and standards to define and permit the new cements.

37

Solutions for Industry and Manufacturing

The eco-effective future of industry is a world of abundance that celebrates the use and consumption of products and materials that are, in effect, nutritious — as safe, effective, and delightful as a cherry tree.

— Bill McDonough

All around the world, as you read these words, coke ovens are blazing, electric motors are whirring and gas boilers are heating water as industries work to manufacture the goods we demand.

In doing so, they use nearly a third of the world's energy and produce 36% of the CO_2 emissions, mostly in the manufacture of chemicals, petrochemicals, iron, steel, cement and paper.[1]

The challenge for industry is to reduce this to zero by increasing efficiency, using non-carbon fuels, recycling wastes and redesigning the way products are made. The challenge for the world's consumers is to use and waste less so that industry can produce less, for we are currently exceeding our planetary ecological footprint three-fold.

In 1981, when Ron Nelson was energy manager of Dow Chemical's Louisiana Division, with 2,400 employees at 20 locations, he set up a contest to find energy-saving projects. By 1993, 575 projects

An Interface fabric loom. The global carpet company Interface is on track to achieve total sustainability by 2020.
interfaceglobal.com

INTERFACE INC

had averaged a 204% return on investment, saving Dow's shareholders $110 million a year, while reducing Dow's emissions.[2]

In 2007, when the International Energy Agency did a study to see if the manufacturing sector could reduce its emissions, they found that it could reduce its CO_2 emission by 19–32% if it adopted advanced technologies already in use, made systematic improvements to its motor and steam systems, and recycled more materials.[3] It's a beginning, but nowhere near enough.

DuPont: 72% Reduction

DuPont shows what is possible. They are a $27 billion chemical company with operations in 75 countries. In the late 1980s and 1990s, they changed course to focus on reducing their footprint on the planet, while increasing their shareholder value. Aiming to reduce their greenhouse gases by 65% below 1990 by 2010, they had achieved a 72% reduction by 2008, consisting of a 4% reduction in CO_2 and an 82% reduction in methane, N_2O, PFCs and HFCs.[4]

They did this by setting up a solid governance structure with a board-level environmental policy committee, an executive climate change steering committee, an energy team in each business unit and a compensation plan that rewards employees for environmental performance. Every month the energy teams calculate how much energy they are using per pound of product and how much steam and electricity per building, enabling them to plan more efficient lighting, heating, cooling, compressed air and cogeneration. Since 1990, they have reduced their energy use by 6% while increasing production

- *Beyond Neutrality — Moving Your Company Toward Climate Leadership:* bsr.org
- Climate Savers Smart Computing: climatesaverscomputing.org
- *Cool Companies* by Joseph Romm: cool-companies.org/book
- Industrial Efficiency: oee.nrcan.gc.ca/industrial
- Industrial Efficiency: ase.org/section/program/industry
- *Natural Capitalism: Creating the Next Industrial Revolution,* by Paul Hawken, Amory Lovins and Hunter Lovins, Back Bay Books, 2008
- The Climate Group: theclimategroup.org
- WWF Climate Savers: worldwildlife.org/climatesavers

by 41%, saving $2 billion in costs. DuPont's goals for 2010 include holding energy use flat at the 1990 level, increasing the use of renewable energy to 10% and achieving a further 15% reduction in greenhouse gases below the 2004 level.

Catalyst Paper: 71% reduction

The Catalyst Paper Corporation, which makes paper for newspapers and phone books around the world, is another good example. Chasing a 70% reduction in their emissions by 2010, they achieved a 71% reduction by 2007 by reducing their energy use, switching from fossil fuels to renewable biomass fuels (wood wastes and sawmill leftovers) and improving the energy efficiency of their equipment. As with DuPont, employee engagement has been critical.[5]

These are inspiring examples, but industry is being very slow to act. In Britain, the National Audit Office found that only 12% of large businesses had worked with Britain's Carbon Trust to reduce their emissions, and only 40% of the identified potentials had been implemented, mainly because of competing investment priorities and a lack of commitment from senior management. Overall, only 5% of the potential reductions were being acted upon.[6]

Don't Delay

Appoint a team, get board-level support, set goals and develop a plan. You will find great resources through The Climate Group, the WWF Climate Savers and Business for Social Responsibility. Don't let your industry association support those who are trying to slow the larger agenda. This is serious business, and we all need to get on board.

The Carbon Reducing Business Leaders[7]			
DuPont	chemicals	72%	1990–2007
Catalyst	paper	71%	1990–2007
Astrazeneca	pharmaceuticals	63%	1990–2005
Bayer	chemicals	60%	1990–2004
Interface	carpets	53%[a]	1996–2007
STMicroelectronics	electronics	50%	1994–2004
Polaroid	electronics	50%	1994–2004
IBM	computers	40%	1990–2005
BASF	chemicals	38%	1990–2002
3M	chemicals	37%	1990–2004
BT	telecoms	35%[b]	1991–2006
British American Tobacco	tobacco	34%	2000–2005
Unilever	consumer goods	34%	1995–2004
Pfizer	pharmaceuticals	29%	2000–2004
Alcoa	aluminum	25%	1990–2007
Johnson & Johnson	health care goods	17%[c]	1990–2007

a: Interface's goal is to achieve total sustainability by 2020.
b: BT's goal is 80% below 1996 by 2016.
c: Sales increased by 372% during this period.

38

Solutions for Forest Companies

Let the land dictate the practices, not the mill's need for logs. Be patient, be prudent, be good stewards — of the whole forest ... the watersheds, the birds, the plants, the animals.

— Kane Hardwood,
Collins Pennsylvania Forest

Ever since they evolved some 370 million years ago, Earth's forests have been harnessing the Sun's light to convert carbon dioxide into carbon and storing it away in their trees, underbrush, forest floor and soil — a miracle of transformation that has enabled a million species to evolve and flourish in their shade.

Paradoxically, it is the sunlight captured by these ancient forests that is causing us so much grief today, as we release their ancient fossilized carbon to power our fast-moving civilization. Can we call on the forests to rescue us today by increasing the amount of CO_2 they store? It will take a small revolution in the way we practice forestry, but it will bring a thousand other benefits, and we

do need to be alarmed, since the steadily rising temperatures are already bringing more stress, insects, disease and fire to Earth's forests. The reasons for Earth's forest companies to contribute to the solution are many.

In general, the older a forest, the more carbon it stores; so the more forestry methods mimic the way old forests work, the better things will be. Conversely, the more our forestry methods tear at the forest, the worse things will be. In the North American Pacific Northwest, when an old-growth forest is clearcut and converted to a managed forest, its capacity for carbon storage is reduced by up to half,[1] and it will be 200 years before it stores as much carbon as it did, if it is never touched again.[2] One study found that even including the storage of carbon in timber products, the conversion of 5 million hectares of mature forest to plantations in the Pacific Northwest over 100 years resulted in a net release of 1.5 billion tonnes of carbon to the atmosphere.[3]

This is a strong argument for leaving the forest alone as wilderness and parks, but we do need lumber, and it is possible for us to have our forest and cut it too. In Northern California, the 38,000-hectare (94,000-acre) Almanor forest has been managed by the Collins Companies in an ecological manner since 1943, using selective logging practices in a multi-aged stand to harvest the annual growth. Over six decades of continuous logging, the forest has produced more than two billion board feet of commercial timber — much more than if it had been clearcut — and yet its inventory of timber and carbon is higher than it was in 1943, much of it in mature trees that provide

COLLINS PINE

The Collins Almanor Forest in Chester, California.

- Clearcutting the Climate: forestclimate.org
- Forest Stewardship Council: fsc.org
- Forest Stewardship Council (Canada): fsccanada.org
- Forest Stewardship Council (UK): fsc-uk.org
- Forest Stewardship Council (US): fscus.org
- The Collins Companies: collinswood.com

habitat to wildlife. The longer the tree harvesting rotation, the more carbon is stored. In the Pacific Northwest, at its maximum, a 160-year rotation cycle will store 590 tonnes per hectare, compared to 363 tonnes in a 40-year cycle.[4]

When the Pacific Forest Trust worked with MacMillan Bloedel, a Canadian forest company, they found that compared to clearcutting, the practice of variable retention silviculture stored 14 more tons of carbon per acre in the timber zones, 23.6 more tons in the habitat zones, and 111 more tons in the old-growth zones. As soon as the emerging carbon markets are able to acknowledge this, companies that use ecological methods of forestry will be able to earn additional income from the increased carbon storage.[5] If the industry is to benefit from carbon pricing, it will be essential to develop sound measurement systems.

The Forest Stewardship Council (FSC) has built a solid reputation for certifying sustainable managed forests, and by 2008, 41 million hectares of forest had been certified in 79 countries, including 20 million hectares in Canada, 17 million in Russia, and 9 million in the USA. At the time of writing, there was a simmering controversy about the FSC's approval of logging in some ancient forests, which will hopefully soon be resolved.

The biggest factor that discourages companies from adopting ecological forestry practices is corporate competition to maximize short-term share-values, which the market believes can best be achieved by clearcutting the forest and putting the money in the bank. It is essential that governments intervene to bring carbon regulation and wisdom to the otherwise mindless markets and

COLLINS PINE

FSC-certified CollinsWood cedar and pine in a residence in Northern California.

use carbon pricing to create an incentive for good practices and penalize carbon loss.

Forest companies use a lot of energy to process their lumber, which can be a further source of emissions. One of the ways that the pulp and paper company Catalyst has reduced its emissions is by using forest and factory biomass wastes to replace coal and gas. (See #37.) The Austrian town of Güssing is also using forest-based fuel and energy in its hugely successful carbon footprint reduction. (See #27.)

39

Solutions for the Media

> The climate change issue must morph from the environmental and science desks to the entire newsroom.
>
> — Bud Ward, Co-Founder, Society of Environmental Journalists

Compared to other industries, the media has a relatively small carbon footprint, even including the explosions Hollywood loves so much. It has an enormous mental footprint, however, because the public is still quite confused about climate change and expects the media to explain what's happening.

The media's relationship with climate change has passed through four phases. In Phase One, (1980s and 1990s) it was just one among many environmental issues, receiving only occasional reporting. In Phase Two (1990s–2006) it started to receive regular coverage but by reporters who sometimes had little grasp of the issues. The scientific consensus was increasingly solid, but faced with a PR campaign financed by the coal and oil industry to spread confusion, many journalists used the ideals of "balance" and "neutrality" to treat the story as a conflict between two competing views, using phrases such as "some scientists believe"

and "skeptics contend." This works for politics but not for science. The damage that has been done by giving equal attention to professional climate scientists and unqualified amateurs is enormous.

By Phase Three (2006–2008) most media outlets had accepted that the science was real, but many journalists and reporters were still poorly informed, and their coverage tended to focus on dire warnings, technical discussions about carbon trading and offsets or lists of how we can supposedly save the world by changing our light bulbs. On the positive side, climate coverage broke out of its environmental corner and started to appear in business, health, energy, agriculture, politics, and even in sports, education, entertainment and fashion stories.

By Phase Four (2009 onward) most climate journalists were able to provide the depth of coverage that their colleagues expect for politics, sports and entertainment. They understand the science and know how to report it; they understand the major solutions; and they have an up-to-date knowledge of what businesses, communities and governments around the world are doing. They are as familiar with the climate leaders as their sports colleagues are with the leaders in the hockey divisions and their entertainment colleagues with the Hollywood film rankings.

Here's how you can flourish in Phase Four:

Percentage of 923 peer-reviewed articles dealing with climate change sampled in scientific journals (1993 – 2003) that expressed doubt as to the human cause of global warming.[1]

Percentage of 636 articles about global warming in the mainstream media (1990 – 2004) that expressed doubt as to the human cause of global warming. [2]

- Don't worry about feeling ignorant. Take the time to do in-depth reading, and keep asking questions. Read the best news and science reporting, the best blogs and the best email forums, both locally and globally.

- On the science, quote only from reputable peer-reviewed scientists and sources, such as the IPCC, *Science, Nature,* Proceedings of the National Academy of Sciences, AGU journals, *New Scientist* and realclimate.org.

- Spend more time researching the solutions than the problems. The public is weary of bad news, but their appetite for solutions is inexhaustible. The focus on solutions also provides a wide variety of hooks, heroes and local angles. Your community may not have a great cycling network, but someone may be going on holiday to Paris or Copenhagen, allowing you to turn their trip into an inspiring story. (See #25.)

- Build relationships with people who are developing the solutions — the eco-activists, green builders, solar companies, green investors and low-carbon lifestylers. Join the Society of Environmental Journalists and use their resources and their list of climate experts. If there is a Green Drinks night in your community, where business people mingle with activists, make yourself a regular.

- Educate your fellow journalists, and offer to be their mentor, so that climate stories can be integrated into transport, lifestyles, cooking, gardening, business and technology stories.

- If you work at a newspaper, radio or TV station, encourage your company to become carbon neutral, so that you can cover the solutions with integrity. The role the media can play in helping us understand how to reduce our carbon footprint is enormous.

- Climate Change Media Partnership: climatemediapartnership.org
- Climate Progress: climateprogress.org
- *Communicating on Climate Change — An Essential Resource for Journalists, Scientists, and Educators,* by Bud Ward, Metcalf Institute: metcalfinstitute.org/dl/CommunicatingOnClimateChange.pdf
- DeSmogBlog: desmogblog.com
- Environment Writer: environmentwriter.org
- Fairness and Accuracy in Reporting: fair.org
- Filmmakers for Conservation: filmmakersforconservation.org
- Green Drinks: greendrinks.org
- Media that Matters: mediathatmatters.org
- Network of Climate Journalists of the Greater Horn of Africa: necjogha.org
- Planet Ark Reuters: planetark.com
- Rural Journalism: ruraljournalism.org
- Society of Environmental Journalists: sej.org
- *The Carbon War*, by Jeremy Leggett: Penguin, 2001
- The Heat is Online: heatisonline.org
- TVE Asia Pacific: tveap.org
- Yale Forum on Climate Change and The Media: yaleclimatemediaforum.org

Film and TV companies could make films and TV programs about climate change freely available beyond their initial broadcast, designating them a "copyright free zone," as TVE Asia Pacific has been urging. When it supplied a TV series called *Climate Challenge* to Vietnam Television in 2007 it was the first time that climate change had received in-depth coverage in Vietnam, and marked a turning point in the country's public understanding of the issue.

40

Form New Partnerships

In order to avoid getting out-maneuvered politically, green-economy proponents must actively pursue alliances with people of color, and they must include leaders, organizations and messages that will resonate with the working class.

— Van Jones

When we reduce our private emissions, it's good. When we work with others, the sky becomes the limit. This is why it's important to be creative and work together in new teams. In North America, the coal and oil industries have cooperated to obstruct progress. We need to cooperate to accelerate progress and give people the confidence that a climate-friendly world is possible.

Take the greatest challenge of all — how to get the CO_2 out of the atmosphere before it melts Greenland and Antarctica. Trees and oceans do it very slowly, but no one knows how to do it in a hurry. In 2007 Richard Branson, chair of Britain's Virgin empire, set up a partnership with Al Gore, James Lovelock, James Hansen, Tim Flannery and Sir Crispin Tickell to create the $25 million Virgin Earth Challenge for the first person or team to come up with a commercially viable way to remove greenhouse gases from the atmosphere.

In 2007 several hundred of the world's top corporations, industry associations and environmental organizations teamed up to create the Global Roundtable on Climate Change. As well as calling on the world's governments to set targets for GHG reductions, the organization is engaged in projects on carbon capture, reducing emissions from deforestation, solar-power LED lighting, enhanced soil sequestration of carbon, energy efficiency and policy analysis.

Also in 2007, HSBC bank launched a $100-million climate partnership with WWF, The Climate Group, the Smithsonian and the Earthwatch Institute to help some of the world's great cities — Hong Kong, London, Mumbai, New York and Shanghai — respond to the challenge of climate change. They are also supporting climate champions worldwide who are doing field research and taking valuable knowledge and experience back to their communities; conducting the largest ever field experiment on the world's forests to measure carbon and the effects of climate change; and helping to protect some of the world's major rivers — including the Amazon, Ganges, Thames and Yangtze — from the impacts of climate change, benefiting the 450 million people who rely on them.

That same year, leading US companies, including Duke Energy, DuPont and Ford teamed up with major environmental groups to form the US Climate Action Partnership (USCAP), calling on the US government to quickly enact strong national legislation to require a significant reduction in greenhouse gases and supporting the need for car-

- Business for Innovative Climate and Energy Policy: ceres.org/bicep
- Business for Social Responsibility: bsr.org
- Climate Smart: climatesmartbusiness.com
- Global Round Table on Climate Change: earth.columbia.edu/grocc
- Green Chamber of Commerce: greenchamberofcommerce.net
- HSBC Climate Partnership: hsbccommittochange.com
- US Climate Action Partnership: us-cap.org
- Vermont Businesses for Social Responsibility: vbsr.org
- Virgin Earth Challenge: virginearth.com

As business leaders, are we taking due responsibility for our children, and their future?

bon taxes. In 2009 a new alliance, Business for Innovative Climate and Energy Policy, which includes eBay, Gap, Nike, Starbucks, Levi Strauss and Sun Microsystems challenged USCAP's goals by calling for the US to embrace a 25% reduction in CO_2 emissions below the 1990 level, and other progressive policies.

When we combine our efforts, there is no limit to what we can tackle. In 2000, 12 major US businesses teamed up with the World Resources Institute to form the Green Power Market Development Group to develop corporate markets for new green energy, with the intention of speeding up the market's growth. In 2008, Stoneyfield Farm, an organic yogurt company, teamed up with Clean Air-Cool Planet to found Climate Counts, which labels companies according to their carbon reduction performance and encourages consumers to shop accordingly. (See #31.) Around the world, 250 leading businesses work together as members of Business for Social Responsibility.

Chambers of Commerce have a critical role to play, reversing what has often been a traditional conservatism, sometimes extending to climate denial. In 2008 a number of mostly Californian businesses teamed up to form the US National Green Chamber of Commerce to strengthen the voice of businesses united to create a green public policy and a sustainable economy. Chapters are now forming across the US.

Local partnerships are equally important. In Oregon, businesses and organizations teamed up to create the Zero Waste Alliance. In British Columbia, more than 50 small businesses have worked with EcoTrust to become Climate Smart,

reducing and offsetting their emissions by 64%. In Vermont, more than 650 businesses, employing 35,000 people, are members of Vermont Businesses for Social Responsibility, using the power of business to change the world. As well as educating their members about green and socially responsible business practices, they employ a full-time policy advocate who lobbies for progressive energy and environmental legislation with Vermont legislators and government officials.

What else is possible, if we put our minds to it?

- Joint purchasing agreements for breakthrough technologies, such as Plug-In Hybrid Electric Vehicles, LED lighting systems and solar installations.

- Joint research funding for critical technologies that individual companies and governments seem reluctant to tackle, such as hot rocks geothermal, cement and tidal power.

- Shared local commitments to critical policy goals such as zero waste, green power, green vehicle standards, public transit and locally grown food.

41

Become a Carbon Farmer

There is nothing more difficult to take in hand, more perilous to conduct, or more uncertain in its success, than to take the lead in the introduction of a new order of things.

— Niccolò Machiavelli

For many farmers, the winds of change are already blowing hard.

In some areas, climate change will cause devastating crop losses. For every 1°F rise in the heat, crop yields fall by 3–5%[1] at a time when the world needs more food. India could lose 30–40% of its agricultural output.[2] By 2100, extreme drought will affect a third of the planet.[3] Many agricultural regions are already heading toward water shortages. Genetically modified crops, meanwhile, even if you like them, are failing to deliver.[4]

- Carbon Coalition Against Global Warming: carboncoalition.com.au
- Carbon Farmers of America: carbonfarmersofamerica.com
- Carbon Farmers of Australia: carbonfarmersofaustralia.com.au
- Conservation Technology Information Center: ctic.purdue.edu/CTIC
- Cornell Soil Fertility Management: css.cornell.edu/faculty/lehmann
- Diary of a Carbon Farmer: envirofarming.blogspot.com
- Farmers Union Carbon Credit Program: carboncredit.ndfu.org
- Holistic Management: holisticmanagement.org
- Keyline Designs: keyline.com.au
- Manitoba Zero Tillage Research Association: mbzerotill.com
- Soil Carbon Australia: soilcarbon.com.au
- Soil Carbon Coalition: soilcarboncoalition.org
- Soil Conservation Council of Canada: soilcc.ca

At the same time, the fluctuating price of oil is hurting many farmers; there's a growing awareness that meat and dairy cause 14–18% of global warming; and there's an increasing consumer demand for organic food. It's a complex world, in which farmers must plan for both financial and ecological sustainability.

A handful of organic soil contains a billion carbon-based organisms, and globally, the world's soils store about 2500 gigatonnes (Gt) of carbon. US farmers have lost 30–50% of their original soil carbon, however, due to plowing and other poor farming practices. For the Earth as a whole, we have lost between 42 and 78 Gt of carbon.

It used to be thought that rising CO_2 would increase carbon accumulation in plants, but in reality the warmer soils are causing organic material to decay more rapidly.[5] In England and Wales, from 1978 to 2003, the soil lost carbon at 0.6% a year, releasing 47 million tonnes of CO_2 a year, equivalent to 8% of the emissions from fossil fuels.

Abandon the Plow

What can a farmer do to increase the soil's carbon content? When the soil is not plowed, it retains its store of carbon. In Argentina, a third of the farmlands never see a plow — farmers get rid of their weeds by planting off-season crops that kill them. In southern Brazil, zero-tillage has become a social movement, reducing costs and increasing yields by two-thirds, while accumulating carbon.[6] In Canada, farmers are turning to conservation tillage, cover crops and direct seeding to store up to half a tonne of carbon per hectare, depending on the crop.[7] In the US and Canada the National

The corn on the left has been grown with a biochar/charcoal soil supplement; the corn on the right without, at Eprida's Lab at the University of Georgia Bioconversion Center, Athens, US.

Farmers Union is helping farmers to bundle their saved carbon into credits that are sold on the Chicago Climate Exchange under a five- to six-year contract. Other successful carbon-storing practices include organic farming (#42), and the use of management intensive grazing (#43), holistic soil management, cover crops, woodland regeneration and agroforestry.

Rattan Lal, an acknowledged soil expert, estimates that with improved farming practices, US farmers could recapture as much as 1 tonne of carbon per hectare and restore 75% of their soil's carbon in 50 years. Globally, he believes farmers could recapture 0.4 to 1.2 Gt of carbon a year.[8]

Create Biochar

More than 500 years ago, the native people of the Amazon turned their domestic wastes into charcoal by smoldering it and letting it work its way into the earth. Today this "black earth," known as terra preta, biochar or agrichar, is up to half a meter deep and storing 150 grams of carbon per kilogram of soil compared to 20–30 grams in the surrounding soils, while adding enormous fertility to the soil.

The discovery has caused a global interest in biochar as a way to capture carbon and in the use of pyrolysis (burning without oxygen) to burn farm and forests wastes. Many people are experimenting with ways to turn organic wastes into biochar, which can be buried in the soil as a carbon store while producing useful bio-oils and biogas that can be used to generate electricity. In New Zealand,

- Biochar Discussion List: terrapreta.bioenergylists.org
- Biochar Fact Sheet: csiro.au/files/files/pnzp.pdf
- Biochar Fund: biocharfund.com
- Biochar Manufacturers: biochar.pbwiki.com/BiocharManufacturers
- Canadian Biochar Initiative: biochar.ca
- Carbonscape: carbonscape.com
- Eprida: eprida.com
- Gardening with Biochar: biochar.pbwiki.com
- International Biochar Initiative: biochar-international.org
- UK Biochar Research Centre: geos.ed.ac.uk/sccs/biochar

Carbonscape has developed a microwave machine (the Black Phantom) that turns organic waste into charcoal, producing a tonne of charcoal a day.

Johannes Lehmann, at Cornell University, has estimated that if we were able to convert all agricultural, forest and urban wastes to biochar by pyrolysis, we could store more than 160 MT of carbon a year, and if we were able to convert all slash and burn that happens in the world's forests to slash and char, we could store a further 210 MT, for a total of 370 MT a year,[9] which is 3.7% of the world's annual carbon emissions of 10 Gt.[10] Others have suggested creating large-scale biochar plantations on marginal lands. Individual farmers can learn about biochar equipment through the links above.

42

Become an Organic Farmer

21st Century regenerative farming is the brightest hope for our planet to reverse the effects of global warming, and to protect and improve the health of farmers, global citizens and future generations.

— Timothy LaSalle, CEO, Rodale Institute

If you are a regular farmer, raising pigs, grains or vegetables in a conventional manner, it might seem like a risky business to consider going organic. Around the world, however, 700,000 farmers have decided it makes sense, including 12,000 in North America.

For most organic farmers, it is a combination of factors. They look at the rising price of pesticides and fertilizers, and they see trouble ahead. They read about the public's concern over pesticides and the continuing incidence of cancer, and they worry for their health and that of their families. Since 1975, studies have consistently shown that farmers develop and die of more cancers than does the general population.[1] Then they look at the size of the market, and they realize that it can make financial sense. Once they have been farming organically for a few years they discover other reasons:

- Organic farming is better for wildlife. Compared to a regular farm, an average organic field has five times more wild plants, 57% more species, 44% more birds and twice as many butterflies.[2]

- Organic food is better for animal reproduction. Compared to those fed conventional food, female organic rabbits have twice the level of ovum production, and organic chickens have a 28% higher rate of egg production.[3]

- Organic food contains more nutrients. Compared to 50 years ago, conventionally grown crops have lost 50% of their iron, calcium, copper and magnesium. Meats and cheese have lost 63% of their iron; broccoli has lost 63% of its calcium; and potatoes have lost 100% of their vitamin A.[4] None of this applies to organic food. A study of 41 comparisons of nutrient levels found that in every single case, organic crops had higher nutrient levels.

Organic soils also store more carbon, which feeds carbon-based worms, organisms and

SHARI MACDONALD

Nikki Spooner gathering organic salad greens at Varalaya organic farm, Mayne Island, BC. varalaya.ca.

microbes. In 1998 the Rodale Institute, based in Pennsylvania, completed a 15-year research project alternating corn with soybeans and other legumes on an eight-acre plot, enriching the soil on some fields while plowing under immature plants on others. Compared with conventional crops grown with fertilizers on adjacent fields, the soil-carbon level in the experimental plots soared, and nitrogen losses were cut by half, thanks to the use of manure, compost, cover crops and mulch, while yields were just as good. Overall, the organic fields stored 15% to 28% more carbon, drawn down from atmospheric CO_2.[5]

The average draw down was 1.66 tonnes of CO_2 per acre-foot per year. On a 130-hectare organic farm, the soil absorbs the amount of CO_2 produced by 117 cars. If organic farmers were paid for this at $25 a tonne, the land would earn $13,000 a year. If all 175 million hectares of US cropland were converted to organic production, they would absorb as much CO_2 as 217 million vehicles produce — that's 88% of the cars in the US.[6] When the Institute for Science in Society (ISIS) looked at organic soil, they concluded that if the whole world changed to organic methods on 1.5 billion hectares of cropland, more than 11% of global emissions could be reduced.[7]

Because organic farming does not use nitrogen fertilizers, which are made with natural gas, it also requires 30% less use of fossil fuels. It takes two to three years for yields to develop as the soil restores its store of carbon, but once established, yields are generally competitive with conventional crops — except in drought years, when they are 25–75% higher, because organic soils store more water.

- Carbon in Earth's forests: 610 Gt (billion tonnes)
- Carbon in Earth's atmosphere: 800 Gt
- Carbon in Earth's soils: 1580 Gt
- Carbon that could be absorbed through carbon farming, per year: 6–10 Gt
- Total carbon that could be absorbed through carbon farming: 400 Gt
- Canadian Organic Growers: cog.ca
- Hero Farmers: hero-farmers.org
- International Federation of Organic Agriculture Movements: ifoam.org
- Organic Farming Research Council: ofrf.org
- Organic Trade Association: ota.com
- Rodale Institute: rodaleinstitute.org
- Soil Association: soilassociation.org

Because the use of nitrogen fertilizers produces nitrous oxide, which is 298 times more powerful a greenhouse gas than CO_2, eliminating its use would reduce global greenhouse gas emissions by a further 5%. ISIS concluded that when all factors are combined, compared to conventional farming, localized organic farming systems could reduce global greenhouse gas emissions by more than 30%.[8]

If you want to think about changing to organic methods, go to the Rodale Institute's Hero Farmers website, where you'll find a 15-hour online program, a crop conversion calculator and an organic price report. Visit organic farmers and talk to them in person. Then consider becoming a Hero Farmer — and an essential part of the solution to global warming.

43

Become a Green Rancher

The most powerful tool we have to heal our climate is locked up in feedlots. We just need to open the gate.

— Abe Collins, Vermont Carbon farmer

For millions of years, huge herds of buffalo, deer and antelope roamed Earth's grasslands, grazing in tightly bunched herds to avoid their predators. Wherever they grazed, their manure and urine fertilized the land, and their hooves trampled native grass seeds into the soil. Their impact was sudden, but when they moved on, the grasslands quickly recovered, and the tall native grasses stored copious amounts of carbon, their roots reaching as deep as six meters.

When modern settlers arrived, their first instinct was to kill the predators. In North America, two million wolves were shot or poisoned during the 19th century — and the same thing happened on the Tibetan grasslands and elsewhere. Without their normal predators, the cattle and sheep wandered freely. The native grasses were no longer trampled or fertilized, and instead of being buried, the seeds were blown away on the wind. The soil became more gravelly. The native grasses, once belly high to a horse, were replaced by invasive weeds with shallower roots. The grasslands eroded and turned to desert, and their enormous store of carbon was released to the atmosphere. The wolves and other predators, by corralling the grazing animals so tightly, had been the unwitting guardians of the grasslands' carbon.

Today, more than 120 million pastoralists tend 3.5 billion cattle, sheep, goats and camels on 3.4 billion hectares of grassland (a fifth of the Earth's land surface), from the plains of North America to the pampas of Argentina and the grasslands of New Zealand. The ranchers are learning, however. In Vermont, Abe Collins practices "mob stocking" on 130 acres at Cimarron Farm, moving his 80 dairy cows up to eight times a day to replicate the way wild animals used to graze. As a result, his pastures have become so thick with carbon storing plants that he has been able to eliminate grain inputs, saving tens of thousands of dollars a year. In Bethesda, Maryland, where Martha Holdridge has been using management-intensive (rotational) grazing at West Wind Farm, soil testing by West Virginia State University demonstrated a five-year carbon content increase in 14 tested paddocks from 4.1% to 8.3%, storing an additional 9 tonnes of carbon per hectare (4 tons per acre), or 1.8 tonnes per hectare per year.[1]

In Australia, farmers call it "carbon farming." Instead of letting their sheep graze everywhere, they use high-density, short-duration grazing and direct-seed their cereal crops into the native perennial grasses. Their sheep graze on small 20-

Summer on Cimarron Farm, where the cattle have rotated into a new pasture.

Methane reduction in grazing animals	
Cystein food supplement + nitrates[2]	<100%
Feed additive based on fumaric acid[3]	<70%
Garlic[4]	<50%
Early season grazing[5]	29–45%
Grinding/pelleting low quality forages[6]	20–40%
4% canola oil mix[7]	33%
Enzyme inhibitors[8]	30%
Alfalfa grass, instead of grass only[9]	25%
Rotational grazing, high quality forages[10]	22%
Organic sugars and special bacteria[11]	20%
Vaccination[12]	20%
Legume lotus/tannins[13]	16%

- American Sahara: wildflowers-and-weeds.com/sahara.htm
- *The Carbon Fields*, by Graham Harvey, Grassroots, 2008
- Conservation Reserve program: nrcs.usda.gov/programs/crp
- Grass-Fed Beef: csuchico.edu/agr/grassfedbeef
- Grass-Fed Food: eatwild.com
- Holistic Management International: holisticmanagement.org
- Grassfarming: eatwild.com/environment.html
- Grassroots Food: grassrootsfood.co.uk
- Rowett Research Institute, Aberdeen: rri.sari.ac.uk
- Ruminant Livestock Efficiency Program: epa.gov/ruminant
- West Wind Farm: westwindfarm.biz
- Wild Farm Alliance: wildfarmalliance.org
- *Wolf Totem*, by Jiang Rong, Penguin, 2008

hectare paddocks and are moved every four to six days, giving the land 70 to 90 days to recover before being regrazed. The hooves push the manure and native grass seeds into the soil once again, and the perennial native grasses return, storing carbon as they grow. Since the carbon-rich soil can hold more water, the yields increase, erosion ceases, and the land stores up to five times more carbon than before, gaining a tonne of carbon per hectare per year.

If all of Earth's ranchers were to adopt these practices, they could capture 3.4 Gt of carbon a year, contributing immensely to the solution we so urgently need.

The Methane Problem

While they are grazing, cows and sheep belch methane, which traps 25 times more heat than CO_2 over 100 years. (See p. 12.) Around the world, 1.3 billion cows each produce 250 to 400 liters of methane a day. Between them, cows and sheep produce 5% of the world's greenhouse gas emissions. In New Zealand, they produce 43% of the country's emissions.

Agriculture research scientists are engaged in a quest to reduce these emissions. While some solutions are still experimental, others show promise, including a diet that is richer in high-quality grasses. Methane can be reduced per kilogram of meat produced when cattle are raised intensively in feedlots and fed grain, because the animals are pushed to grow so rapidly, but this has to be balanced against the acidosis, rumenitis, liver abscesses, bloat, bovine respiratory disease, higher feed costs and nitrous oxide (N_2O) emissions from the feedlot manure lagoons, which trap 298 times more heat than CO_2, because of which the livestock industry produces 65% of the world's emissions.

When a cow digests more easily, it produces less methane. In the wild, when herds of grazing animals wandered constantly in search of more tasty grasses, the native grasslands offered a huge diversity of grasses, wildflowers, mosses, lichens and liverworts. The Yunnan Mountain Grasslands in southwest China host more than 15,000 plant species, providing a delicatessen and a medicine chest for the grazing animals. Restoring cows to organic pastures rich in medicinal herbs would appear to be the best way to reduce their methane emissions.

44

Become an Energy Farmer

Each of us has the ability to act powerfully for change; together we can restore that ancient and sustaining harmony.

— David Suzuki

Farmers use a lot of energy — so they should try sustainable energy. A typical chicken house can be lit with compact fluorescent lamps, refreshed with large, slow fans instead of small fast ones, insulated and weather-stripped to keep out the cold and heated with an air-to-air heat exchanger or a solar wall. Solar heat can generate hot water; solar electricity can power lights, electric fences and water pumps.

If you create value-added farm products, you can use a glazed solar dryer for fruits, vegetables, grains and herbs. In California, Garry Vance, who farms 62 acres of pecans at Korina Farms, built a drying facility with a large solar wall in the roof.

If you need to dry grain, you can learn from the late Bill Ward, of Kansas City, who bored a

A wind farm in western Canada.

hole in the top of his silo and added a small windmill with hollow blades that sucked up a draft of air, drying and cooling his grain.[1]

If you have a dairy farm, you can reduce your energy bills by two-thirds by switching to variable frequency vacuum pumps and using heat exchangers to capture the heat thrown out during chilling.

If you grow under glass, you can build a solar greenhouse with thermal mass and an insulated north face or a winter-hardy bubble-insulated greenhouse that will protect your crops in temperatures as low as –30°C, as Kat and Ross Elliot are doing in Perth, Ontario.

If you have a creek or irrigation ditch with a good hydraulic head, you can install a Pelton wheel to generate hydropower.

If you have lagoons of animal manure, you can capture the biogas using an anaerobic digester, burning it to produce heat and/or electricity using an Integrated Manure Management System, as they do at Highmark Farm, Alberta, where 25,000 cows provide enough biogas to produce 4 MW of power, enough for 2500 homes — ten cows per home.[2] If you raise pigs, you can use a hoop house, yielding dry manure ready to spread on fields.

If you have wind, a developer may have already knocked on your door, offering a royalty if you agree to a turbine on your land. A large wind turbine, including access roads, will require less than half an acre and earn you $2500 – $5000 a year, while you farm around it. But don't sign a doorstep contract before you speak to neighboring farmers and consult a group such as Windustry. You could become a developer yourself or form a

Courtesy CANWEA

cooperative with your neighbors, but it takes a lot of time, money and technical expertise.

There's also the potential to grow energy crops for biodiesel or ethanol, or to generate electricity. It's controversial, because every hectare lost to food production has a knock-on effect and indirectly causes a hectare of rainforest to be cut down to grow the missing food. Most ethanol crops also have a low to marginal net-energy gain. (See p. 52.) Ethanol can also be made from agricultural cellulosic wastes, however, including damp and below-grade grain, nutshells, peach-pits and cotton-gin trash. Biodiesel can be made from soy-bean cake, a co-product of soy farming, which does not compete with food production, though it does with animal feed.

One sustainable possibility is ethanol made from native perennial switchgrass grown on marginal lands, which can produce 540% more renewable energy than the energy used to grow it.[3] Its roots, which can reach 2.5 meters (8 feet) down into the soil, store 15 tonnes more carbon per hectare than corn or wheat, while providing excellent habitat for wildlife.[4] Agricultural wastes and crops such as switchgrass can also be burned to generate electricity. In Australia, the horticultural company Growcom is planning to generate electricity and fuel for cars from banana wastes.

In the small German village of Juehnde, 200 families and nine farmers use the manure to make their village self-sufficient in heat and electricity. Financed by the villagers, the farmers have built a biogas cogeneration plant and woodchip central heating plant, using 25% of their farmland and 10% of their annual wood growth for the energy they need.[5]

- Biomass Research: nrel.gov/biomass
- Bubble Insulated Greenhouse: tdc.ca/bubblegreenhouse.htm
- Carbohydrate Economy: carbohydrateeconomy.org
- Dream Farm: i-sis.org.uk/DreamFarm.php and i-sis.org.uk/DreamFarm2aWorkofArt.php
- Farm Energy Audit: tinyurl.com/5ruyy8
- Farm Energy: farmenergy.org
- Harvesting Clean Energy: harvestcleanenergy.org
- Highmark Renewables: highmark.ca
- Methane to Biogas: epa.gov/agstar
- Methane to Markets: epa.gov/methanetomarkets
- Renewable Energy on Farms: farm-energy.ca
- Solar Greenhouses: builditsolar.com/Projects/Sunspace/sunspaces.htm
- Wind Energy for Farmers and Ranchers: nrel.gov/learning/fr_wind.html
- Windustry — Harvesting the Wind: windustry.org

In the Umbria region of Italy, the Castello Monte Vibiano Vecchio farm and vineyard is cutting its CO_2 emissions to zero by using electric cars, bikes and golf-carts; a solar-powered battery recharging center; a fleet of mini-tractors that burn biofuel made from crop wastes; boilers that burn local woodchips; storage tanks painted white to reflect back the heat; and 10,000 newly planted trees.

In England, Dr. Mae-Wan Ho, Director of the Institute of Science in Society, has developed plans for an abundantly productive Dream Farm that puts all the pieces together. Powered by waste-gobbling bugs and human ingenuity, it will have zero inputs and zero emissions.

45

Keep On Farming!

> If you think about your and my grandchildren, this is what really worries me.
> I don't want them — if I'm still alive by then — to say, why didn't you do something about it, when you could have done?
> — Prince Charles

Farmers are becoming critical players as the world searches for solutions to the climate crisis, the energy crisis, the food crisis, the water crisis and the health crisis — all at the same time.

Many young people want to become organic farmers, but very few can afford the land. The children of existing farmers, who once wanted a different career, are increasingly changing their minds as they realize the importance of farming.

The farms of the climate-friendly future will be very different from the operations we have grown used to, with their oil-consuming machinery, methane-producing animals and eroding soils. In growing numbers they will be organic, to benefit from the market, the restored local ecology, the absence of chemicals and health risks, the increased carbon storage and the greater personal control.

The new farms will grow a wider variety of crops to benefit from the increased yields and decreased diseases that polyculture brings. Out of

- American Farmland Trust: farmland.org
- Community Supported Agriculture: sare.org/csa
- GRACE Factory Farm Project: factoryfarm.org
- Land Trust Alliance: lta.org
- Maryland Agricultural Land Preservation Foundation: malpf.info
- Pocket Markets: foodroots.ca/pocket_market.htm
- SPIN Farming: spinfarming.com
- The Land Institute: landinstitute.org

a need to reduce their energy bills, they will use solar, wind, geothermal, microhydro and biomass energy systems. Out of a need to store carbon in their soil, and attracted by carbon pricing, they will adopt holistic, ecosystem-based methods and rotational grazing.

To achieve all this, we will need more farmworkers — and there will be plenty of young people eager to return to the land, where they will build zero-energy cob and straw bale houses, and regenerate the rural culture.

Some farmers will also earn a good living in the cities and suburbs, using Small Plot INtensive (SPIN) farming to practice market gardening in the gardens and backyard lots of their friends and neighbors, selling the produce at street markets and pocket markets.

The farmers of the future will probably end their use of genetically modified crops, disappointed by the falling yields,[1] the rising cost of pesticides and fertilizers, the ecological and health concerns and the shrinking demand for non-organic food.

N2O Emissions

Their farms will also be organic to avoid the carbon-taxed cost of nitrous oxide (N_2O) emissions produced by unused nitrogen fertilizers and by animal manure. Globally, farming is responsible for 65% of the world's nitrous oxide emissions. It is possible to reduce the N_2O by up to 35% by using soil tests to apply the fertilizer only when it's needed, so that it doesn't get washed away and released, but it's easier to go organic, so that you no longer need the fertilizer at all.

Farmland next to a housing development in upstate New York.

JIM NEWTON/ AMERICAN FARMLAND TRUST

N_2O is also produced from animal manure. In Japan, Toyota has developed an environmentally friendly manure composting process called resQ45 that uses an enzymatic agent to cut the time needed to convert wastes into compost to 45 days, instead of the normal 90 to 180 days, reducing the N_2O by 90% and the methane entirely.[2]

Perennial Grain Crops

In Kansas, the Land Institute is developing edible grain varieties of perennial native plant species, such as intermediate wheatgrass, which have roots up to two meters deep and do not need plowing. In the next 25 to 50 years, the large-scale development of high-yield perennial grain crops may be possible.[3]

Protecting our Farmland

Finally, it is essential to protect good farmland from being turned into subdivisions with names like Green Meadows and Owl Acres. It is mainly in the US, where the belief in private property rights has been allowed to pre-empt the wisdom of retaining farmland, that this is happening. Between 1982 and 1997, 25 million acres of rural land were converted into subdivisions, malls and parking lots. Every minute, the US loses two acres of farmland, with the prime land disappearing 30% faster than the non-prime land.[4] In Canada and Europe, when land is zoned agricultural, it has some legal protection.

One solution, where there is local support, is for farmers to donate or sell their development rights in exchange for a permanent conservation easement that guarantees that the land will never be developed. By 2007, the Maryland Agricultural Land Preservation Foundation had preserved 266,000 acres of productive agricultural land and woodlands by this means, funding it with 0.5% of their local property transfer tax, making it the most successful program of its kind in the nation.

This is not the time to give up farming. The Earth needs you. We are entering a time of crisis, but as the Chinese know every time they say the word "crisis," as well as being a time of danger it is also a time of opportunity and new beginnings.

46

Solutions for the Auto Industry

We need to design transportation solutions that overcome our reliance on fossil fuels.

— Toyota, 2007[2]

When auto executives meet to plan their future, they must consider two compelling facts — global warming, which demands the rapid development of zero-emissions transportation, and the price of oil, as production passes its peak and begins to fall. If oil costs $4 a liter ($16 a gallon) in 2020, the demand for gas-powered vehicles will be zero.

In 2007 the global auto industry produced 73 million vehicles — 87% cars and light trucks and 13% commercial vehicles, trucks and coaches. The global fleet is approaching 700 million vehicles.[1] Road vehicles produce 10% of global CO_2 emissions, 5% of the total greenhouse gas emissions.

100% Climate-friendly Vehicles by 2030

To achieve a climate-friendly world by 2030, we need every vehicle to be a zero-emissions vehicle (ZEV), powered by sustainably sourced electricity and/or sustainably produced biofuel. We should also assume a major shift to other means of travel such as transit, railways, and cycling.

- Better Place: betterplace.com
- California Cars Initiative: calcars.org
- Converting Gas Guzzlers: calcars.org/ICE-conversions
- EV World: evworld.com
- Hypercar: hypercar.com
- Green Car Congress: greencarcongress.com
- Plug In America: pluginamerica.org
- Plug-In Vehicles Tracker: pluginamerica.org/plug-in-vehicle-tracker.html

When the first mass-produced electric (EVs) and plug-in hybrid electric vehicles (PHEVs) arrive on the market, the pent-up demand will astonish automakers. But can the industry deliver the vehicles needed by 2030? It takes six to seven years to bring a new vehicle to market, and the average life of a vehicle in North America is nine years. If this applies globally, there is time for two complete fleet renewals before 2030. Climate friendly air-conditioning is also essential. (See #97.)

To achieve the goal, starting immediately, an increasing share of each year's new vehicles should produce less than 100g/km of CO_2. After 2014, no more fossil-fuelled cars should be planned, and after 2020 all new vehicles should be ZEVs.

In 2008, the average US passenger vehicle produced 180g/km (170g/km in Canada). The European Union's requirement is 120 g/km, starting in 2015. (See #76.) From 2020, global regulations should require 0g/km for *all* new vehicles. Nissan/Renault has prepared its business plan on this basis, knowing that zero-emissions vehicles are the future.

Auto companies and independent businesses should also establish conversion factories to retrofit the existing stock of vehicles into EVs and PHEVs. (See #76.)

The key to the transition will be partnerships between governments and businesses such as Better Place to begin the mass deployment of electric cars by installing thousands of EV charging posts, and battery exchange stations. Governments can also help with incentives, feebates, carbon taxes and grants. The public can help by buying the ZEVs as soon as they arrive.

A colleague in Seattle drove 2500 miles on one tank of fuel in his converted Toyota Prius Plug-In Hybrid, similar to this early conversion.

Fuel Efficiency Improvements

The biggest change for cars is to displace the liquid fuel with cleaner, cheaper, domestically produced electricity. There are also ways to reduce CO_2 emissions by increasing the efficiency of gasoline and hybrid gasoline cars, such as stopping the engine when the car is not moving; installing tire-pressure and fuel-efficiency indicators; and using power-saving air conditioners, dual-speed superchargers and hydraulic hybrid powertrains. Toyota is adopting four intake valves per cylinder, low-friction materials and a variable-valve-lift mechanism, promising a 5–10% improvement. Teijin, the Japanese textile manufacturer, has produced a concept car, the PU_PA, that uses carbon fiber and bioplastic, resulting in a 50% reduction in the vehicle's weight.

Renault has reduced the emissions from its Logan Eco2 Concept car to 97g/km by means of aerodynamic improvements (2g/km), Michelin Energy Saver tires (2g/km), a re-calibrated injection system (5g/km), improved running gear (3g/km), battery charging only as required, reducing energy-hungry parts (4g/km), lengthening the final drive ratio by 8% (4g/km) and optimizing the play between moving parts while using low-viscosity lubricants (2g/km). When driven by a good eco-driver, its emissions fall to 71 g/km. Its materials are 95% reusable, by weight.

Zero-Emissions Auto Assembly Plants

Though it is only a tiny portion of the greenhouse gases it produces, the energy used to build cars is also important. Volvo is planning to make all its plants CO_2 free, starting in Ghent, Belgium, where it uses solar, wind and biofuel energy to produce electricity and heat. In Canada, by 2007, GM had reduced its emissions to 40% below the 1990 level, and Ford had reduced its by 40% below 2000 level. Toyota — which is working toward zero emissions at every level — wants to reduce its manufacturing emissions by 27% per vehicle below the 2002 by 2011 and has implemented more than 250 *kaizens* (small productivity improvements) in its northern California assembly plant. It also has a policy of zero landfill waste and a target of no more than 30kg of paid recyclable waste per vehicle.

This may seem like a big stretch today, but by tomorrow it will all seem normal.

47

Solutions for Trucking

A rising carbon price is essential to "decarbonize" the economy — to move the nation towards the era beyond fossil fuels.

— James Hansen

Every day, millions of tonnes of materials are trucked to factories, retail stores and landfills around the world in 70 million trucks, and with every 100 gallons of diesel burned, a tonne of CO_2 is released into the atmosphere.[1]

The average 18-wheeler, being driven 100,000 miles a year at 6.5 mpg, will burn 15,000 gallons and produce 150 tonnes of CO_2. The US has 3 million commercial trucks. Canada has 455,000.

For the planet, the problem is the carbon footprint. For the truckers, the problem is also the future price of diesel. How will we truck our goods when climate change demands that we cease burning fossil fuels and diesel costs $16 a gallon?

The most obvious solutions are high-speed electric rail (see #48), and electric and hybrid electric trucks. The more goods we can produce in strong local economies, the fewer goods we'll need to ship around.

In the short-term, there are many ways to cut fuel use. Idling trucks use more than 2 billion gallons a year in the US, producing 20 million tonnes of CO_2.[2] Some 800 million gallons of diesel are used annually in the US just to keep overnight sleeper cabs warm or cool. Solutions

Fuel Saving Innovations for Existing Trucks	Saving
New trucks. The Mercedes-Benz Actros, a 40-ton tractor-trailer, gets 12 mpg (14 times more efficient than a 2.3 ton Hummer at 10 mpg).	Up to 85%
Trailers and containers shipped on railway flat cars instead of by road for trips over 1,000 miles	Up to 65%
Transportation management systems that allow for better load matching, real-time tracking and optimized dispatch and delivery. Empty haul movements account for up to 15% of the heavy goods miles traveled. Air Products has increased its fleet mileage efficiency by 50% since 1975 by using logistics scheduling software.	Up to 50%
Diesel-electric hybrid and hybrid reefer (refrigeration) trucks.	Up to 35%
Smart driving habits, assisted by driver training, on-board software systems and fuel economy indicators.[3]	Up to 33%
Fuel economy drops significantly above 55 mph. Speed can be limited with electronic speed governors, and incentive programs that reward drivers for fuel efficiency. Slowing down also reduces engine repair costs.	16–27
Hydrogen fuel-injection engines. These use power from the electrical system to generate hydrogen from water injected directly into the combustion chamber, creating ultra-lean homogeneous combustion.[4]	10–15%
Improved aerodynamics, roof fairings, side fairings, cab extenders, trailer tails[5]	5–15%
Single wide-base tires	3%
Low-viscosity drive-train lubricants	Up to 3%
Lightweight wheels, axle hubs, roof posts, upright posts and floor joists.	1–3%
Deflecktor covers, to reduce turbulence on wheels	2%
Automatic tire inflation system. Also extends the tires' life by 8%.	0.5%

The Smith Newton Electric Truck can carry up to 16,000lbs (7,400kg) , with a range of 100 miles (160km) per charge and a top speed of 50 mph (80kph).

- American Clean Energy Systems: acesfuel.com
- Bison Transport: bisontransport.com
- Blue Fuel: energysynergy.ca
- BlueCool Truck: bluecooltruck.com
- Deflecktor: deflecktor.com
- EnviroTruck: cantruck.com/envirotruck
- EPA Smartway Partnership: epa.gov/smartway
- FleetSmart Canada: fleetsmart.nrcan.gc.ca
- Freight Best Practice: freightbestpractice.org.uk
- GreenFleets BC: greenfleetsbc.com
- Green Road: greenroadtech.com
- Green Trucker: greentrucker.com
- Hybrid Truck Users Forum: htuf.org
- Modec: modeczev.com
- Rail/Truck Carbon Calculator: tinyurl.com/d9z6gu
- Schneider National: schneider.com
- Smartidle: smartidle.com
- SmartWay: epa.gov/otaq/smartway
- Smith Electric Vehicle: smithelectricvehicles.com
- TrailerTail: atdynamics.com
- Transformational Trucking: tinyurl.com/nwwswq
- Trucks Deliver a Greener Tomorrow: trucksdeliver.org

include auxiliary power units, commercial transponders for bypassing weigh stations, cab cooling technologies, and rest areas that provide heat, air conditioning and electrical power.

Taken together, these initiatives could save more than 100% of the fuel, proving that with the right statistics, trucking needs no fuel at all. In the US, the EPA's SmartWay Partnership helps trucking companies measure their environmental performance, identify goals for improvement and develop a plan. In Canada, FleetSmart, SmartDriver and Greenfleets BC offer the same.

Schneider National, North America's largest truckload carrier, is aiming to operate the most energy efficient fleet in the industry,[6] while Bison Transport, based in Winnipeg, Manitoba, Canada, is pursuing the same title with its Sustainable Transportation strategy. So there's your challenge!

The truck of the future may be a tribrid, combining electric, hydrogen and sustainable algae-diesel or blue fuel.

Electric trucks are already on the road. The Dutch company TNT has more than 100; Unicell and Purolator are developing Quicksider electric delivery vans; Coca-Cola is using Zap Xebra electric trucks in Montevideo, Uruguay, and had 327 hybrid trucks on US roads by 2009. The 5.5 tonne Modec has a range of 160 km (100 miles), carrying a two-tonne payload at 80 km/hr. Amsterdam is planning an electric cargo tram for local deliveries.

Blue fuel is DiMethyl Ether, which can be made from fossil fuels or biofuel, or by combining green hydrogen with waste CO_2 from industrial facilities to produce methanol, synthesized to blue fuel. Where there is ample but remote wind energy that is not useful for the grid, blue fuel could be manufactured for use as a diesel substitute.

48

Solutions for Railways

Building a new national high speed train network is the single most important thing we can do to get us off oil, and change the direction of the nation for the better. It is the backbone of a truly sustainable society.

— Andy Kunz

Most people outside Europe have no idea that a high-speed 300 km/h train can be such a comfortable and easy way to travel. In France, most high-speed trains are so tight to their timetables that they can run every four minutes on the same track. In Japan, the average delay is no more than 24 seconds. In North America, the Amtrak Acela Express from Boston to Washington comes closest, with speeds up to 150 mph, but it is still a 10-hour journey. With dedicated, European-style, high-speed rail, its journey of 440 miles could be done in three hours.

High-speed railways are the future. For inter-city trips, we need high-speed electrified trains at over 200 mph. For regional journeys connecting local destinations, we need electrified commuter trains at up to 125 mph. For local journeys within our cities, we need electrified metro, light rail, bus rapid transit, trucks and streetcars.

In Europe this requires the rapid expansion of the already booming rail network, which is 80% electrified. In North America it requires an entirely new railway network, linking every city. It will be an enormous investment, but no more so than the highways and airports we built for our cars, trucks and planes. If we do not make the investment, we risk losing our modern civilization as the price of oil pushes cities, businesses and industries into bankruptcy.

Compared to cities that only have buses, cities that also have railways have 400% greater transit ridership, 36% fewer traffic deaths, 21% less vehicle ownership and mileage, and 14% lower consumer expenditures on transport — for a means of travel that uses three times less space than cars.[1]

For external costs such as accidents, noise, air pollution and impacts on wildlife (not including congestion), rail-freight has five times less cost than roads and 16 times less than flying, while passenger travel has 3.3 times fewer external costs than road, and 2.3 times fewer than flying.[2]

With electrification, trains can use regenerative braking to gather useful energy, and with power sockets and good food they become a very civilized way to travel — but we must begin investing immediately on a very large scale, as Andy Kunz, one of North America's leading new urban planners, has urged since 2000.[3]

There are some small things that a railway company can do right now:

- For electric trains, buy green power, as Calgary Transit's "Ride the Wind" C-Train does.

- For freight, invest in the Green Goat, a hybrid drive locomotive made by RailPower Technologies Corp that replaces a 900- to 3500-horsepower diesel generator with a 70- to 250-horsepower system and a large pack of electric batteries, cutting fuel use by 40–60%.

European Freight Transport kg CO_2 per 100 tonne/km	
Railway	2.94
Inland waterway	3.44
40-tonne truck	8.63
Flying	67.14[4]

European Long Distance Passenger Traffic kg CO_2 per 100 person/km	
Railway	4.79
Car, diesel	15.78
Car, gas	17.62
Flying	23.00[5]

The Velaro E high-speed train takes 2.5 hours to travel the 625 kilometer from Madrid to Barcelona.

SIEMENS AG

- For cranes, replace on-board diesel generators with RailPower's ECO Crane hybrids, cutting fuel use by 70%.
- Teach eco-driving skills, as Germany's Deutsche Bahn is doing, reducing energy use by up to 10%.
- Harmonize timetables with local transit services, and provide safe, sheltered bicycle parking.

The greater challenge is to the public's vision of a completely new railway system, with high-speed tracks connecting cities, ports and transport hubs. The most likely strategies are for the track to be owned by a new National Railway Agency and the trains and stations licensed out to private companies, or for the entire system to be publicly owned. In the US a nation-wide system might cost $1 trillion, in Canada $50 billion, using money that would otherwise have been spent on highways, airports and bailing out bankrupt airline companies.

California has already started down this route with a $40 billion plan for an 800-mile electrified track, with trains traveling at 220 mph from Sacramento and San Francisco to Los Angeles,

- American Public Transportation Association: apta.com
- Amtrak: amtrak.com
- California High-Speed Rail Authority: cahighspeedrail.ca.gov
- EcoTransIT: ecotransit.org
- Energy Efficiency Technologies for Railways: railway-energy.org
- High Speed Rail Canada: highspeedrail.ca
- Keep Kyoto on Track: railway-mobility.org
- National Association of Railroad Passengers: narprail.org
- National Corridors Initiative: nationalcorridors.org
- New Trains: newtrains.org
- Rail Transit in America — The Benefits: vtpi.org/railben.pdf

Irvine and San Diego, potentially powered by solar, wind and geothermal facilities alongside the track.[6]

Railway companies need to form partnerships with cities, states, businesses, groups like the New Apollo Project and the climate change movement, so that we can create the political will to make it happen.

49

Solutions for Shipping and Ports

> The only certainty is that we have to act. How could I look my grandchildren in the eye and say I knew about this thing and did nothing?
> — David Attenborough

Some 100,000 civil ships roam Earth's oceans, including 10,000 tankers, 4,300 container ships and 1,500 passenger ships, burning 280 million tonnes of bunker fuel a year. In 2007 this produced 1 billion tonnes of CO_2 emissions,[1] representing 2.76% of the world's CO_2 emissions and 1.2% of the cause of global warming, plus the warming influence of the black carbon that is produced by dirty bunker oil. Before the 2008 financial crisis, the industry was growing by 5% a year. There is — as yet — no international agreement to reduce greenhouse gases from shipping.

Compared to other forms of transport, shipping is highly efficient. A ship can transport a tonne of merchandise 241 km on a liter of fuel, compared to 95 km by train and 28 km by truck.[2] In 2008, at $550 a tonne, bunker oil was 50–60% of a ship's operating costs. Burning 217 tonnes a day, a 28-day round trip for a container vessel cost $3.35 million. For a ship with 7,750 six-meter containers, that's 2.8 tonnes of CO_2 per container.[3]

The MS Beluga SkySails.

Slow Down

Hapag Lloyd, the fifth largest shipping line in the world, reduced its use of fuel by 25% by cutting its speeds from the standard 23.5 to 20 knots, and by 50% for ships that had been speeding up to 25 knots.[4] By going two knots slower, the trip from Asia to Europe takes four days longer, but it uses 17% less fuel and produces 21% less CO_2. The Danish Maersk Line, named Sustainable shipping Operator of the Year in 2009, has been utilizing super-slow steaming since 2007.

Add SkySails

In 2008 the cargo ship *Beluga SkySails* sailed 11,952 nautical miles from Germany to Venezuela and back pulled by a 160-square-meter kite, cutting fuel use by 20%. Using a figure-of-eight flight pattern, the kites produce two to three times more power than a regular sail. SkySails operate best at a speed of 10 knots and a wind of 25 knots. With bunker oil at $550 a tonne, the payback is less than two years.

Future Ships

In 2008 the Japanese shipping line Nippon Yusen K.K. equipped its cargo ship *Auriga Leader* with a 40 kW solar system, generating 0.2% of the ship's power. Nippon Yusen's goal is to reduce its fuel use and CO_2 by 50% by 2010. The *E/S Orcelle*, planned by Sweden's Wallenius Marine for production in 2025, will carry 10,000 vehicles; it will be powered by the sun, wind, waves and hydrogen, using a pentamaran, five-hulled design that needs no ballast or stern propeller. Other developments include using counter-rotating propellers, stern flaps, sharkskin coating made from polyolefin film to reduce water-friction; creating an air pocket

SKYSAILS

between the ocean and the hull;[5] hydrogen fuel cell ships; and fixed-wing sails.

More Solutions

By simply polishing the propellers once a year, which is done as an underwater operation by a Canada's All-Sea Enterprises, emissions and fuel-use can be cut by 2%. Shipping is also responsible for 133,000 tonnes of particulate pollution, including a significant amount of black carbon (see p. 14), 90% of which could be eliminated if ships moved to ultra low sulfur diesel, enabling particulate filters to be installed.[6] Future ships could also operate on sustainably managed biodiesel, or blue fuel. (See #48.)

Solutions for Ports

In 2008, 55 ports signed the World Ports Climate Declaration in Rotterdam, committing themselves to solutions such as installing electric plugs to power ships while in harbor, as Seattle and California are doing; limiting the tonnage that shippers can move by truck within a port, moving more by train and barge; replacing diesel trucks with hybrid vehicles to move containers, as New York and California are doing; using electric hybrid gantry cranes (EcoCranes, EcoRTG); and giving docking fee rebates to the cleanest ships, while charging more to the dirtiest. The Swedish port of Göteborg is enabling some ships to reduce their portside emissions by 94–97% by using wind energy to power ships at berth.[7] The fact that all

- All-Sea Enterprises: all-sea.com
- E/S Orcelle: tinyurl.com/6cb6d4
- EcoPorts: ecoports.com
- Electric Boat Association: electric-boat-association.org.uk
- Electric boats: electricboats.ca
- FellowSHIP: fuelcellship.com
- Green Marine: green-marine.org
- International Association of Ports and Harbors: iaphworldports.org
- Kiteship: kiteship.com
- Live Ships Map: marinetraffic.com
- Ports — Best Practices: c40cities.org/bestpractices/ports
- Ports and Nature: newdelta.org
- Project GreenJet: tinyurl.com/3wjp4y
- Shark Skin Coating: sharkskincoating.com
- *Shipping, World Trade, and the Reduction of CO_2 Emissions:* marisec.org/co2.htm
- SkySails: skysails.info
- Solar Navigator: solarnavigator.net
- Solar Sailor: solarsailor.com
- Sun21: transatlantic21.org
- Sunboat: sunboat.com
- Wingsails: shadotec.com
- World Ports Climate Conference: wpccrotterdam.com
- Zemships: zemships.eu

ports are — by definition — at sea level might also exercise some port managers' minds in view of rising sea levels.

Global Solutions

The pressure is on for the entire shipping industry to accept mandatory energy-efficient design and operational standards, leading to an 80% reduction in emissions by 2050. We also need a global levy on shipping fuel, with the funds being used to help developing countries adapt to climate change and adopt zero carbon technologies.

50

Solutions for Aviation

Sooner or later, aviation will have to shoulder the burden it imposes on the planet.

— *The Economist*, June 2006

There are almost 23,000 commercial aircraft, and they all burn kerosene, producing 750 million tonnes of CO_2 in 2007, or 2% of the world's CO_2 emissions.[1] CO_2 causes 44% of global warming (see 10), but flying also produces nitrogen oxides that contribute to tropospheric ozone formation and water vapor that causes condensation trails and cirrus clouds; so the IPCC recommends that we multiply the impact by 2.7%, increasing aviation's responsibility for climate change to 2.4%.[2] Eighty percent of the impact comes from long-haul trips, though short-haul flights up to 500 km produce proportionally more CO_2 per kilometer because more fuel is used during take-off.

In 2007, 2.2 billion people flew, one in four for business, the rest for personal reasons. 40% of the world's air travel happens in North America. The airlines also ship 44 million tonnes of freight a year, and are projecting rapid growth for the future. Aviation enjoys freedom from fuel taxation and benefits from many loans and subsidies.

The International Air Transport Association, representing 230 airlines, published *Building a Greener Future* in 2008, in which it set goals to reduce aviation's emissions by 18% by improving infrastructure and flight operations; to meet 10% of airline fuel needs with biofuels by 2017; and to build a zero-carbon emissions aircraft within 50 years.

Boeing has pledged to reduce its CO_2 emissions by 15% with each new generation of aircraft and by 25% for its worldwide fleet by 2025, and to work with its customers to reduce their emissions by 25% by 2023. Air Canada's goal is to improve its fuel efficiency by 25% by 2020. Virgin Atlantic's goal is a 30% improvement by 2020. How can these goals be achieved?

- **Buy lightweight airplanes.** Europe's Airbus A380 uses 12% less fuel than its competitors; the Boeing Dreamliner will use 20% less. Friction drag can potentially be reduced tenfold by covering a plane's surface with tiny pill-like discs, which reduce turbulence.[3]

- **Reschedule night flights.** Night flights are 25% of total air traffic, but their contrails cause 60–80% of aviation's climate effect, due to the colder night atmosphere.[4] Slowing down from a cruising speed of 800 to 775 km/hr also saves fuel.

- **Operate with full flights.** In 2007 British Airlines flew 80 billion empty seat kilometers. Reducing the number of business and first class seats would improve efficiency; so would charging for luggage by weight.

Flying over Alaska's mountains and glaciers.

- **Improve air-traffic management.** The industry says it could reduce emissions by up to 12% with optimized traffic-control procedures. Each minute of flying time saved reduces fuel consumption by 62 liters and CO_2 emissions by 160 kg.[5]

- **Stop ground idling:** Delta's reduction of engine idling has cut its ground emissions by up to 40%. Virgin estimates it can save up to two tonnes of fuel per flight (five tonnes of CO_2) by using electric tractors to tow planes to and from the runway.

- **Become carbon neutral.** Include the price of offsets in all tickets on a voluntary basis, prior to making it mandatory. In 2009 Harbour Air and Helijet in British Columbia and NatureAir in Costa Rica offset all their emissions.

- **Support plans to include aviation in carbon trading.** When carbon pricing becomes mandatory, this will help airline initiatives to reduce their emissions.[6]

Biofueled Flying

The airlines are placing their hopes in second- and third-generation biofuels, testing biofuel mixes made from the jatropha plant, coconut and babassu nut oil, camelina oilseed and algae. *New Scientist* magazine has calculated that if all of the world's aviation fuel (five million barrels of oil a day in 2007) came from jatropha nuts, growing them would require 1.4 million square kilometers — twice the size of Texas. If the biofuel were grown from algae, however, it could be done on 66,000 sq km, about the size West Virginia or

- AeroNet: aero-net.org
- Aeroscraft: aerosml.com/ml866
- Air Cargo World: aircargoworld.com
- Air Transport Action Group: atag.org
- Algae Jet Fuel: algaelink.com/jet-fuel.htm
- Algal Biomass Organization: algalbiomass.org
- Aviation and the Global Atmosphere (IPCC): grida.no/climate/ipcc/aviation
- Aviation Environment Federation (UK): aef.org.uk
- Aviation Global Deal Group: agdgroup.org
- Boeing and the Environment: boeing.com/environment
- Clearer Vision, Cleaner Skies: enviro.aero
- *Climate Change and Aviation: Issues, Challenges and Solutions,* by Stefan Gossling and Paul Upham, Earthscan, 2009.
- Green Flight International: greenflightinternational.com
- International Air Transport Association: iata.org
- International Civil Aviation Organization: icao.org
- Solar Impulse: solarimpulse.com
- Virgin Atlantic's Eco Flight Plan: (tinyurl.com/5wgpx6)

Ireland, which is just 0.13% of the world's 50 million square km of farm and pastureland.[7]

What else is possible?

The solar-powered Solar Impulse has flown for 25 hours, but carrying only one person at 27 mph. A French test pilot managed a 50-km electric flight around the Alps in a single-person plane. Boeing is testing a small slow-speed hydrogen airplane, but hydrogen's water emissions form more contrails and cirrus cloud. Looking into the future, Richard Branson dreams about making a larger version of his Virgin Galactic spaceship that could lift passengers out of the atmosphere and convey them from New York to Sydney in 2.5 hours. Or maybe we just need to stop flying so much.

51

Invest in the New Apollo Program

*It's not five minutes to midnight.
It's five minutes after midnight.*

*— Angela Merkel, Chancellor of
Germany, on the urgency of combating
climate change*

Most people are happy as long as the electricity arrives when they flick the switch. For the world's power utilities, however, a huge amount of effort and investment is required to produce this power in a safe, reliable, affordable manner.

In 2008 most of the world's electricity came from coal (41%), gas (20%), hydroelectric (18%) and nuclear power (15%). Only 2% came from geothermal, wind, solar and biomass.

That year, the world's 50,000 fossil-fuelled power plants produced 11.4 billion tonnes of

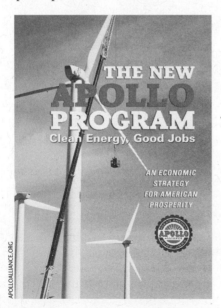

The Apollo Alliance is a coalition of labor, business, environmental and community leaders who work to catalyze a clean energy revolution that will put millions of Americans to work in a new generation of high-quality, green-collar jobs. apolloalliance.org

CO_2, representing 31% of the world's CO_2 emissions, but the urgency of global warming requires an end to most fossil-fueled power unless its CO_2 emissions can be captured — a prospect that is still very far off. (See p. 60.) There is also understandable resistance to the construction of more nuclear power plants and hydroelectric dams. This points to enormous change and a restructuring of the way we generate electricity.

How much power do we need? In 2009 the world consumed 20,000 terawatt hours (TWh) of electricity,[1] which has been forecast in a "business-as-usual" analysis to rise to 34,000 TWh by 2030, with most of the increase coming from the developing world.[2]

How much could we save if the world's buildings, appliances, factories and equipment were far more efficient? Estimates range from 25% to 70% with known technologies and policies. (See #52.) If we set a 33% efficiency increase as our goal, our future demand could be reduced to 23,000 TWh.

North America's Future Electricity

In the US, electrical demand in 2006 was 4,000 TWh. Based on a business-as-usual estimate, the Energy Information Agency suggests we'll need 5,800 GWh by 2030. With a 33% reduction for energy efficiency, this falls back to 4,000 TWh, but we also need to provide power for electric cars and railways.

In 2005 American cars and trucks drove 4.8 trillion kilometres.[3] If we assume 25% less driving in 2030 because of far greater use of cycling, transit and high-speed trains, that falls to 3.6 trillion kilometers. If we also assume an efficient electric

	World[6]		USA		Canada	
	2006	2030	2006	2030	2004	2030
Coal	8200*	0	1990[7]	0	95[8]	0
Coal CCS	0	500	0	100	0	14
Gas	4000	2000	813	250	30	10
Hydro	3000	4000	289	289	337	337
Nuclear	2600	2000	787	250	85	0
Wind	284	7500	32	1250	5[9]	137
Solar thermal	0	5000	0	1000	0	14
Geothermal	50	3000	15	1000	0	109
Solar PV	5	3000	0.6	500	0	34
Ocean	0	2000	0	250	0	34
Biomass	227	1000	38	100	7	14
Total	20,000	30,000	4,000	5,000	576	685
* Gigawatt-hours per year						

Top Ten Fossil-Fuel-Powered Nations[10] (CO$_2$ emissions in million tonnes/year)	
China	3100
USA	2800
Russia	661
India	583
Japan	400
Germany	356
Australia	226
South Africa	222
UK	212
South Korea	185

vehicle that uses 20 kWh per 100 km, the electricity needed comes to about 720 TWh — increased to 950 TWh for heavier vehicles. The high-speed electric railway network will need a further 50 TWh (see #75), giving us a 5,000 TWh goal for 2030.

Can it be done? Yes, if we invest in a huge New Apollo program to develop wind farms, solar-thermal plants, geothermal plants, tidal and wave energy, solar PV and small-scale biofuel and microhydro plants, and if we continue use the existing hydroelectric dams and nuclear plants that still have life in them.

The challenge is enormous, but at our current rate of consumption the world's oil may effectively be gone by 2030, the gas by 2068,[4] and the coal by 2138;[5] so even without global warming, we know we're going to have to solve this problem.

The task is to mobilize the political will and investment incentives needed to build it, using Europe's feed-in tariffs (see #69) while overcoming local opposition from people who have bumper stickers that say, "I want to save the world — but please do it somewhere else."

We need to build a supergrid to deliver the new power to the cities, which will require cooperation among many competing jurisdictions (see #70). We will also need to close down the existing coal-fired power plants, compensating the companies for their stranded assets and assisting the coal-mining communities with green economic development and the ecological restoration of scarred landscapes.

For some power utility owners and managers, this might seem scary or just plain impossible. For others, it is what they have been saying for years. Many utilities tend to be quite conservative, because the public demands reliable, affordable power, which has traditionally been provided through a few large hydro, coal or nuclear power plants rather than a mass of small, distributed generators. The problems of intermittency, baseload power and power storage all have solutions, however (see #57), including the fact that the more renewable energy we produce, interconnected through a supergrid, the better the prospects look for a reliable zero-carbon power supply.

52

Maximize Energy Efficiency

Energy efficiency is the cheapest, quickest and cleanest way to extend energy supplies, while also taking great strides toward addressing global climate change.
— Kateri Callahan, Alliance to Save Energy

It was Amory Lovins, from the Rocky Mountain Institute, who coined the word *negawatts*, meaning "saved energy." You can't see a negawatt the way you can see a wind turbine, but the electrons that result are identical to those produced by the wind — and they are three times cheaper.

When a utility invests in a program that saves energy, it costs only 1 to 3 cents/kWh,[1] compared to 5–7 cents for wind and up to 20 cents for nuclear. The potential is huge — estimates suggest that we could save 25–70% of all the power we use. Here are some of the most effective ways a utility can harvest this sea of latent power.[2] (See also #66.)

1. Lobby for the Best Policies:

- Decouple a utility's profit from the power it sells, giving it a financial incentive to sell less power, as California has done so successfully.

- Establish a public benefit fund to finance efficiency programs by means of a surcharge on every utility bill, as many states do.

- Allow the use of "white certificates" that give a market value to negawatts so that they can be sold contractually, as New England is doing.

- Establish the highest efficiency standards for appliances, buildings and equipment, as Japan is doing in its Top Runner program.

- Establish an independent efficiency utility, as Vermont has done with Efficiency Vermont, which in 2007 achieved a statewide load reduction of 1.8% at a cost of 2.6 cents/kWh, versus 10.7 cents to supply electricity. Delaware's goal, through its Sustainable Energy Utility, is a 30% reduction in the use of all fuels by 2015,

funded by a 36-cent-per-month electric bill surcharge and a $30 million private bond issue, to be repaid through savings.

2. Do Integrated Resource Planning

This is the industry name for long-term planning, to show how future demand can be met by integrating efficiency, renewable energy, stored energy and smart grid development.[3]

3. Develop a One-Stop Efficiency Program

Develop a one-stop package for everyone who owns or rents a building, including a free energy audit; a costed list of recommended changes; access to grants, loans, rebates and tax credits; and access to a competent company to do the work. Develop special programs for low-income households, landlords, farmers, small businesses, industry, schools and colleges.

4. Engage the Public

Use creative ways to tell people how to save energy, such as America's Six Degree Challenge, Canada's Flick-Off campaign and Project Porchlight.

5. Build an Efficiency Partnership

Set up a partnership to improve and then market the most efficient appliances. Between 1998 and 2008 the Northwest Energy Efficiency Alliance (NEEA) saved enough power for 100,000 homes; in 2006 alone they sold 10.7 million efficient light bulbs, four times the national average. NEEA also collaborated with Oregon's Department of Energy to create new standards for Oregon's residential building code that are reducing new home

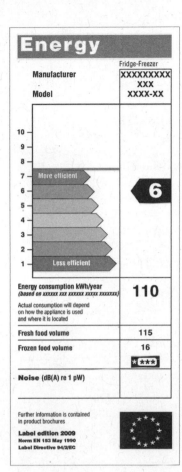

Europe's efficiency label for a fridge-freezer.

- Closing the Efficiency Gap: rt.rmi.org/research/cgu.html
- Demand Response links: enernoc.com/links.html
- Efficiency Programs Best Practices: eebestpractices.com
- Flex Your Power (California): fypower.org
- Flick Off: flickoff.org
- Fort Collins: fcgov.com/ZILCH
- Good Cents: goodcents.com
- Information Gateway: reegle.info
- Northwest Energy Efficiency Alliance: nwalliance.org
- Project Porchlight: projectporchlight.com
- Regulatory Assistance Project: raponline.org
- Six Degree Challenge: sixdegreechallenge.org
- SMART 2020: smart2020.org
- Smart Meters Ontario: smartmetersontario.ca

energy use by 15%. California's Flex Your Power is another great partnership exercise.

6. Install Smart Meters and Pay-As-You-Go

When unmetered condo dwellers get their own metering and electricity bills, consumption falls by 20–40%. When the residents of Woodstock, Ontario, switched to a pay-as-you-go service, combining advance payment with smart metering, their consumption dropped by 15%.[4]

Smart meters bring electricity management into the 21st century, enabling people to see how much power they are using in which space, at which time and (with time-of-use pricing) at what price. The meters cost $150 to $250 to buy and install, and can reduce demand by up to 30%, depending on incentives. In Italy, which has had smart meters in 30 million homes since 2005, energy use has dropped by 5% a year.[5]

7. Build a Demand Response Network

Demand response networks use smart technology to pool the voluntary power savings of numerous customers, enabling a utility to meet its peak demand without building expensive new power plants. Southern California Edison has 12 such programs, and EnerNOC's Negawatts Network does the same.

8. Invest in Smart Technology

In China, industrial motors will use 34% of the country's power by 2020, producing 10% of China's carbon emissions. The best electrical motors, with electronically controlled variable drives, use 40–70% less energy.[6] Worldwide, according to SMART 2020, we could reduce global electrical demand from motors by 22% by 2020, saving $107 billion a year. The same applies to many other technologies.

53

Develop the Best Low — Income Programs

By skimping on design, the owner gets costlier equipment, higher energy costs, and a less competitive and comfortable building; the tenants get lower productivity and higher rent and operating costs.

— Amory Lovins, Hunter Lovins
and Paul Hawken.

Carbon taxes and public benefit surcharges on utility bills are excellent public policies, but because they increase the price of power they can be a sour pill to swallow for anyone who lives on a low income. It costs money to replace inefficient light bulbs and old fridges, and if you live from month to month, facing rising food and gas prices, where does the money come from? Furthermore, if you rent your home, what incentive does your landlord have to pay for better insulation when it's you who pays the bills?

Utilities across North America have spent 20 years developing programs to help people on low incomes save energy and reduce their bills, so they have learned what works. The best programs educate their customers; provide comprehensive energy audits; replace inefficient lights and appliances; undertake weatherization, air-sealing, insulation and caulking; install thermostats; replace old furnaces and boilers; replace windows; and provide financing while making the most of state and federal grants.

Most of what constitutes a "best program" can be found in *Meeting Essential Needs: Exemplary Utility-Funded Low Income Energy Efficiency Programs*, published by the American Council for an Energy-Efficient Economy (ACEEE) in 2005.

1. The perfect project will provide one-stop shopping for all low-income families, whether they live in a single- or multi-family building or a mobile home. It will be a single program that applies across the state or province, creating synergy with government funding programs, and will be advertised widely and colorfully in several languages.

2. It will be comprehensive, applying to the whole house, and cover oil and gas as well as electricity ("fuel-blind"). Its staff will spend up to two hours per household on customer education and offer a full menu of energy-saving measures at zero or close to zero cost. They will work with manufacturers and distributors to benefit from the bulk purchase of efficient appliances.

3. The program will be delivered through a partnership with private and community-based contractors, using creative marketing and

Patty Larocque, who works with Warm Up Winnipeg, a low-income energy efficiency program of B.U.I.L.D. - Building Urban Industries for Local Development.

- ACEEE: aceee.org
- Energy Bucks: energybucks.com
- Energy Partners: pge.com/energypartners
- Green Landlords Project: bcsea.org/greenlandlords
- Massachusetts: masssave.com
- New Hampshire: nhsaves.com
- New York State: getenergysmart.org
- Pay As You Save America: paysamerica.org
- Vermont: efficiencyvermont.com
- Warm Front: warmfront.co.uk
- Warm Up Winnipeg: warmupwinnipeg.ca
- Wisconsin: focusonenergy.com

working through existing social and religious groups as well as social networking websites such as Facebook. Local businesses are usually best at the physical side of the work, while non-profit societies do a better job at customer education and outreach.

4. The contractors who deliver the program will be well trained and ideally from the same low-income communities it is targeting. They will have excellent technical support and infra-structure, a wide range of diagnostic tools, a generic priority list based on known cost-sav-ings, a simple energy-savings calculator, and a real-time database that allows administrators to track what's happening, backed by quality assurance and evaluation. As a result, the pro-gram will deliver reliable negawatts that a utility can build into its long-term planning.

5. The program will reduce energy use by an average 30% and save more than it costs. As solar prices fall, the same program can be used to deliver solar PV, solar hot water and other zero-carbon options. It may be financed through a public benefit fund or an energy efficiency fund, depending on the local situa-tion. Organizationally, it will be run with help from many partners and collaborators, includ-ing government agencies and community organizations.

6. In addition to negawatts, it will provide other benefits including lower credit and collection costs and less bad debt for the utility, improved health and safety for the customers and lower energy bills.

Solving the Split Incentive

Even the best programs get bogged down by the notorious "split incentive," where the landlord owns the building and major appliances but the tenant pays the bills. Overcoming this barrier is imperative. In British Columbia, the Green Landlords Project has developed an integrated solution that involves nine separate but integrated actions, likened to a Rubik's Cube because they all need to work together. The actions include developing an Energy Efficiency Loan Fund, providing landlords with reliable information about the paybacks from different efficiency investments, improving financial incentives, establishing a single-door agency that can make everything easier, implementing tenant education programs, and changing the billing so that a loan can be attached to a utility meter or a building's property tax, instead of to an individual.

If utilities do not address the needs of people on low incomes, they will meet understandable resistance to anything that increases their bills. This is therefore an essential component of the road-map to a climate-friendly world.

54

Invest In Wind Energy

An acre of windy prairie could produce between $4,000 and $10,000 worth of electricity per year — which is far more than the value of the land's crop of corn or wheat.

— Denis Hayes

In 2008 the US government showed that wind has the potential to deliver 20% of the US's power by 2030, using 300 GW of turbines. This certainly underestimates what's possible, as it did not assume the use of European-style feed-in tariffs to accelerate progress; or the potential for greater efficiency, rising gas prices or the arrival of carbon pricing for coal- and gas-fired power.

The goal laid out in Solution #51 of generating 25% of the US's energy from the wind (1,250 TWh) would require the construction of 625 GW of capacity,[1] using 5% of the 12,000 GW potential. It would create 2.5 million high-paying jobs,[2] while requiring a density of turbines on the land

A wind farm in western Canada.

similar to Denmark's. Germany produces 34 times more wind energy per square mile than the US, and enjoys 80,000 high-paying wind energy jobs.

By the end of 2008, the US had developed 25,300 MW, representing 0.2% of its potential, which is to produce far more, mostly in the open farmland of the mid-west from the Canadian border to the Gulf of Mexico.[3] This would bring hundreds of thousands of jobs to the region and enormous income to the farmers, at prices that would be consistently cheaper than electricity from both coal and gas with a carbon price attached. It would also preserve the enormous volume of water that coal-fired electricity requires.

At $2 million per MW, the $1.25 trillion investment would be similar in the long term to the cost of new gas-fired or nuclear power plants, both of which will see rising fuel prices during their operational lives, while wind will always blow for free. It is for this reason that utilities such as Xcel Energy like the wind, because it provides a useful hedge against future fuel price volatility.

Studies show that wind energy can integrate well into the grid, despite being an intermittent source of power. In 2008 Denmark received 25% of its energy from the wind, chasing a goal of 50%. Spain, which gets 9% of its energy from the wind, had a period in March 2007 when the wind was producing 27% of its power. At 20% there is almost no additional cost for firming up the power, and even at 25% the cost for storage and firming is only half a cent per kWh.[4]

The greater the geographic distribution of the turbines, the smaller the chance that they will be

- 20% Wind Energy by 2030: 20percentwind.org
- American Wind Energy Association: awea.org
- Canadian Wind Energy Association: canwea.ca
- Public Renewables Partnership: repartners.org
- Utility Wind Integration Group: uwig.org
- Wind Powering America: windpoweringamerica.gov
- Windustry: windustry.org

windless at the same time, and changes in the wind are never instantaneous, unlike the sudden forced outages with large conventional plants. Modern wind forecasting tools warn system operators about likely changes up to an hour ahead, giving them ample time to prepare.

In 2008, a study found that the US could theoretically replace 100% of its carbon-emitting pollution (assuming the use of electric vehicles) with up to 645,000 giant 5 MW wind turbines, requiring just 13 sq km for their footprints, and 3% of the US landmass for their spacing area.[5]

Canada's Windy Future

In its 2008 report, *Windvision 2025*, the Canadian Wind Energy Association found that wind energy could generate 20% of Canada's electricity demand by 2025, as called for in Solution #51, adding 55 GW of capacity and generating 52,000 full-time green-collar jobs. With such a large landmass, Canada could easily become a major wind energy exporter if the US moves too slowly in its own development. In Quebec alone, a 2004 study found more than 100 GW of wind potential in sites within 25 km of Hydro Quebec's transmission lines.[6]

US Wind Potential			
State	**Wind Potential MW[7]**	**Installed, 2008 MW**	**% Potential Installed**
North Dakota	138,400	344	0.25%
Texas	136,100	7,118	5%
Kansas	121,900	465	0.4%
South Dakota	117,200	98	0.08%
Montana	116,000	165	0.14%
Nebraska	99,100	73	0.07%
Total USA[8]	12,000,000	25,300	0.21%

Lobby for the Best Policies

For this much energy to be produced, wind-producing companies, utilities, farmers and cooperatives must work with local advocacy organizations and politicians to overcome the obstacles and win approval for policies that support the goal, including legislated targets, feed-in tariffs, federal agency purchasing requirements, investment in R&D, and the grid extensions that will be needed to bring the power to the cities.[9]

Wind Power Development around the World			
	Square miles	**MW wind 2008**	**KW wind per square mile**
Denmark	16,627	3,125	188
Germany	137,742	22,000	160
Spain	194,883	11,000	56
Texas	262,017	5316	20
USA	3,615,123	25,300	7
South Dakota	75,952	98	1.3

55

Invest in Solar Thermal and Geothermal Power

> Humankind evolved to be active when the sun was up, which is why human activity and energy usage correlate significantly with the energy delivery from direct solar systems.
>
> — David Mills, Ausra

Earth's deserts receive an astonishing amount of energy, much of which could be harvested through concentrating solar thermal power. And there is an even greater amount of energy in the rocks beneath us. Both of these sources can provide firm, base-load power, making them very important to our future.[1]

The challenge is very specific: can we produce enough solar thermal and geothermal power, combined with other kinds of electricity, to allow us to phase out carbon-producing power plants by 2020 or 2030?

The Solar Thermal Future

The world's 33 million square kilometers of desert receive 80 million TWh of potential power a year, which is 4,000 times more than the 20,000 TWh the world used in 2009. Solar thermal power generation is a proven technology, with 6,000 MW of capacity expected by 2012. A 354 MW solar plant has been operating in California's Mojave Desert since the late 1980s, and new plants are being constructed or planned in Florida, Arizona, Spain, France, Greece, Portugal, China, Australia and nine other countries.

Solar thermal plants produce electricity from the sun's heat, using a steam turbine. By storing the heat in molten salt or water they can extend their hours of operation to 13–16 hours a day. This is mostly daytime power, so the correlation with the grid's load is 92–96%, making it a good substitute for coal and gas-fired power plants.[2] If solar-thermal were to produce 90% of US power needs, the land needed would be around 15,000 square miles, which is 13% of Arizona, and less than the land being used today for coal mining and coal-fired power plants, or the flooded land needed to generate the equivalent hydro-electric power.[3]

In the US, we would need to build 500 GW of solar thermal plants to produce 1,000 TWh a year of solar thermal power by 2020, as proposed in Solution #51. By 2007 the US Bureau of Land Management had received right-of-way requests on more than 300,000 acres of California desert for the development of 34 large solar thermal plants, totaling 24 GW, by companies including Ausra, BrightSource and Solel.

In the southwestern US, electricity from solar thermal plants costs 13–17 cents kWh, which is expected to fall as production is scaled up. When you add carbon taxes, the rising price of natural gas and feed-in tariffs (see #69), it makes sound economic sense. In Spain, where ten solar thermal plants are being planned, operators receive 40 cents/kWh under Spain's feed-in tariff; in Israel, they receive 19 cents. The US Department of Energy hopes to see the cost fall to 5–7 cents/kWh

- Ausra: ausra.com
- BrightSource Energy: brightsourceenergy.com
- Desertec Australia: desertec-australia.org
- Desertec Foundation: desertec.org
- Desertec USA: desertec-usa.org
- NREL Solar Thermal research: nrel.gov/csp
- Solar Millennium: solarmillennium.de
- Solel: solel.com

Possible Solar Thermal & Geothermal Development to 2020					
US	Solar thermal GW	Solar Thermal TWh	Hot Rocks Geothermal GW	Hot Rocks Geothermal TWh	Coal TWh
2010	0	0	0	15	2000
2014	125[5]	250	30[6]	250	1500
2016	250	500	60	500	1000
2018	375	750	90	750	500
2020	500	1000	120	1000	0

by 2020, making it among the cheapest forms of energy available.

Solar thermal can achieve a similar substitution for fossil fuels elsewhere, using the deserts of India, China, Central Asia, North Africa, the Middle East, Southern Africa, Australia and South America. Europe consumes 4,000 TWh of electricity a year, while the deserts of the Middle East and North Africa are bathed in 630,000 TWh — 150 times more than Europe needs. The technology exists — it is up to us to make it happen.

The Geothermal Future

The second big source of baseload power is hot rocks geothermal energy, using the heat stored in granite 6–10 km underground. According to a team at the Massachusetts Institute of Technology, this has the potential to deliver 2000 to 20,000 times more baseload power than the US needs, and a similar amount for Canada. (See p. 50.)

In 2007 the MIT team suggested that the investment needed to deliver 100 GW by 2050 was $1 billion.[4] The scenarios presented here require 120 GW of geothermal for the US by 2020, and 13 GW for Canada — far larger quantities that are needed far sooner, and call for immediate development. The policy commitments needed are similar to solar thermal — a clear goal to drive the development and a feed-in tariff set at the level needed to attract investors. Partnerships with the oil and gas industry make sense, because that industry has the drilling expertise and the investment capital. Hot rocks geothermal power has much higher front-end risks,

because the cost of drilling is so high, so high tax write-offs for capital costs would be helpful.

- Canadian Geothermal Energy Association: cangea.ca
- Geothermal Energy Association: geo-energy.org
- Geothermal Energy in California: energy.ca.gov/geothermal
- Geothermal Resources Council: geothermal.org
- IEA Geothermal: iea-gia.org
- International Geothermal Association: geothermal-energy.org

Europe's first commercial concentrating PS10 solar power tower near Seville, Spain. The 11 MW tower produces electricity using 624 large movable heliostat mirrors.

DESERTEC-UK/POLLY HIGGINS

56

Invest in Solar Energy

All of the world's energy could be achieved by solar many thousands of times over.

— Paul Booth, Head of Renewable Energy Supply and Marketing, Shell.

Solar thermal's better-known sister is solar photovoltaics (PV), which is growing at a rapid 40% a year. By 2007 the installed capacity had reached almost 10,000 MW globally, heading toward 27,000 MW by 2010.

In Germany, thanks to the government's feed-in tariffs, the cost of installed solar PV was cut in half between 1997 and 2007, and continued price reductions are certain. In 2008 the German city of Freiburg got 3% of its electricity from PV, and some analysts predict that PV could meet 25% of Germany's demand by 2050, up from 1% in 2007.

In the US the 2008 Utility Solar Assessment Study found that with the best policies, the US could produce 8% of its electricity (388 TWh) from solar PV by 2025, putting it on track to achieve the needed 500 TWh by 2030. (See #55.) Somewhere before 2020, solar PV looks likely to become competitive with other kinds of electricity in most regions of the US, making it acceptable to require its use in new buildings, and as a point-of-sale upgrade in existing buildings.

From a utility perspective, solar PV has many advantages. It can be used as a hedge against expensive summer peak power and it can be built to scale as needed. It does not need a lengthy approval process, and because it can be built close to the demand it has lower transmission and distribution costs. It can also be controlled independently, helping to prevent costly blackouts from the loss of centralized thermal or nuclear generation. It complements wind well, as wind is strongest at night, solar in the day.

As well as larger utilities such as PG&E and Southern California Edison, smaller public utilities are also showing leadership, including Austin Energy, the Tucson Electric Power Company and the Sacramento Municipal Utility District. As members of Solar America Cities, each of these cities has a comprehensive plan to transform the solar market, as does San Francisco.

Seven Steps to Solar Success

Step 1: Appoint an internal champion who can take the lead and educate utility staff and engineers to become excited by solar PV instead of resisting it.

Step 2: Form a partnership with energy advocacy groups and lobby the government to adopt

- Byron Bay: beyondbuildingenergy.com
- California's solar incentive program: GoSolarCalifornia.org
- Database of State Incentives: dsireusa.org
- Interstate Renewable Energy Council: irecusa.org
- PG&E's Solar Schools Program: need.org/pgesolarschools/ec.htm
- Power Purchase Agreements: mondial-energy.com/howitworks.htm
- Solar America Cities: solaramericacities.energy.gov
- SolarBC: solarbc.ca
- Solar Electric Power Association: solarelectricpower.org
- Utility Solar Assessment Study: solarelectricpower.org
- Utility Solar Water Heating Initiative: www1.eere.energy.gov/solar/ush2o

A large solar array on a barn in Germany, where solar installers receive 62 cents kWh with a 20-year contract. (€0.47/kWh, 2008).

Europe's feed-in tariff, which has replaced net metering and the Renewable Portfolio Standard as the best policy. (See #69.)

Step 3: If there is any delay in introducing a feed-in tariff, lobby to include solar in your state or provincial Renewable Portfolio Standard.

Step 4: Similarly, if there is any delay, ensure that all residents have net metering with no upper limit on size, enabling them to sell surplus solar energy back to the grid. This should be linked to a pro-solar utility rate that combines time-of-use rates with no standing demand charges, enabling solar customers to earn the best prices on summer afternoons.

Step 5: Work to establish standardized state and national interconnection standards, using New Jersey's model as the best practice. The Interstate Renewable Energy Council publishes an excellent newsletter that has all the relevant details.

Step 6: Form partnerships with local businesses to install solar PV on commercial roofs, so that the utility owns the panels and sells the power to the owners through a Power Purchase Agreement, with the excess solar energy going to the grid.[1] Using this model, Nevada Power has partnered with the Las Vegas Valley Water District to install a 3.1 MW solar system on six rooftops.

Work with community organizations to reduce costs through mass installations in solar neighborhoods, as Beyond Building Energy is doing in Byron Bay, Australia.

Step 7: Train the next generation of workers, as PG&E and Southern California Edison are doing with their solar training workshops.

Solar Hot Water

By 2008, 40 million Chinese homes were using solar hot water systems. In Israel, Spain and Hawaii solar hot water is required in all new buildings. With its ability to reduce power consumption for water heating by 50–70%, it acts as a demand-side resource that can reduce a household's yearly greenhouse gas emissions by 1 to 3 tonnes. In Oregon, the Eugene Water and Electric Board has installed more than 1,000 systems through its Bright Way to Heat Water program, using a cash discount and zero-percent financing, coupled with Oregon's $1,500 tax credit.

The key to solar/thermal take-off is the development of a regional partnership including a non-profit organization to take the lead, industry, government (for incentives), colleges (for training), building and plumbing inspectors (to remove needless obstacles), and the use of bulk purchases to reduce the price.

57

Learn from Austin, Texas

I get to be the mayor of the capital city of the most polluting state of the most polluting country on the planet.... I see truly a non-carbon economy. It's cleaner. It's healthier. We're about out of alternatives, so it's going to be easier and more cost effective to start to do the right thing.
— Will Wynn, Mayor of Austin[1]

Austin, Texas, has set itself the goal to become the clean energy capital of the world and the leading city in the nation in the fight against global warming. It wants its buildings to be the most energy efficient in the nation, and the city's public utility, Austin Energy, aspires to have the most aggressive utility greenhouse gas reduction in the US.

In 2008 the utility owned 2,630 MW of generating capacity, consisting of two natural gas plants (1400 MW), a shared stake in a 600 MW coal-fired plant, a 400 MW nuclear plant, 214 MW of wind farms and 13 MW of landfill gas. This makes it a typical player on the power scene, facing the same challenges with regard to global warming and peak oil that all utilities face.

Being owned by the city, the utility has been able to set some clear goals for 2020:

- achieve 15% savings through efficiency and conservation
- meet 30% of its needs from renewables, including 100 MW from solar power
- achieve peak demand savings worth 700 MW
- achieve carbon neutrality for all new generating units
- establish a cap and reduction plan for its existing emissions

Austin Energy's Power Saver Program invites its customers to become Power Saver Volunteers, offering them low-interest loans, cash incentives

to recycle old refrigerators and more than $2000 in rebates for efficient appliances and retrofits. For those on a low to moderate income it offers a free home improvement service. To help control the summer peak load, it invites customers to become Power Partners, giving them free programmable thermostats that the utility can cycle off for up to ten minutes at peak demand.

For multi-family buildings it has provided cash rebates for efficiency improvements to more than 48,000 apartment owners, and to its commercial customers it offers rebates and incentives up to $200,000. It will also pay 15 cents/kWh when commercial customers agree to reduce their power during peak periods. As the reward for its efforts, Austin Energy saved 8.2% of its demand between 2003 and 2007, halfway to the 2020 goal, suggesting there is room for far more than 15% savings.

Next is their Green Building Program, rated #1 in the US, which offers consultancy, advice and training to help people build in a green, energy-saving manner. As a result of the goodwill established in the building community, the city has been able to write a building code that requires all new houses to be "zero-energy ready" by 2016, which means being 70% more energy-efficient and being prepared to obtain 30% of their energy from solar and other clean technologies.

To stimulate more use of renewable energy, Austin Energy has a GreenChoice Program through which (by 2008) 10,000 households and 500 businesses had signed up to buy 100% green power (mostly wind) at a fixed price for 10–15 years, instead of the variable fuel charge that rises or falls with the price of natural gas. Because they have marketed their program so successfully to businesses, GreenChoice has been the US's top-performing green power program ever since it started in 2001, selling 578 million kWh of green energy in 2007.

To encourage more use of solar power, Austin Energy offers a $4.50 per watt rebate and a solar loan from a credit union, which led to 2 MW of solar being installed by 550 homes and businesses by 2008 — still a long way from its 100 MW goal. By 2009 Austin Energy was integrating 12% renewable energy into its grid, en route to its 30% target by 2020.

Looking ahead to the arrival of electric vehicles, Austin Energy is building support for Plug-In Hybrid Electric Vehicles (PHEVs) by establishing Plug-In Partners as a national grass-roots initiative with more than 750 city, business, non-profit and governmental members. They are generating "soft" fleet orders for future vehicles to demonstrate that there is a huge pent-up market for PHEVs, and to this end they have set aside $1 million for PHEV rebates as soon as the cars are available. The utility's logic is that since most electric cars will charge up at night when there's a surplus of wind energy that's hard to store, their widespread use will help with grid stability.

Austin Energy's initiatives are market leading, but even more is needed if we are to achieve a climate-friendly world by 2030 and hopefully forestall some of the climate tipping points from kicking in before it is too late. We need all utilities to adopt Austin's ideas and ramp up their goals dramatically to achieve 33% energy saving and 100% renewables by 2030.

58

Build a Smart Grid

We need a new grid capable of networking millions of distributed energy devices such as solar panels, wind turbines, electric vehicles and smart appliances. We need an Internet of energy that employs the latest in digital technologies. We need a Smart Grid.
— Patrick Mazza

The world's power grids are a wonder of mid-20th century technology, and grid managers pride themselves on the reliability of their systems. All around the world, however, grids are stuck in the past, powered by large, centralized and often rather old power plants.

In North America, utilities have been spending only 0.1% of their revenue on research and development,[1] which is one of the causes of the power disturbances that cost US electricity consumers $79 billion a year.[2] At the same time, three demanding visitors are knocking at the door:

1. The smart grid, offering digital two-way communications and a whole new world.

2. Global warming, bringing the urgent need to replace coal- and gas-fired power with green

and build thousands of miles of new transmission lines in addition to the 60% of the grid that needs upgrading because of aging power lines and substations.[3]

3. Peak oil, bringing the need to change to electric and plug-in hybrid vehicles, with battery storage capacity linked to the grid.

Today's grid was designed to send power from A to B. In the smart grid, substations and control rooms will talk to each other digitally, and air conditioning systems will conspire with system operators to reduce peak power. Smart meters will flash their digital eyebrows at homeowners to say, "You want more power *now*? If you wait an hour, it'll be 5 cents cheaper," and "Wowee! We're selling solar for 35 cents!" In a smart grid, if one part of the grid is in difficulty, it will use its smart sensors to tell the other parts, enabling rapid diagnosis and action, creating a self-healing grid.

Some of today's grid managers — many of whom are close to retirement — do not like the millions of solar panels, wind turbines, electric vehicles and smart appliances that are lining up to tackle global warming. They want reliable, baseload power from coal, gas, hydro and nuclear power that they can manage from one control room.

Tomorrow's smart grid managers, by contrast, are excited by the vision of an Internet-style grid that encourages local micro-grids, stores electricity in electric cars parked in smart garages and pools customers in demand response networks that smooth out peak power loads, reducing the need for expensive spinning reserves and gas-fired peaker plants.

GE's "Scarecrow" ad for the smart grid is a modern take on the classic song, "If I Only Had A Brain," from The Wizard of Oz, imagining what can happen when old technologies become smarter.

The smart grid's peak-shaving ability alone could eliminate the need for $46 to $117 billion in new plants and lines over the next 20 years. A $600 million investment in smart appliances could remove the need for a $6 billion investment in new power plants.[4]

In a smart grid demonstration in Washington State, homeowners saved 10% on their power bills and cut their power use by 15% during peak hours. A grid-friendly appliance project found that if every major household appliance were fitted with peak-shaving equipment, the entire US electricity use could be reduced by 20%, saving $120 billion on new power plants and transmission lines.[5]

Step 1: Appoint a smart grid team and join the GridWise Alliance, which is bringing stakeholders together to move the grid into the digital age.

Step 2: Learn from Xcel Energy, which chose Boulder, Colorado, to demonstrate how the smart grid works, where it has installed 50,000 smart meters, a high-speed communications network and smart grid sensing equipment.

Step 3: Choose a community for a local pilot project. We need 50 such pilot projects, so that we can build the technical know-how and political support to move to a full North American smart grid.

Power Storage

The fundamental solution for power storage is the supergrid (see #70). For short-term storage, there are several additional options:

- Pumped storage — using low-cost off-peak power to pump water from a lower to a higher

- Galvin Energy Initiative: galvinpower.org
- GE Plug Into the Smart Grid: plugintothesmartgrid.com
- Global Smart Energy: globalsmartenergy.com
- GridWise Alliance: gridwise.org
- Illinois Smart Grid Initiative: ilsmartgrid.org
- IntelliGrid: intelligrid.epri.com
- Next Generation Utility: ert.rmi.org/research/next-generation-utility.html
- Night Wind: nightwind.eu
- Pioneer City 2030: www.newenergynexus.org
- *Powering Up the Smart Grid,* by Patrick Mazza: climatesolutions.org
- SF6 Emission Reduction Partnership: epa.gov/electricpower-sf6
- Smart Electric News: smartelectricnews.com
- Smart Grid News: smartgridnews.com
- The Modern Grid Strategy: netl.doe.gov/moderngrid
- V2Green: v2green.com

elevation reservoir, where it can be used to generate hydropower.

- Flow batteries — using large vanadium-based rechargeable batteries. When used on King Island, Tasmania, wind penetration increased to 40%. Other battery technologies are also being developed.

- Demand response, such as storing night-time energy in refrigerated warehouses, in order to reduce peak-time demand, which is being demonstrated in Holland's Night Wind Project.

- Vehicle to grid — using the storage capacity of electric vehicle batteries.

- Compressed air — using off-peak surplus power to store compressed air in underground aquifers, and blending it with natural gas to fire combustion turbines (being developed in Iowa).

- 10 kW lithium ion batteries — by investing in these for short-term power storage, utilities will build the market to reduce their price for all purposes.

59

Solutions for the Gas Industry

> There is a parallel universe in operation out there. Politicians try to negotiate a reduction in greenhouse gas emissions, while business executives lay plans to expand their carbon footprint.
>
> — Terry Macalister, *Guardian*

The thrust of this book is that our civilization must move to a climate-friendly world as rapidly as possible and that we can and must meet 100% of our global energy needs with renewables. We have to be very proactive and adopt far more sustainable lifestyles, but the lights will not go out, transport will not grind to a halt, and we will not freeze in winter or fry in summer.

This puts the fossil fuel companies in a dilemma. Should they dig in their heels and fight for a dying past or plan for an orderly shut down and build a new portfolio from renewables?

Natural gas is a non-renewable fossil fuel, and as North America's gas runs out, we will need to import it as liquefied natural gas from Russia and Iran — those paragons of democracy — at a much higher price. Natural gas is also a huge contributor to global warming.

When it is burned, natural gas produces fewer carbon emissions than coal or oil. Globally, the industry produced about 3,000 billion cubic meters of gas in 2008, creating 6 billion tonnes of CO_2. This represents 16% of the global CO_2 emissions and 8.5% of the cause of global warming.[1] As a rule of thumb, 1000 cubic meters of natural gas produce 2 tonnes of CO_2.

That's not the whole story, however. Natural gas is 90% methane,[2] which has a climate impact over 100 years that is 25 times greater than CO_2. And nor is *that* the whole story, for methane's natural life in the atmosphere is not 100 years — it is 8.4 years, and its impact over this period is up to 125 times greater than CO_2. (See p. 12.)

This would not matter if the gas were all burned — but it is not. As it travels from the well-head to the boiler, about 1.4% escapes through valves, pipes, compression stations and by deliberate venting to blow and purge the system.[3] A 2005 German study of Russian pipelines found an average 1.4% leakage rate;[4] US government and industry suggest a 1.5% loss rate;[5] a 1990 British report suggested a leakage rate of 5.3 to 10.8%.

If 1.4% escapes as methane, this means that 38 billion cubic metres[6] of natural gas are released as raw methane every year. If we measure the impact over 8.4 years, however, which is both the natural life of methane and the time-scale of the emergency we face, those 38 billion cubic meters produce 10 billion tonnes of CO_2e a year, increasing gas's carbon footprint to 16 billion tonnes of CO_2e per year, revealing the disastrous nature of what's happening as being equivalent to half the CO_2 emissions from all fossil fuels.[7]

There are also "incidents," such as when backhoe operators rupture a gas line, or the automatic venting that occurred one night in June 2001 at a compressor station in Langley, British Columbia, when 10 million cubic feet of gas was released in 28 minutes, producing 1,414 tonnes of CO_2e, or 7,000 tonnes over methane's lifetime.[8] You'd need to drive 770,000 cars for those 28 minutes to create that much climate damage.[9]

Immediate Solutions

The best government solution is a methane tax to increase the incentive to clean up the leaks, known as "fugitive emissions".[10] This can be done with handheld remote detectors, infrared video scanners that use laser beams to show the leaks in

> • Leak Detection & Measurement of Fugitive Methane
> Emissions: epa.gov/gasstar/documents/heath.pdf

real time, or airborne detectors that can be used from a helicopter at 1,000 feet. The possibility of airborne imaging from balloons or circling drones at 80,000 feet is also being discussed.[11] Ninety percent of the leaks can be fixed at a net saving to the company; even a small leak that could be fixed for $500 might be losing gas worth $10,000 a year.[12]

For the venting that occurs in compression stations, the gas can be captured as fuel for a natural gas engine, or a turbine can be used to generate electricity from the unwanted pressure. The utility Enbridge found that if turbines were used in combination with fuel cells in all its Ontario operations instead of the gas being vented, they could generate up to 40 MW of clean electricity.

Diversify into Renewables

If not gas — then what? One possibility is the geothermal industry, because the gas industry has the expertise to drill at depth. Another is biogas harvested from agricultural wastes, sewage, manure, algae and organic landfill waste. In Germany, a Green Party study found that biogas has the potential to replace 100% of the European Union's natural gas imports, and operators are already putting biogas directly into the country's natural gas pipelines.

During the transition phase, gas companies can offer their customers carbon offsets to make the gas carbon neutral, as Direct Energy, Russia's Gazprom and MXEenergy are doing — though I doubt that they include the fugitive emissions in their carbon calculations, as they should.

A gas pipeline crossing Texas. Gas pipelines have an average 1.4% leakage rate, by which raw methane escapes to the atmosphere.

© INSPHOTOS DREAMSTIME.COM

60

Solutions for the Oil Industry

> Unless we redesign our civilization in numerous ways, all of the science in the world won't save us.
> — William Calvin, neurobiologist, author of *Global Fever — How to Treat Climate Change*[12]

In 2004 Lord Oxburgh, chair of Shell, said that the threat of climate change made him "really very worried for the planet," and that carbon capture technologies were urgently needed.[1]

In 2008 BP's most senior risk manager, Jan-Peter Onstwedder, said that the world's known fossil fuel reserves contained a quarter more carbon that we could emit without causing dangerous climate change, putting the purpose and value of new oil exploration in doubt. "A quarter more" is clearly understating it; three quarters is closer to reality.

How are we to deal with an industry that earns such enormous wealth by extracting and selling a product that puts the whole planet in peril? As a

Felix Kramer of CalCars in his Plug-In Hybrid at San Francisco's Step It Up action in 2007. If 50% of cars and light trucks were electric and 50% were plug-in hybrid electric, we would need 90% less liquid fuel for personal transport.

world, we burn 32 billion barrels of oil a year, each of which produces 457 kg of CO_2, pouring 14.6 billion tonnes of CO_2 into the atmosphere every year. This represents 40% of the global CO_2 emissions and 20% of the cause of global warming.[2]

The need to stop burning fossil fuels is as urgent as the need to stop slavery was 200 years ago. Some oil companies are diversifying a tiny slice of their assets into solar, wind and biofuels, but compared to their continued investment in oil and gas, it is like removing the leg-irons on a slave ship. Exxon Mobil is investing $30 million a year in new technologies, some of which relate to renewables, and $12 billion a year — 400 times more — in new oil and gas developments.

In 2008 Exxon Mobil, Shell, BP, Chevron, Total and ConocoPhillips' 2nd-quarter profits hit a record $51.5 billion, thanks to the high price of oil. Projected for a year, this is more than $200 billion. At least until 2009, the US government was still giving $7 billion a year in tax breaks to the oil and gas industry.[3]

In addition, the 12 oil-exporting nations that are members of OPEC (United Arab Emirates, Qatar, Saudi Arabia, Kuwait, Iraq, Iran, Libya, Algeria, Venezuela, Nigeria, Angola and Ecuador) earned $645 billion in oil exports in the first half of 2008, suggesting earnings of $1,300 billion for the year as a whole.

Why does this matter? It matters because

• $1 billion could open up the world's enormous hot rocks geothermal resources, for which the oil industry already possesses the drilling expertise.[4]

• $10 billion a year could develop concentrating solar power in the world's deserts to the point where it is competitive.[5]

These investments would be only 5% of the profits that the six oil companies earned in 2008, and less than 1% of the global oil industry's combined profits and export earnings. Instead of realizing these opportunities, however, the oil industry has invested $100 million in Alberta's tar sands (with a further $125 billion projected by 2015), where the oil is so mired in dirt and difficult to extract that the entire boreal forest landscape must be destroyed to access it, releasing all the carbon stored within it, and where every barrel requires four to six times more energy to extract than is needed for conventional oil,[6] producing 23% more CO_2 per barrel.[7]

The potential size of the future Alberta tar sands is 140,000 square kilometers, larger than the whole of England, every inch of which would be turned to a blackened mess in order to extract the estimated 180 billion barrels of crude — and all this to feed our planetary addiction for only another five to six years. The World Wildlife Fund has warned that developing North America's shale and tar-sand oil reserves could increase atmospheric CO_2 by up to 15%.[8]

What should the oil industry do?

• Accept the need for obligatory carbon pricing, and for all oil to become carbon neutral, so that oil's climate costs are internalized into its price.

• Global Gas Flaring Reduction Initiative: worldbank.org/ggfr
• Oil Depletion Protocol: oildepletionprotocol.org
• Return to the Tar Sands: tothetarsands.ca
• *Stupid to the Last Drop: How Alberta is Bringing Environmental Armageddon to Canada (and Doesn't Seem to Care)* by William Marsden, Vintage Canada, 2007
• *TarSands* by Andrew Nikiforuk, Greystone, 2008
• Tar Sands Time Out: tarsandstimeout.ca
• Tar Sands Watch: tarsandswatch.org

• Accept the need for a moratorium on all new tar sands developments until an agreement on carbon neutrality has been achieved.

• Accept the need for a global Oil Depletion Protocol, to plan the decline in the world's oil supply in a rational manner.

• Accept the need for a windfall tax on oil industry profits, to invest in sustainable energy and climate solutions.

• Stop gas flaring from oil wells. In Africa alone, 40 billion cubic meters of gas are burned as waste every year — enough to generate half of Africa's electricity needs.[9] Globally, up to 170 billion cubic meters are flared each year, some as raw methane, which is far worse.[10] In 2007 oil industry flaring produced 400 million tonnes (MT) of CO_2,[11] the worst offenders being Iran (14 MT), Algeria (8 MT), Indonesia (8 MT), Malaysia (6 MT), Mexico (5 MT), Britain (5 MT), USA (5 MT) and Canada (4 MT).

61

Prepare for the Deluge

Dad, I'm scared and angry. Your generation created this problem. What are you going to do to fix it?

— **Mary Doerr, 15-year old daughter of John Doerr, Greentech Network**

Many governments are developing climate action plans, but few have focused on the need also to prepare for the storms, floods, hurricanes, droughts, forest fires, heat waves, diseases, crop failures and sea-level rise that climate change will bring, plus the impact of rising energy prices.

In Illinois a report by the University of Maryland found that if left unchecked, global warming could cost the state $43 billion a year by the 2030s because of flooding, farming losses from soil erosion and heat waves, dairy industry losses because of heat, the need for more water treatment and so on. Shared among Illinois's 13 million people, this is $3,300 for every citizen, or $13,000 for a family of four, per year.[1] In California a study by Next 10 estimated that the potential economic damage and loss of assets caused by climate change ranged from $7 to $47

billion a year ($184 to $1,237 per person a year, for 38 million people.)[2]

These are not imaginary costs dreamed up by academics with time on their hands. Climate change will cause increasingly expensive impacts. Coastal areas need to prepare for a possible two-meter rise in sea level by 2100, with huge implications for low-lying lands, drinking water, sewage treatment, underground electrical services and the saltwater contamination of farmlands — and the sea level will continue to rise unless we achieve a rapid reduction in emissions. The last time the world was 3°C (5.4°F) warmer, sea levels were 25 meters (80 feet) higher.[3]

Cities will need to prepare for more heat waves and smog alerts, at least until electric vehicles assign smog to the history books. Health services will need to prepare for an increase in West Nile virus, malaria, Lyme disease and dengue fever. Many regions will need to fight more forest and bush fires. All regions will need to brace for more sudden storms and deluges. Pipes, roads, bridges and settlements in flood plains that have been built to handle 100-year storm events will find them happening every ten years. Regions that depend on mountain snow-pack for water and power will need to prepare for its disappearance. Ecologists and conservationists will need to prepare rescue strategies for threatened species. All who depend on food from California will need to prepare for the reality that the state's agriculture may cease by 2050, due to the loss of 90% of the Sierra Nevada snowpack.[4]

We must also prepare for the growing energy crisis. Once global oil supplies have peaked,

- Climate Change Economics: climatechangeecon.net
- Climate Impacts Group (Pacific Northwest): cses.washington.edu/cig
- Future Sea Level Rise (San Francisco): futuresealevel.org
- ICLEI Climate Adaptation: iclei-usa.org/programs/climate/Climate_Adaptation
- Pew Center on Global Climate Change: pewclimate.org
- Sea level Rise Maps: flood.firetree.net
- *The Economic Impacts of Climate Change and the Costs of Inaction:* cier.umd.edu/climateadaptation
- US State Impacts: nextgenerationearth.org/usstates/statelist

Flooding in Ironbridge, Shropshire, England, known as the birthplace of the industrial revolution because it is near the place where Abraham Darby perfected the technique of smelting iron with coke (made from coal). This enabled him to build the world's first cast-iron bridge, in 1779. It might be renamed Coal's Revenge due to its frequent flooding.

whether in 2010 or 2020, gas may cost $5 a liter ($20 a gallon) and natural gas $20 a gigajoule, causing enormous disruptions for everything from asphalt to commuting to food.

There are very few examples of best practice to learn from. In 2008 the Pew Center on Global Climate Change found that only Alaska, Washington, Oregon, California, Florida and Maryland had any adaptation planning in progress, and many of their efforts were in the earliest stage.[5] ICLEI, which helps local governments with sustainability planning, has a Five-Milestone Program to help communities prepare for climate resiliency. In Seattle, King County has prepared a countywide flood control plan, funded by an increase in property taxes. New York has a task force to protect vital infrastructure. Toronto is incorporating a long list of short-term adaptation actions into its planning and budgeting.

There is a good reason why this tops the list of government solutions. Many governments are meeting resistance from citizens who don't understand why climate matters. Being confronted with the stark evidence is like getting a soaking with ice-cold water — which is what happened in Chicago when a team of climatologists hired by the city to look at Chicago's likely climate in 2095

predicted that if emissions weren't halted, Chicago would have a climate like Houston's. As a result, Chicago's commitment to climate action is among North America's strongest.

Actions

- Commission a study to examine the economic costs of coping with climate change by 2095 if the world follows a business-as-usual path, and the assets at risk.

- For coastal states, request a sea-level-rise assessment report to consider a range of scenarios for 2050 and 2100, and advise on how to plan for future sea level rise, as California is doing.[6]

- Undertake comprehensive climate adaptation and energy crisis planning.

- Require every city and region to prepare a fully costed climate adaptation and energy crisis plan looking 100 years ahead, with an update every three years, supported by a common planning framework and grants.

- Require listed companies, financial professionals and investment banks to report their potential impacts of climate change, future carbon costs and energy price increases, and to provide full disclosure of their risks and opportunities.

62

Plan for a Climate - Friendly World

Make no small plans, for they have no power to stir the soul.

— Niccolo Machiavelli

Norway, New Zealand, Iceland and Costa Rica are working to become carbon neutral nations. Costa Rica aims to get there by 2021, using tree planting as one of its methods. Norway has pledged to get there by 2030 — but mostly by financing environmental projects abroad. New Zealand says it will get there first, but this requires finding a solution to the methane burps of its 5 million cows and 40 million sheep.

Most countries are nowhere near what's needed. In 2009 the Climate Change Performance Index left the top three places in its annual ranking of

- Apollo Alliance: apolloalliance.org
- BC Community Energy and Emissions Inventory: env.gov.bc.ca/epd/climate/ceei
- Climate Change Performance Index: germanwatch.org/klima/ccpi.htm
- Climate Neutral Network: climateneutral.unep.org
- CO_2 Reduction Targets: pewclimate.org/what_s_being_done/targets
- F-Gas Phase Out: mipiggs.org/govaction
- Green New Deal: neweconomics.org
- Repower America: repoweramerica.org
- State Action Maps: nextgenerationearth.org/maps
- The Big Ask: thebigask.eu
- The Climate Registry: theclimateregistry.org
- The Next Industrial Revolution: thenextindustrialrevolution.org
- *Plan B, 3.0: Mobilizing to Save Civilization:* earth-policy.org

the world's nations empty, to emphasize this underperformance. Scotland is chasing 42% below the 1990 level by 2020. Germany's goal is 40% below. Britain is aiming at 30%. The European Union's overall goal is 20%. Sweden is aiming at 40% below 1990 by 2020, and to be 100% carbon neutral by 2050. The US goal, through the American Climate and Energy Security Act, is 4-7% below 1990.

In North America, Ontario is pursuing a 15% reduction, and British Columbia,[1] Oregon, eight US Northeast states and Canada's eastern provinces have committed to 10% below 1990 (Minnesota 7%[2]). Other US states, including Washington and California, are aiming only to reduce their emissions to the 1990 level by 2020. In 2009, after 20 years of warnings from climate scientists, 33 US states and six Canadian provinces still had no reduction targets, and 14 states had no climate action plans.[3]

This is not good. Most of the commitments are neither sufficient to keep Earth out of the extreme climate danger zone nor fast enough to replace oil and gas before global scarcity sets in.

Learn from World War II

We can achieve remarkable things when the public will is there. During World War II, when a house cost $6,950 and a loaf of bread 9 cents, the US spent a third of its $1,235 billion GDP in a shared effort to save humanity from the scourge of Nazism — and this on the heels of the Great Depression, when the US public debt had already risen to 45% of its GDP. As well as ensuring victory, the investment laid the foundation for the unparalleled prosperity of the post-war years.

The Dell-Winston School Solar Car Challenge teaches high school students how to build and safely race road-worthy solar cars. winstonsolar.org

The US's GDP in 2007 was $13.84 trillion. If we were to invest a third to overcome this crisis ($4.6 trillion), it would re-circulate as incomes, jobs and nation-wide improvements. Spread over 25 years, it would be $184 billion a year. In 2007 the US military budget was $440 billion, and Americans spent $150 billion just on advertising. Spending $184 billion a year is $12 per person per week. Using another approach, in 2008 two European researchers calculated that for a 90% chance of limiting the temperature rise to no more than 2°C, we would need to invest 2% of our GDP every year from 2005 to 2100. That's $276 billion a year, or $18 per person per week.[4]

We need to act with far greater vision if we are to prevent this catastrophe and redesign our world for a sustainable, climate-friendly future. Al Gore's campaign to Repower America, with two million supporters, is pushing for 100% clean electricity by 2018. The Apollo Alliance has proposed investing $313 billion over ten years in the US in renewable energy, retooling manufacturing,

Actions

- Appoint top-notch climate science and climate solutions advisers.
- Establish citizen-based climate solutions teams.
- Declare all greenhouse gases as listed pollutants, and require all industries to report their emissions to the Climate Registry. Include imported electricity, flying, shipping and an estimate of imported goods.
- Pass a climate action law with legally binding targets that reflect the urgency of the crisis, with annual goals, budgets and progress reports (25–40% below 1990 by 2020; 80–100% by 2030–2040; rapid carbon drawdown thereafter).
- Require all communities to report their greenhouse gases and to develop targets, policies and actions to reduce them, as the government of British Columbia is doing.
- Establish a single government department to address climate change, with combined responsibility for climate change, energy and transport.
- Set congruent targets for renewable energy, zero-energy buildings, transit, rail, cycling, organic farming, zero waste, tree planting and other solutions.
- Legislate the phase-out of all F-gases, as Denmark, Austria and Ireland have done.[5] (See #97.)
- Require all government ministers and departmental and agency heads to develop cross-sectoral carbon reduction plans and to factor a shadow price for carbon into all policy and investment decisions covering transport, housing, planning and energy, as Britain is doing priced at $51/tonne, rising to $120 by 2050. Include farming, forestry and grasslands decisions.

green buildings, urban renewal, high-speed rail, public transit and a smart grid. The Earth Policy Institute's "Plan B — Cutting Carbon Emissions 80%" by 2020, and Britain's Green New Deal lay out similar sets of solutions.

63

Lead by Example

We must quickly mobilize our civilization with the urgency and resolve that has previously been seen only when nations mobilized for war. These prior struggles for survival were won when leaders found words at the 11th hour that released a mighty surge of courage, hope and readiness to sacrifice for a protracted and mortal challenge.

— Al Gore

© STANICAT DREAMSTIME.COM

If 100 million US rooftops each had a 5 kW solar PV system, they would generate 900 TWh of electricity a year, meeting 23% of the US electrical demand.

What use are big plans if they are not accompanied by big actions? People are hardly going to follow a government if its leaders and bureaucrats drive gas-guzzling cars and leave the office lights on at night.

In British Columbia, every government agency has been told to become carbon neutral by 2010, reducing what it can and buying offsets for the remaining emissions. The government allocated $100 million to fund energy upgrades but no money for carbon offsets. With this kind of pressure, staff quickly found opportunities to carpool, teleconference and turn the lights off. In New Zealand the government wants all its operations to be carbon neutral by 2012.

In California every state department and agency must use 20% less energy by 2015. Every new building must be built to the LEED Silver standard, and all leased space of 465 square meters (5,000 sq. ft) or more must carry an Energy Star rating. In British Columbia, the requirement is for LEED Gold.

Not to be outdone, President Bush signed an executive order in 2007 that required every US federal agency to increase its energy efficiency and reduce its greenhouse gas emissions to 30% below the 1990 level by 2015 — a rare example where the White House was ahead of California.

New York State has gone even further. In 2001, under the Governor's Executive Order 111, "Green and Clean" State Buildings and Vehicles, each of the 14,000 buildings that the state owns, leases or operates must be 35% more efficient by 2010 than it was in 1990. 25% of their electricity must come from renewable sources by 2013, and 100% of all new light-duty vehicles must be alternative fuelled, except for police, emergency and specialty vehicles.

Green Purchasing

With their huge purchasing budgets, governments have an enormous opportunity to show leadership and develop the market for green products. As well as insisting on energy-efficient appliances, they can require carbon-reducing items such as 100% post-consumer waste recycled paper, as the US Congress and many universities are doing.

- California's Best Practices Wiki: bestpractices.ca.gov
- "Environmental Purchasing Guide": richmond.ca/services/Sustainable/environment/policies/purchasing.htm
- Environmentally Preferable Purchasing: epa.gov/opptintr/epp
- Environmentally Preferable Purchasing (California): green.ca.gov/EPP
- Green California: green.ca.gov
- Green Suppliers Network: greensuppliers.gov
- National Governors' Association Center for Best Practices: nga.org
- Responsible Purchasing Network: responsiblepurchasing.org
- State Clean Energy Programs: epa.gov/cleanenergy
- US Communities/Green: gogreencommunities.org
- Working Green (California): workinggreen.dgs.ca.gov

Government Action — Best Practices		
Carbon Footprint	US Federal	30% reduction below 1990 by 2015
	Vermont	50% below 1990 by 2018
Carbon Neutral	BC	100% carbon neutral by 2010
Buildings	New York	35% more efficient than 1990 by 2010
	BC	All new owned or leased buildings to be LEED Gold or equivalent
Electricity	Pennsylvania	50% clean energy by 2010
Fleet	Montana	30 mpg fleet average by 2010
	New York	100% of light-duty vehicles to be alternative fueled by 2010
Vehicle Fuel	California	40% of fuel used to be California produced renewable by 2020
Pension Funds	California	$1.5 billion invested in Green Wave initiatives (out of $250 billion total).
Paper	US Congress	100% post-consumer recycled

They can also require a priority for healthy, locally grown food, as Northumberland County Council is doing in Britain, recycling $3 million a year into local farms and businesses. Governments are always providing food for meetings and conferences. Serving local, organic vegetarian food with organic pasture-raised meat would send a powerful message about meat's carbon footprint, protecting our health and supporting local farmers.

Leadership is essential. Green purchasing needs senior management involvement and clear targets. In New York State, the government's Green Procurement and Agency Sustainability Program promotes green purchasing by all government agencies and provides training for state employees. In Oregon, a Sustainability Supplier Council is developing recommendations for sustainable state purchasing policies.

Think what cooperation could do to bring prices down if governments made collective green purchase agreements for efficient appliances, solar PV and plug-in hybrids — as Austin Energy is doing with its Plug-In Partners. (See #57.) Nationally, the group known as US Communities helps governments pool their purchasing power and acts as a one-stop source for a range of environmentally certified products and services.

Actions

- Set up a green purchasing program.
- Write environmental clauses into all contracts.
- Establish a carbon reduction target for all government operations.
- Require all government operations to become carbon neutral by a set date.
- Set targets for energy efficient appliances and equipment.
- Set energy efficiency standards and LEED design goals for government buildings.
- Buy or generate green energy for all government operations.
- Launch a green fleet program and a low-carbon vehicle fuel standard.
- Launch a green ride program to encourage cycling, transit and ridesharing by government staff.
- Establish an energy efficiency fund to finance the retrofit of government buildings.
- Allow agencies to use energy service companies (ESCOs) for energy saving projects.
- Encourage government pension funds to invest in green businesses.
- Link up with other governments to make bulk purchases of green supplies.

64

Get Everyone Engaged

Each and every one of us should stop playing small and license ourselves to become one of the giants of the new century. We will need champions by the truckload.

— Van Jones

Opinion polls show that although most people are concerned about global warming, they are still confused by the science and don't know what to do once they've changed the light bulbs and wondered about buying an electric car. We need to get everyone engaged as we did in World War II, when almost everyone volunteered to serve either in the armed forces or on the home front.

Bring in the Public

The non-profit world is doing an amazing job with its concerts, teach-ins, walks, bike-rides, demonstrations and study groups (see #11–#20), but they reach only a fraction of the public. How can we reach *everyone*? There are three keys that might unlock a wave of engagement.

The first is that we must frame what's happening not as "Change the light bulbs and try not to despair" but as "This is amazing: we are pioneering a new, climate-friendly world." Fear without hope is paralyzing. Back this up with prime-time

A Spitfire and a Flying Fortress. During World War II, ordinary people raised the money to pay for planes like these.

TV spots, as the UK government did with its *Today's Climate is Tomorrow's Challenge* video.

Secondly, we need to bring new community leaders out of the woodwork, as happened in World War II, when ordinary heroes appeared everywhere. Because we can't identify them in advance, we need a willingness by governments to fund different community and public engagement programs until the winning approaches emerge. In World War II Britain, it was the Women's Voluntary Service (see #22). Today, it might be the Climate Action Circles (see #11), Britain's Green Neighbourhoods initiative, or something no one has dreamed up yet. Once a winner has emerged, we should roll it out to every household across the nation.

Thirdly, we must establish green bonds and public share offerings so that people can contribute financially, as they did in World War II. In 1940, when the British government asked people to buy savings certificates to help pay for the war, the small town of Marple in Cheshire raised £128,360 toward the cost of a Lancaster Bomber, four Spitfires and four Hurricanes. In total, the bonds raised £1,754 million for the war effort — 3% of Britain's GDP.

Bring in the Schools

Twenty years after the climate alarm bells first rang in 1988, there are still very few schools where climate change is being taught in the curriculum. We *must* bring it into the classroom, framed not in grim forecasts but with action learning and a passion for the solutions. While curriculum is being developed, we must fund schools to bring in speakers and resources, so that

Actions

- Set up a major fund for non-profit public engagement activities.
- Integrate climate change into the school curriculum.
- Provide funding for climate speakers and resources in schools.
- Advertise! Learn from Britain's Act on CO_2 campaign, and 1010uk.org
- Establish green bonds, so that the public can contribute financially.
- Require every business to report its CO_2 emissions in its annual accounts.
- Require every business to submit a costed plan for CO_2 reductions and to implement every measure with a payback under three years.
- Invite businesses to join sectoral task forces.
- Apply green strings to all government grants and funds.

- Act on CO_2: campaigns2.direct.gov.uk/actonco2
- Carbon Trust: carbontrust.co.uk
- Fostering Sustainable Behaviour: cbsm.com
- Green 500, London: green500.co.uk
- *New Rules: New Game:* futerra.co.uk/downloads/NewRules:New%20Game.pdf
- Seattle Climate Partnership: seattle.gov/climate/partnership.htm
- Ten Tips for Sustainability Communications: futerra.co.uk/downloads/10-Rules.pdf
- The Rules of the Game: futerra.co.uk
- *Today's Climate is Tomorrow's Challenge:* tinyurl.com/buyypx
- Tools of Change — Proven Methods of Promoting Health, Safety and Environmental Citizenship: toolsofchange.com

time is not lost. It was children who mobilized their parents to recycle in the 1980s, and children who mobilized the village of Ashton Hayes, England, to such great efforts. (See #16.) Schools must also be included in programs like Britain's Carbon Reduction Commitment. (See #79.)

Bring in the Business Community

Some great things are happening, but very few businesses are involved. To get them all engaged, every business must be required to integrate CO_2 into its annual accounting, with write-offs and incentives for footprint-reducing investments. Businesses over a certain size must be required to submit costed plans for CO_2 reduction and to implement every measure with a payback under three years, or be taxed on the capital that would have been used.

A climate solutions loan fund should be established to help with the investments needed. Initiatives such as Seattle's Climate Partnership,

London's Green 500 and Britain's Carbon Trust, which advise organizations on ways to reduce their emissions, should be funded to expand their work. Business leaders should be invited to participate in sectoral task forces to develop programs and measures for further emissions reductions, as British Columbia has done.

Apply Green Strings to all Government Funding

To widen the process of engagement, whenever the government cuts a check to a person, business, city, university or non-profit society, a percentage of the money should come with green strings attached, requiring evidence of CO_2 reduction planning before it is signed (also known as carbon-conditionality), as the Green Party of Canada and Greg LeRoy, executive director of Good Jobs First, have proposed.[1]

All these initiatives should be approached with a positive "Build a climate-friendly future" approach, alongside "Avoid disaster." We need to sing together as we work to change the world.

65

Inspire People

> Humans are motivated by syntropic dreams of life and wholeness, not by entropic nightmares of death and collapse.
>
> — Guy Dauncey

How can we motivate people to join the challenge of building a climate-friendly world? When it happens, the results can be amazing. (See #11 to #20.) For every household, business, church or school that has become engaged, however, there are 1,000 that have not. We have yet to find the formula that can motivate large numbers of people.

It is all about the "framing." After 20 years of communications about global warming, we have learned what does *not* work, but we are slow at discovering what *does* work.

Messages that dwell on scientific arguments, call for personal sacrifice; pass the responsibility to industry or government; emphasize complex policy solutions such as cap-and-trade or high-tech solutions such as nuclear or clean coal; dwell on awful warnings about the future; urge us to change our light bulbs or ride our bicycles; or show us pictures of polar bears and remote islanders who are losing their homes because of sea level rise — none of these messages works.

Some succeed in making people distressed, but none makes people feel motivated and confident that their actions will make a difference. The whole climate-change discussion has been framed in the negative, rather than as a positive challenge to get engaged with the solutions.

So what *does* work? There is one particular framing that works both for climate change and the peak oil energy crisis. It talks in historical terms about a global transition to a green, post-carbon future and the birth of a new civilizational era. It conveys five core messages:

1. Fossil fuels have been amazing.

If we had not discovered the power of fossil fuels, we could not have harnessed steam, built railways, learned how to fly or built the world we know today. Fossil fuels are a gift from the past that enabled us to build the scientific and technological capital needed to design solar systems, electric cars and geothermal heat pumps. They are the launch ramp that will enable us to achieve lift-off into the solar age of permanent renewable energy.

2. The age of fossil fuels is almost over.

Regardless of whether oil peaks in 2010 or 2030, it *will* peak, followed by a rapid decline, and likewise for natural gas. The age of fossil fuels is a very brief, 200-year period of history, sandwiched between the past and the future. In the journey of human civilization, it is a stepping-stone between the pre-carbon past and the post-carbon future.

3. Climate change is Gaia's message, saying "Time's Up!"

Fossil fuels carry the solar energy gathered by ancient trees, plants and sea creatures over a period of 200 million years. When you release that ancient carbon into the atmosphere all at once, it is bound to have a destabilizing effect. If we don't stop almost immediately, the rising temperature will put us in extreme peril.

4. A climate-friendly world is doable, desirable and will generate a gazillion new jobs.

What makes us doubt that we can do this? We are *good* at this kind of thing. A hundred years ago we were all riding horses. The sun, wind, tides,

Hoping for a green, peaceful, sustainable Earth, with many new green collar jobs.

© MGIVIAN DREAMSTIME.COM

biomass, geothermal energy and energy efficiency offer us many thousand times more energy than we need. The zero-carbon revolution will bring innovation and a host of green-collar jobs that can stabilize the middle class and lift people out of poverty. If we stick with fossil fuels, our economies will collapse because of rising prices and the impacts of climate change. If we change, a whole new world will become possible. The carbon years are the launch ramp for the journey to the solar age.

5. The Future Needs YOU!

Historically, when changes like this happen, they are led by the pioneering efforts of individuals and groups. We need to reinvent our buildings, suburbs, cities, transportation, farming, forestry, industry — everything that uses energy. There are so many opportunities. Those who succeed will be those who can combine a good idea with the skills and the determination to see it through. Think big. Dream big. Act big!

In Britain the Big Green Challenge, funded by the government agency NESTA, has challenged groups to achieve big CO_2 reductions in their communities, putting £1 million on the table for the winning proposal. Of the 350 groups entered, ten were each given £20,000 to test their ideas during 2009. (biggreenchallenge.org.uk; community.wwf.ca/livingplanetcity)

- Continue to pass on the urgent warnings that the climate scientists are giving us and the warnings about the declining oil and gas supply.

- Reframe the government's climate messaging in a way that combines the very real urgency with a positive vision of a climate-friendly world, placing the shift in a historical context.

- Don't be shy to present this message to the public. If you don't frame it this way, the climate deniers will frame it their way.

- Support exciting challenges that draw out people's creativity.

66

Build a Super - Efficient Nation

> The generations living today get to retrofit, reboot, and reenergize a nation. We get to rescue and reinvent the US economy.
>
> — Van Jones

We're smart, bright and innovative, but compared to where we'll be ten thousand years from now, we're just beginners in the world of technology. We also have a deep tendency to conservatism, especially when comfortable.

The incandescent light bulb was invented 200 years ago and fine-tuned by Thomas Edison 130 years ago — yet many people still use it, even though 90% of its energy is wasted as heat. The same goes for most fridges, electric motors and other appliances. We pay good money to power this inefficiency. So what does it take to become a super-efficient nation?

In 2008, when an avalanche cut the power lines in Alaska, consumers achieved a 30% reduction through their "Turn off, turn down, unplug" campaign. In California, thanks to a decades-long emphasis on efficiency, the average household power use since 1974 has increased only from 7,000 to 8,000 kWh/year, compared to 14,000 kWh in the rest of the US. The European Union has a goal for all member nations to be 20% more efficient by 2020, saving $130 billion a year.

Our goal should be a 33% reduction in energy use by 2030, eliminating the forecast increase in demand and going further. Don't be shy of electricity prices rises: the states and provinces with the lowest price (Alabama, British Columbia, Kentucky, Wyoming) use the most energy. If the income from a 15% price-rise is used to help consumers reduce their use by 33%, we'll all be better off. The American Council for an Energy-Efficient Economy has calculated that if US utilities were required to reduce their electricity demand by 15% and their natural gas demand by 10% by 2020, the nation would enjoy bill-savings of $169 billion a year.[1]

These are the 10 key policies that are needed, plus action on buildings. (See #67.)

1. Write a National Energy Plan that makes efficiency the top priority, as Italy has done since 1991. Require all gas and electric utilities to practice integrated resource planning, and legislate a 2% annual increase in energy efficiency (33% by 2030), with four-year goals and sanctions for non-compliance. In 2007 Vermont reduced its electricity use by nearly 2% and was likely to exceed 2% in 2008. Several studies show that 33% savings by 2030 is feasible and cost-effective.[2] Give regulatory utility boards the freedom to innovate. Allow them to make large investments in efficiency, such as the California Public Utilities Commission's $2 billion, and BC Hydro's $500 million in British Columbia.

2. Adopt "decoupling" legislation and incentives that enable private gas and electric utilities to

Smart meters enable people to control their power consumption and reduce their costs.

PICTURES OF THE FUTURE, SIEMENS AG

- ACEEE state energy policies: aceee.org/energy/state
- Alliance to Save Energy: ase.org
- California Center on Energy Efficiency: californiaenergyefficiency.com
- *Curbing Global Energy Demand — The Energy Productivity Opportunity:* mckinsey.com/mgi/publications/ Curbing_Global_Energy/index.asp
- Database of State Incentives: dsireusa.org
- Japan's Top Runner Program: eccj.or.jp/top_runner
- National Action Plan for Energy Efficiency: epa.gov/eeactionplan
- State Energy Program: eere.energy.gov/state_energy_program
- Utility Decoupling: tinyurl.com/4adqmy

earn a profit while selling less power, as California has done.

3. Write national appliance standards based on Japan's Top Runner program, which takes the most efficient appliance in a category and requires all others to match it. Require all major appliances to be built with smart chips that allow for demand response. Make energy-use labeling mandatory for everything that uses energy, with an easy five-star system. Limit stand-by power use to 1 watt per device.

4. Phase out the incandescent light bulb, as Australia (2010), Britain (2011), Canada (2012), and the US (2014) are doing. Extend the phase-out to other inefficient lights such as low-efficiency fluorescent tubes and mercury discharge lamps used for street lighting. Ban clothesline bans. Make mass bulk purchases of efficient appliances and light bulbs to reduce the price.

5. Allow utilities to pay up to 10 cents/kWh for saved energy. Most saved power costs only 2 cents/kWh, so this will allow far deeper penetration while still averaging less than the price of new power.[3] Create a market for certified energy efficiency savings, monetized as "white certificates" or "white tags," as the UK, Italy, France, Connecticut, Pennsylvania and Nevada are doing.

6. Create a low- or zero-interest energy efficiency loan fund that can be used by governments, industry, organizations, schools and consumers to finance efficiency investments, as Northern Ireland is doing in partnership with

the Carbon Trust. Allow paybacks of up to 20 years.

7. Create a 100% tax deduction for investments in energy efficiency by those on moderate incomes (50% for the rich) and for the interest on associated loans.

8. Establish a public benefits fund that provides 50:50 matching grants for state or provincial benefit funds that support efficiency programs through a small rider on utility bills.

9. Adopt tiered-rate pricing, time-of-use pricing, pay-as-you-go and smart metering, creating price signals that persuade consumers to save energy. Install meters and controls that give customers information in real time, so that they can adjust their usage and see an immediate impact.

10. Work with industry. Create energy-saving agreements with big consumers, as China is doing. Provide tax credits for manufacturers who improve their efficiency. Invest in leading-edge efficiency research. Establish organizations such as Britain's Carbon Trust to help industries save energy and money while reducing their carbon footprints.

67

Make All Your Buildings Zero Carbon

The road to energy independence, economic recovery and greenhouse gas reductions runs through the building sector.

— Ed Mazria

In the beginning, we lived in caves and grass huts. Later we built houses from stone, brick and timber, heated with wood. Today we heat them with coal, oil and gas, pouring CO_2 into the atmosphere. They are often very inefficient — buildings consume 48% of the US energy and 76% of its electricity.[1]

Viewed today, this seems like an enormous challenge. Viewed from a future where all buildings are smart and super-efficient, powered by the sun and earth, it's a natural transition.

1. Require all New Buildings to be Zero Carbon by 2016

There are zero-carbon buildings going up today, using solar orientation, super-insulation, heat-exchange, solar energy, biomass cogeneration and other well-accepted green building techniques. Larger developments are being heated and powered with district heating, using energy from biomass, solar and sewage. The resistance comes from the building industry, not the public or the architects. The solution is political determination coupled with support, incentives and training for the industry. In Britain all new buildings must be zero carbon by 2016. Wales wants this by 2011. In France new buildings must produce more energy than they use by 2020.

In the US, architect Ed Mazria has created The 2030 Challenge, calling for all new and renovated buildings to be carbon neutral by 2030 (60% by 2010, 80% by 2020). The US Conference of Mayors, Illinois, Massachusetts and the US government (federal buildings) have all signed onto the goal. In California, all new residential buildings must be net zero energy by 2020, commercial buildings by 2030.

2. Adopt Green Building Codes

The 2030 Challenge has a Building Codes Guide that city governments can tack onto their existing codes, allowing the drive for carbon reductions to get underway immediately. (See #26.)

3. Adopt Energy Labeling

In Britain the owner of every house for sale must provide a Home Information Pack, including an Energy Performance Certificate completed by an accredited energy assessor.

4. Plan Zero-Carbon Towns

The British government is planning 10 new Eco Towns of up to 15,000 zero-carbon homes. This helps the building industry get ready and sets the standard for all new developments. Governments should require all new subdivisions of ten or more houses to be zero-carbon by 2016 (50% by 2012).

- Architecture 2030: architecture2030.org
- Britain's Eco Towns: tcpa.org.uk/ecotowns.asp
- Energy Efficient Codes Coalition: thirtypercentsolution.org
- Home Information Packs: homeinformationpacks.gov.uk
- Passive Housing in Europe: tinyurl.com/66j6f2
- US Weatherization Assistance Program: eere.energy.gov/weatherization
- Warmfront: warmfront.co.uk

In the Drake Landing solar community in Okotoks, Alberta, 90% of the heat for the 52 homes comes from solar thermal energy collected in summer from 800 flat plate solar panels, which is stored underground in boreholes and used to heat the homes in winter. dlsc.ca

5. Adopt Green Heat Policies

In Germany all new homes must use renewable energy to meet 14% of their heating needs (20% in Baden Wurttemberg). In Spain, Israel and Hawaii, all new buildings must include solar hot water systems.

6. Retrofit All Existing Buildings by 2030

Existing buildings make up 99% of the supply. One solution is to apply San Francisco's point-of-sale energy upgrade requirement (see #26) nationwide. Germany is paying $2.25 billion a year for the systematic upgrade of all pre-1978 buildings, tackling 5% a year for 20 years. Encourage all building owners to invest in energy-saving measures by offering low- or zero-interest loans. In Britain, where every house in the country is to have a green makeover by 2030 (25% by 2020), six million low-income homes will get free insulation and five million will receive half-price energy saving measures.

7. Adopt Low Income Policies

The best approaches are laid out in #53. The task for governments is to ensure that cities and utilities carry them out, assisted by programs such as Britain's Warmfront and the US's Weatherization Assistance Program, which reduces heating bills by 32% and overall energy bills by $358 per year.

8. Solve the Financing Problem

Germany provides subsidized loans for low energy buildings. New York, Maryland and Massachusetts provide tax credits for green buildings. Provide zero-interest or tax-deductible loans for energy upgrades and tax incentives for small buildings, and pay-as-you-save loans attached to utility meters.

9. Support the Building Professionals

Colleges need to provide zero-carbon training for building trades and building inspectors, who need to enforce the new codes rigorously. Building companies and developers need tax breaks for investments in training and for building zero-carbon homes.

10. Teach Your Leaders Well

... and feed them on your dreams. Send your staff and political leaders on a tour of US and European eco-housing to get them inspired.

Architecture 2030 has a 2030 Challenge Stimulus Plan which involves spending $96 billion a year for two years to lower people's mortgage rates in return for their improving the energy efficiency of their homes to 75% below code, the cost of which is more than covered through the reduced mortgage payments. The Plan would create nine million new jobs in two years, and reduce CO_2 emissions by 505 million tonnes.

68

Make All Your Cities Green

We must rebuild our cities around energy efficiency and human needs, rather than the car and wasted energy.

— Jay Inslee and Bracken Hendricks, authors of *Apollo's Fire: Igniting America's Clean Energy Revolution*

Our cities are where most of us live, work, travel and leave our carbon footprints. Some cities are showing great leadership (see #21–#30), but when we look beyond the occasional green initiative, the vast majority of the world's cities and towns are following a business-as-usual track, with only the occasional nod toward the need for action.

Some cities will be hit in a devastating manner by the one- to two-meter sea level rise that is possible by the end of the century if we don't get our act together. Some will be hit by heat waves, water shortages and new diseases. But *all* will be hit by peak oil and the ensuing energy crisis, as first oil and then natural gas become so expensive that people will no longer be able to afford to drive cars or heat their homes. If our cities do not storm-proof themselves against this, businesses will go bankrupt, suburbs will be abandoned and there will be food riots in many neighborhoods.

What can governments do to persuade cities and towns to abandon their traditional business-

as-usual attitudes and get on the fast track to becoming sustainable?

The reward for success looks very enticing, if we look to cities like Portland and Copenhagen for direction. Back in the 1970s, Portland broke the mold when it decided to spend the federal money allotted for freeways on public transit and urban development. Today many of Portland's citizens enjoy quiet electric light-rail transit, cycle paths, relatively dense urban neighborhoods where you can walk to local restaurants, and a thriving economy with skilled workers who are attracted to the city's new urban ambience. Nearly 25% of all trips in Portland are walking trips, and Portland was the first city in the US to reduce its vehicle-miles driven per person each year instead of increasing it. It is also the only city in the US that has been able to reduce its carbon footprint below the 1990 level. It has a long way to go, but it's on the right track.

A climate-friendly city will be more neighborly, enjoy a rich culture, and have a stronger economy than it did before, stimulated by the wave of new green-collar jobs. It will look and feel more like great cities of the past but with the blessings of clean water, clean streets, clean energy, abundant trees, advanced resource-recovery sewage treatment, electric transport, a public plaza in every neighborhood, glorious public parks, thriving arts, green buildings and community democracy. Here are some of the policies that governments can use to get their cities onto the green track:

Climate Solutions

- Work with the Clinton Climate Initiative's C40 Cities Program to create a common

Trainee Mario Francis works at a South Bronx garden in Sustainable South Bronx's BEST program. ssbx.org

GREEN JOBS NOW

- BC Climate Action Charter: livesmartbc.ca/community/charter.html
- C40 Cities: c40cities.org
- Canada's Green Municipal Fund: sustainablecommunities.fcm.ca/GMF
- Washington's Commute Trip Reduction Law: seattle.gov/Transportation/commute.htm

measurement tool that cities can use to calculate their emissions, track their reductions and share best practices.

- Require all cities and regions to reduce their greenhouse gas emissions, and give them the technical and financial assistance to do so.
- Provide funding for cities to hire sustainability staff and form climate action teams.
- Invite all communities to sign onto a climate action charter to reduce their emissions and adopt sustainable policies, as BC has done.
- Tie all urban grants to evidence of carbon reduction progress.
- Work with cities to make bulk purchases of solar panels, electric vehicles, etc.
- Give annual awards to cities that make the most progress.

Transportation Solutions

- Transfer most freeway funding into urban transit, light rail and cycling.
- Change tax codes to encourage low- and zero-carbon travel.
- Adopt Washington State's Commute Trip Reduction Law, requiring employers to develop trip reduction plans.
- Encourage congestion charges and road pricing.
- Provide funding for cities to develop green freight plans, as Amsterdam is doing.
- Provide 50:50 funding to cities that invest in electric vehicle charging networks, as San Francisco is doing. (See #24.)

- Apply Boulder's transit rules to all transit systems. (See #25.)

Smart Growth Solutions

- Require all new developments to adopt transit-oriented, smart growth designs, and better land-use planning to curb sprawl, as California has done.
- Require cities to adopt urban growth boundaries, as Oregon, British Columbia and most European regions do.
- Encourage the adoption of model bylaws/ordinances that remove the barriers to smart growth, green buildings, renewable energy and urban farms.

Energy and Economy Solutions

- Empower cities to develop local renewable energy, as Sweden does.
- Convert urban brownfield sites and depressed neighborhoods into green communities.
- Invest in green infrastructure renewal, as Canada is doing with its Green Municipal Fund.
- Fund a huge green-jobs training program, including urban youth and released prisoners. (See #77.).
- Integrate all this into a green-cities program to help cities become sustainable, protect themselves against the peak oil crisis and kickstart a green urban renaissance.

69

Plan for 100% Renewable Electricity

Nuclear power and fossil fuels are the choices of the past. Renewable energy is the choice of the future that is here today.
— Dr. Herman Scheer, European Member of Parliament, author of *The Solar Economy*

We need to achieve a global shift to renewable electricity as quickly as possible, closing down the coal- and gas-fired power plants and assisting their workers through community economic restoration.

In 2008, after consulting climate solutions experts, Al Gore's Repower America campaign, with two million supporters, called for the US to achieve 100% clean electricity within ten years.

SIEMENS AG

The offshore wind farm Lillgrund in the Øresund, between Malmö and Copenhagen. 48 2.3 MW turbines supply power for 60,000 Swedish homes. Built by Siemens AG for the Swedish utility Vattenfall.

What's the Best Policy?

In North America the policies used so far have been the Renewable Portfolio Standard, which requires utilities to produce a percentage of their electricity from renewables by a certain date, along with tax credits, loans, production incentives and net metering, which allows homeowners who generate power to run their meters backwards. These have sometimes achieved good results, as in Texas, but they are nowhere near as effective as Europe's electricity feed laws, also known as feed-in tariffs or renewable energy payments.

Developed in Denmark in the early 1980s, adopted by Germany in 1991, and widely emulated throughout Europe and elsewhere, electricity feed laws address the core problem of renewable energy, which is that solar, wind and biogas are usually more expensive up front, because of capital costs. The feed laws eliminate this hurdle by giving every farmer, cooperative or private producer of renewable energy guaranteed priority access to the grid and a fixed 20-year contract at the price needed for investors to make a decent return.

As with a conventional power plant, the price is set by what it costs to install, operate and maintain the plant, varying by the type of technology, and size of installation. A Bavarian farmer earns more for solar on a barn roof (75 cents/kWh) than on the ground (57 cents/kWh). Wind earns less for a very windy (8 cents/kWh) than for a less windy site (13 cents/kWh).

The results have been very persuasive. In 2007, Germany installed more solar PV systems than the US has installed in its entire history. The renewable energy revolution has propelled

Actions

- Join the Alliance for Renewable Energy and the International Renewable Energy Agency.
- Adopt Europe's feed laws as rapidly as possible.
- Require regulatory bodies to give the highest priority to 100% renewable energy.
- Work systematically to list and eliminate all barriers to renewable energy and heat.
- Establish a federal renewable energy authority, similar to the Tennessee Valley Authority, with the power to commission and operate projects and build the needed grid.
- Lease sections of the sea-bed to tidal and wave energy companies, to facilitate development.

- Alliance for Renewable Energy: allianceforrenewableenergy.org
- Clean Energy States Alliance: cleanenergystates.org
- *Feed-in Tariffs in America — Driving the Economy with Renewable Energy Policy that Works:* boell.de/climate-transatlantic/index-129.html
- Feed Laws: wind-works.org/articles/feed_laws.html
- Green Power Network: eere.energy.gov/greenpower
- International Renewable Energy Agency: irena.org
- Interstate Renewable Energy Council: irecusa.org
- Policy Action on Climate Toolkit: onlinepact.org
- Ontario Green Energy Act: greenenergyact.ca
- Renewable Energy Policy Network: ren21.net
- Repower America: repoweramerica.org
- State Incentives for Renewables and Efficiency: dsireusa.org

Germany to rapid economic growth, generating 240,000 jobs and 4–5% of Germany's GDP in 2007. Germany has made equally impressive gains in biogas and solar hot water.[1]

The tariffs are paid by a small surcharge on everyone's electricity bill, averaging 1 cent/kWh, or $17 to $27 a year for the average household. For this small an investment, Germany is unleashing a third industrial revolution. By 2009 similar feed laws had been adopted by France, Italy, Spain, China and 36 other jurisdictions. In North America, they had been adopted by Ontario and in a limited form by Washington State, Hawaii and Minnesota, and things look hopeful in Illinois, Michigan, California and Wisconsin.

In 2008 the accounting firm Ernst and Young concluded that Germany's feed laws had delivered five times more renewable energy for 20% lower cost than Britain's market-friendly policies, such as tendering and certificate trading systems.[2]

Governments must also eliminate the numerous barriers to wind, solar, tidal and biogas installations. Germany's nuclear and coal industries fought the feed laws and continue to do so, accompanied by power utilities that are wedded to the old ways. The key to a breakthrough is building strong partnerships so that governments have the support they need to adopt well-designed feed laws and achieve the goal of 100% renewable electricity.

Renewable Electricity Goals for 2020[3]		
	2005	**2020**
New Zealand	60%	100%
Sweden	40%	49%
Ireland	3%	40%
California	10%	33%
Portugal	20%	31%
Denmark	17%	30%
Germany	10%	27%[4]
France	10%	23%
15 US States	Varies	20%
European Union	8%	20%
China	7%	15%
UK	1%	15%
US	7%	15%

70

Build a Supergrid

If the political framework is put into place now, we have the technical abilities to build such a super grid within twenty years. This would not cause any noteworthy economic problems, we just need to commit to this big long-term strategy.
- Gregor Czisch, University of Kassel, Germany[1]

With so much renewable energy available at falling prices, we come to the next great challenge: the need to expand the grid to accommodate the new sources of power.

Supergrid Europe/Middle East

The vision of a renewable energy supergrid has emerged most clearly in Europe, which has enormous solar potential in Spain, North Africa and the Middle East; enormous wind potential in Egypt, Morocco and offshore from Ireland to the Baltic Sea; enormous geothermal potential in Spain, Italy, Germany and Turkey; and enormous hydro potential in Norway that can also be used for storage.

Two versions of the supergrid have emerged, both using high-voltage DC (HVDC) lines, which lose three times less energy over long distances than AC, cost less and can transmit power both ways.[2] Superconducting cables that can conduct large quantities of electricity with virtually no line losses may be practical for local distribution.

The Trans-Mediterranean Renewable Energy Cooperation (TREC) has proposed a land-based network, to bring North Africa's solar potential to Europe, while the Irish wind company Airtricity and the Swiss engineering giant ABB have proposed an undersea network, bringing the enormous offshore wind energy to land.

The two proposals have been merged by Gregor Czisch, an energy systems modeling expert at the University of Kassel, Germany. Including the cost of new power lines, he estimates that the project would cost about $80 billion and deliver energy to Europeans for 7.3 cents/kWh, meeting the energy needs of 1.1 billion people in 50 countries with 100% renewable electricity.[3] It would eliminate Europe's annual 1.25 Gt of CO_2 emissions from electricity generation and enable all electric transport to be CO_2 free.

The merit of this is that when you combine sun, wind, hydro and geothermal in a large, interconnected grid, the problems of baseload power and power storage can be handled by the variety of sources available. The challenges are mostly political, needing support from the 27-member European Union and the North African nations and overcome the power of the coal and gas lobbies, but the process is already underway.

Supergrid North America

In North America, solar thermal, wind and geothermal could each provide 100% of North America's power needs, so linking the renewable resource areas by coastal and overland HVDC power lines makes sense. The Western Renewable Energy Zones project, led by the Western Governors' Association, is mapping renewable energy zones in the US and Canada, to facilitate grid planning and permitting. The regulatory and jurisdictional

- Airtricity Supergrid: tinyurl.com/5t7be8
- Global Energy Network Institute: geni.org
- HVDC Cables: abb.com/hvdc
- Supergrid for Renewables: transnational-renewables.org
- Trans-Mediterranean Renewable Energy Cooperation: desertec.org
- TREC-UK: trec-uk.org.uk

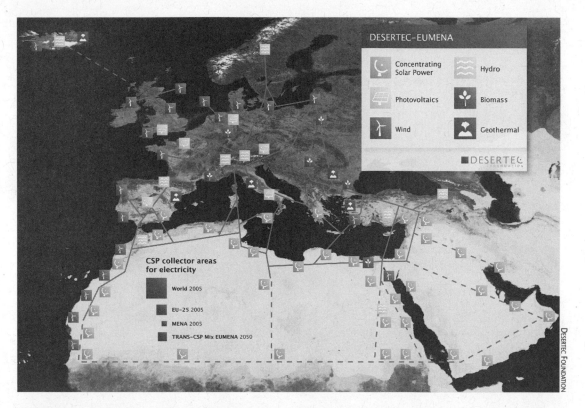

DESERTEC–EUMENA

Concentrating Solar Power

Hydro

Photovoltaics

Biomass

Wind

Geothermal

DESERTEC FOUNDATION

CSP collector areas
for electricity

World 2005

EU–25 2005

MENA 2005

TRANS–CSP Mix EUMENA 2050

DESERTEC FOUNDATION

difficulties that have troubled previous attempts to build a new grid might be lessened by the scale of the vision, which provides a reliable source of renewable power that will never run out. A single publicly owned grid, financed by a small wires charge on every utility bill, would make the most sense, similar to the publicly owned national highway system. The cost in the US has been estimated at $400 billion over ten years.[4]

Supergrid Germany

On a smaller scale, scientists at the University of Kassell have demonstrated that Germany could provide 100% of its energy from renewables by linking 10,000 localized biogas, wind, solar and hydropower plants. When combined, these smaller decentralized grids could meet the needs of the nation-wide fluctuating grid, using biogas and geothermal energy to provide baseload power.[5]

A possible sustainable supergrid for Europe, the Middle East and North Africa.

Actions

- Build a clear picture of a unified, national supergrid, and analyze its advantages, costs and obstacles.
- Build political support for a single, publicly owned supergrid with smart grid capacities into it, enabling energy efficiency functions to operate.

The implications of this big-picture thinking are enormous. In every region of the world, climate activists, renewable energy companies, utilities and political leaders need to work together to realize the vision.

71

Phase Out All Fossil Fuels

> There is something unbelievable about the world spending hundreds of billions of dollars annually to subsidize its own destruction.
> — Earth Council[8]

Sweden is aiming to end its dependency on oil by 2030. Germany will to cease all production of hard coal by 2018. The Swedish region of Kalmar aims to end all fossil fuel use by 2030. This is the future we must embrace, before the impact of fossil fuels overwhelms our planet.

Close Down All Coal-Fired Power Plants

In Britain in 2008, a parliamentary committee called on the government to set a deadline to close down all coal-fired power plants and warned against allowing coal-fired power plants that claimed to be "CCS ready" because of the immaturity of carbon capture technologies. (See p. 60.)

A 7 kW "solar tree" in Gleisdorf, Austria's Solar City.

In New Zealand and British Columbia new coal-fired plants are no longer allowed unless they capture their carbon emissions. In California, utilities can no longer import electricity that produces more CO_2 than a modern gas-fired power plant. In 2008 governor Kathleen Sebelius of Kansas vetoed the development of a new coal-fired plant, hopefully signaling the beginning of the end for the US coal industry.

In 2008 a British jury ruled that the threat of global warming was such that six protesters were justified in doing $70,000 of damage to the Kingsnorth coal-fired power plant in Kent by painting a slogan on its smokestack, using the argument that it is considered legal to break into a house to put out a fire.

Does the coal industry have any alternatives? One possibility is to convert coal-fired plants to biomass combustion combined with carbon capture, producing carbon-negative energy. The Belgian utility Electrabel has retrofitted an 80 MW plant to burn 100% biomass from locally produced pelleted wood, reducing its CO_2 emissions by 500,000 tonnes a year. In England, the Drax Group is building a 400 MW biomass-fired plant in Yorkshire that will burn biomass from a variety of sources, including high-energy crops such as willow. Another option is to diversity into renewables.

End All Fossil-Fuel Subsidies

Behind fossil fuels' global dominance lies the sad reality that most governments still subsidize them with tax breaks and price supports, some dating back to World War I, totalling more than $210 billion a year.[1]

Actions

- By 2020, close down all coal-fired power plants that do not capture their CO_2 emissions, compensating their owners for any stranded assets.
- Establish retraining programs for all affected plant workers and coal-miners.
- Establish renewal programs for affected regions to help them become sustainable.
- Place a moratorium on all new heavy oil sands developments until they are included in a cap-and-trade program.
- Eliminate every fossil fuel subsidy, tax break and incentive.
- Restrict the lobbying registration rights of fossil fuel lobbyists.
- Tax profits from fossil fuel lending at twice and renewables at half the normal rate.
- Reform campaign financing to eliminate the influence of money.
- Collect a windfall oil industry profits tax.

- 700 Mountains: 700mountains.org
- Burning the Future — Coal in America: burningthefuture.com
- Carbon Monitoring for Action: carma.org
- Clean Elections: publicampaign.org
- Clean Energy Action: cleanenergyaction.net
- Coal Moratorium Now: cmnow.org
- Coal-Is-Dirty: coal-is-dirty.com
- Earth Track: earthtrack.net
- End Oil Aid: endoilaid.org
- Exxon's Climate Footprint: foe.co.uk/campaigns/climate/news/exxonmobil_climate_change.html
- Exxon's Greenwash: desmogblog.com/exxons-greenwash
- Moving Beyond Coal: sierraclub.org/coal
- No New Coal: nonewcoal.org.uk
- NRDC Dirty Coal Campaign: nrdc.org/coal
- Oil Change International: priceofoil.org
- Power Past Coal: powerpastcoal.org

The German government has spent more than $200 billion subsidizing its coal industry since the 1960s.[2] Canada gave more than $40 billion between 1970 and 2000 to its fossil fuel industries,[3] including Alberta's tar sands, where oil's carbon footprint is three times greater than conventional oil. In 2006 Canadian subsidies continued at $1.4 billion a year.[4]

In the US from 1992–2002, the oil and gas industry got $26 billion and the coal industry $3 billion in subsidies.[5] In addition, even before the 2003 occupation of Iraq, the US was spending $10 to $20 billion a year defending Middle Eastern oil fields. Globally, the same is happening through World Bank and other agencies, which lent more than $61.3 billion to the oil and gas industry between 2000 and 2007.[6] (See #98.) In 2006 Earth Track estimated that the US oil and gas industry received $39 billion in federal energy subsidies, and the coal industry $8 billion.[7]

Behind the subsidies and tax breaks lies the power of the lobbyists, who continue their lethal game even while the world floods and burns. We should no more allow this than we would allow lobbyists for the enemy during wartime.

End All Corporate Campaign Financing

Behind the lobbyists is the corruption of democracies where political influence can be purchased with campaign donations. We must end this forever. By 2008 Clean Elections guidelines that govern the financing of political campaigns were the law in Arizona, Connecticut, Maine, New Jersey, New Mexico, North Carolina and Vermont. In Canada, only individuals can contribute to a political party's finances, to a limit of $1,100. Instead, parties receive state financing based on their share of the vote at the previous election.

We also need to tax all windfall profits caused by investor speculation (such as the $148 billion that was taken by the five major US oil companies in 2007), returning half to the taxpayer and investing the other half in climate solutions.

72

Stop the Methane Emissions

There is nothing more tragic in the world than to know right, and not to do it.

— Martin Luther King

In the midst of solutions relating to renewable energy and zero-carbon housing, why does methane suddenly appear, like a burp at a garden party, accompanied by its unwanted friend, HFC-134a?

The methane introduction on page 12 is an essential backgrounder. We need a turnaround in greenhouse gas accumulations by 2016, but methane's contribution is being underestimated five-fold due to a quirk in the way global warming potentials (GWP) are measured. The impact of the F-gas HFC-134a is being similarly underestimated.

When we recalibrate GWPs to reflect the impact of greenhouse gases over ten years instead of 100, methane's GWP is 125, not 25, and a new landscape is revealed in which methane reduction assumes far greater significance.

There are 1.3 billion cows in the world, and they each produce 250 to 400 liters of methane gas from their stomachs a day. For solutions, see #43.

Knowing this, what can governments do? We can no longer remain asleep at the switch, assuming methane to be of secondary importance.

Methane emissions from fossil fuels

Ancient buried methane escapes as "fugitive emissions" from coal mines, oil wells, refineries and natural gas pipelines and installations. Most companies underestimate their emissions and do little to capture the escaping methane.

In North America oil refineries use a formula developed by the American Petroleum Institute to estimate their emissions, using technical assumptions and mathematical equations. When the Alberta Research Council measured the actual plume of pollutants escaping from a Canadian oil refinery, they found the methane emissions to be nine times higher than the refinery was reporting to Environment Canada.[1] Because methane's short-term impact is also being underestimated five-fold, its climate impact is 45 times greater than was being reported.

The same problem occurs with gas pipelines, which have an estimated 1.5% leakage rate. With methane's GWP at 125, the escaping methane increases the short-term impact of burning gas by 287%.[2] Methane also escapes when underground pipes are broken by careless back-hoe operators and in industrial accidents, both of which should be taxed to create an incentive to smarten up.

For pipelines, the solution is to tax the methane at an assumed 1.5% escape rate until a gas company can prove otherwise by using laser infrared photo technology to give real time evidence of the emissions, resulting in a tax rebate or increase.

A similar approach should be used for all coal mines, oil wells, coalbed methane facilities and natural gas compression stations.

Methane emissions from farming

The challenge here is twofold, because methane is produced by cattle and other ruminants and by slurry manure lagoons. The solution for the latter is easier, as methane from slurry can be captured and used for heat and electricity, encouraged by a methane tax.

The solution for livestock is more complex because the range of responses is relatively new. (See #43.) A methane tax must be still imposed to send a price signal to farmers and meat-eaters and to accelerate the quest for solutions.

Methane is also produced by rice farming, because flooded paddy fields cause vegetation to break down as methane, not CO_2. Emissions can be reduced by up to 40% by draining paddy fields in mid-season and between winter growing seasons.[3]

Methane from landfills and sewage

Decomposing organic garbage breaks down as methane because it is deprived of oxygen. Landfill gas capture is a mature technology, and it is only cost and indifference that stops operators from using it. If governments impose a methane tax and give tax-inspectors laser infrared cameras to measure emissions, operators will quickly take steps to capture the methane and reduce the flow of organic garbage to landfills by adopting zero waste strategies. (See #28.) The same applies to methane escaping from sewage treatment works.

Actions

- Work with the IPCC to recalibrate the GWPs of all global warming gases to a ten-year time scale.
- Require annual real-time reporting of all methane emissions.
- Phase in a methane tax with varying rates for each source, creating a financial incentive to capture the gas for use as heat, electricity or hydrogen.
- Use the income to help farmers, landfill operators and coal mine operators to reduce their emissions.
- For gas pipelines and compression stations, apply the tax on an assumed 1.5% methane escape rate until proven otherwise.
- Phase out HFC-134a as rapidly as possible.

Methane emissions from power utilities

In 2000 the World Commission on Dams reported that methane is produced by organic matter washed into hydro dams and by rotting vegetation in shallow reservoirs, making the global warming impact of some dams worse than coal-fired power plants. Emissions can vary 500-fold, so assessment is needed for each dam before taxation. Methane is also produced when damp biomass burns incompletely, putting pressure on biomass electricity generators to operate more effectively.

Phase out HFC-134a

The continued production of the coolant HFC-134a is a climatic disaster that merits the most urgent attention. Governments should place an immediate moratorium on all new fluorine industry developments and phase out the production and use of HFC-134a immediately. (See #97.)

73

Reduce the Impact of food and farming

Healing the wounds of the Earth and its people does not require saintliness or a political party, only gumption and persistence.

— Paul Hawken

In 2006, the United Nations report *Livestock's Long Shadow* showed that the livestock industry alone was responsible for 18% of the cause of global warming. (See p. 62.)

Two years later, the Greenpeace report *Cool Farming* found that farming and ranching were responsible for between 17% and 32% of the cause of global warming.[1] A similar study found that food accounted for more than 30% of the European Union's emissions.[2]

That same year, Britain's Food Climate Research Network published *Cooking Up a Storm: Food, Greenhouse Gas Emissions and our Changing Climate*, which found that food and farming produced 18.5% of the Britain's total GHG emissions. Using a wider consumption-oriented approach that included the climate footprint of imported food increased the impact to 24%.[3]

Another report found that 18–20 million tonnes of British food that is wasted every year, including 6–7 million tonnes that is wasted by households, representing 30% of the food people buy.[4] US surveys show that 28% of edible food is wasted by retailers, restaurants and consumers each year, while up to 30 million Americans go hungry.[5] Stopping this would save 18 million tonnes of CO_2 a year, the equivalent of taking a fifth of the US's cars off the road.

If this book allocated its space according to the impact of each cause of global warming, food and agriculture would receive 25 solutions, demonstrating the urgency of the problem.

There are 10 reasons why food and farming have such high emissions:

1. CO_2 from the destruction of rainforests to clear land for cattle and cattle-feed
2. CO_2 from the use of natural gas to make nitrogen fertilizer
3. CO_2 from farm machinery, heated greenhouses and refrigeration
4. CO_2 from the shipping of food and fertilizers across the world
5. Methane emissions from livestock
6. Methane emissions from wasted food rotting in landfills
7. Methane emissions from rice production
8. Nitrous oxide emissions from the application of nitrogen fertilizer
9. Nitrous oxide emissions from manure lagoons
10. The loss of carbon from conventional farm and grassland soils.

- Alternative Farming Systems Information Center: nal.usda.gov/afsic
- Climate and Farming: climateandfarming.org
- Community Alliance with Family Farmers: caff.org
- Food Climate Research Network: fcrn.org.uk
- International Federation of Organic Agriculture Movements: ifoam.org
- No Junk Food: nojunkfood.org
- Organic Farming Research Foundation: ofrf.org
- Sustainable Agriculture Research and Education: sare.org
- Take a Bite Out of Climate Change: takeabite.cc

Alice Waters and students in The Edible Schoolyard, Berkeley, CA. edibleschoolyard.org

Behind these lie four more reasons: the failure of food prices to include the cost of farming's environmental externalities (see #84); the control of the global food industry by a small number of corporations that want to continue business as usual; the widespread belief that a meal must include meat; and the influence that the junk food industry has over our children.

If governments are to achieve their climate reduction goals, they simply must tackle this one, starting with a clear vision of a sustainable future in which farms and ranchlands restore the carbon in their soil and produce bioenergy from their wastes; and they must be included in cap-and-trade schemes, giving them an incentive to reduce their emissions and increase their carbon storage. (See #41–#45.)

Action:

- Establish a sustainable food and farming taskforce, and charge it with developing a strategy to reduce greenhouse gas emissions from food and farming by 80% and achieve 100% organic farming by 2040.

- Place a tax on chemical pesticides and fertilizers, using the income to support new organic farmers.

- Tax methane and nitrous oxide emissions, using the income to help farmers reduce their emissions.

- Establish an agriculture sector carbon market workgroup to develop ways to include farm management in carbon trading, as Washington State has done.

- Use tax incentives to encourage initiatives such as conservation tillage, conservation grazing, tree-planting, riparian restoration, biochar and bioenergy.

- Eliminate the subsidies that support unsustainable farming and livestock production.

- Set up a farmland fund to help states and provinces protect their farmland against development.

- Require agricultural colleges and agencies to teach sustainable climate-safe farming as a condition of funding.

- Require government institutions and schools to provide meat-free menus, as Britain's National Health Service has proposed.

- Ban the sale of junk food in schools, as Latvia, California, Ontario and BC have done.

- Ban all junk food advertising aimed at children, as Sweden has done.

- Establish a meat road map in partnership with the livestock sector, NGOs and the research community to reduce meat's climate impact.

- Write a supermarket climate charter, and offer tax incentives to stores that reduce their carbon emissions, food waste and packaging.

Governments must also help farmers prepare for the disappearance of cheap oil and gas. The food security of entire nations will be at risk if farmers do not shift to organic farming and eliminate their dependency on fossil fuels.

74

Capture Carbon from the Atmosphere

One hundred years ago, visionary political leaders established a system of national forests and parks that are the envy of the world and the treasure of an entire nation. Why not a similar global vision for our generation?

— Eban Goodstein

Before humans started farming, ranching and cutting trees, Earth's lands stored immense quantities of carbon. Grasses, plants and trees sucked CO_2 out of the atmosphere, storing it in their roots, leaves and branches and then in the soil. Some native grassland plants have root systems that go down six meters, storing carbon all the way.

Most of this book's emphasis has been on ways of reducing CO_2, methane and the other greenhouse gases we are adding to the atmosphere. As long as the atmosphere remains overloaded with heat-trapping gases, however, the temperature will continue to rise. We must also draw the carbon overload out of the atmosphere.

As a result of our frenetic consumer lifestyles, we add 10 gigatonnes (Gt) of carbon to the atmosphere every year (36.67 Gt of CO_2). The world's forests and vegetation absorb 1.5 Gt of the additional carbon each year, and the oceans absorb 2.4 Gt, but at the cost of their waters' becoming more acidic and to the detriment of their coral reefs and shellfish. The world's soils, instead of absorbing carbon, have been losing it due to poor farming, ranching and forestry practices. The world's tropical forests, meanwhile, have been working overtime, absorbing 1.3 Gt of the excess carbon we release, pointing to the crucial importance of ending all tropical

deforestation (see #95).[1] The result is a net annual increase of 4.8 Gt of carbon in the atmosphere.

In earlier solutions we saw how improved forestry, farming and ranching practices could absorb large amounts of CO_2. If each of the world's 5.5 billion hectares of farms and ranchlands farmers and ranchers were to store an additional one tonne of carbon per year by zero tillage, organic methods, cover crops, rotational grazing and burying biochar, they could in theory absorb 5.5 Gt a year.

Solution #38 also showed that a forest managed holistically on a 160-year cycle stores 227 more tonnes of carbon per hectare than a conventionally managed forest. Russia, Canada, the USA and China have 1.48 billion hectares of forested land. Some is already protected and growing naturally, but if half (740 million hectares) was managed ecologically on a 160-year cycle, it could sequestrate an additional 168 Gt of carbon, or 1 Gt a year over 160 years, increasing the captured total to 6.5 Gt a year.[2]

In 2008, Earth's atmosphere contained more than 800 Gt of carbon, a dangerous 240 Gt more than is considered safe for a stable climate. If we can eliminate our greenhouse gas emissions entirely by 2030, the annual drawdown of 6.5 Gt into new farmland, ranchland and forest sinks could in theory reduce it to the safe pre-industrial level of 560 Gt within 50 years. I say "in theory" because although 50% of the CO_2 has a short atmospheric life of around 30 years, 20% stays there for 300 years, and 20% stays there for up to 3,000 years. Furthermore, the ability of ocean sinks to store carbon is weakening, while carbon

- Boreal Songbirds Initiative: borealbirds.org
- Canadian Boreal Initiative: borealcanada.ca
- International Boreal Conservation Campaign: interboreal.org

The closer animals graze, replicating their grazing patterns when predators were around, the more carbon is stored in the soil.

and methane emissions are increasing from the melting permafrost and increasing forest fires.

Policies for Large Landowners

- Develop a carbon analysis tool that estimates emissions from the soil, timber and the forest floor for different methods of silviculture, farmlands and grasslands management.
- Require all large landowners to publish annual carbon flow and GHG emissions data.
- Apply a carbon tax to all net carbon losses from the land, and place the money in a carbon solutions fund to finance sequestration initiatives.
- Include farming, ranching and forestry in cap-and-trade schemes, allowing landowners to sell carbon storage credits into the market.

Policies for Forestry

- Prohibit logging in all remaining old-growth forests, which store the most carbon.
- Maximize the protection of parks, ecological reserves and wetlands to minimize soil carbon disturbance and protect biodiversity.
- Ban clearcutting, as several European countries have done, and require forest companies to manage their lands on a 160-year cycle.

- Protect 50% of the Boreal forest — as the governments of Ontario and Quebec have pledged to do — and ensure that the other 50% is managed using eco-certified forest practices. Only 10% was protected in 2008.
- Prohibit the import of illegally and unsustainably harvested timber, as Norway has done.
- Support tree planting on farmlands, ranchlands, institutional lands and wastelands.
- Support global initiatives to protect the world's rainforests (See #97).

Policies for Farmers and Ranchers

- Support riparian restoration to prevent the loss of carbon by soil erosion along creeks, streams and rivers.
- Tie farm support programs to evidence of carbon sequestration and GHG reductions.
- Provide grants and carbon-pricing for the production and burial of biochar from crop and woody wastes.
- Develop a Grasslands Carbon Farming Initiative to help ranchers store more carbon.
- Work to include rangeland management activities in global carbon trading systems.[3]
- Support global initiatives to protect the world's grasslands.

75

Develop a Sustainable Transportation Strategy

Fossil-fueled transportation produces 33% of the US's and 23% of the world's carbon emissions, and oil will soon become scarce and unaffordable. How can we travel and transport goods in a sustainable world? Countries that succeed will move easily into a climate-friendly world. Those that do not will suffer bankruptcies and financial meltdown.

A clear vision is emerging in which cities, towns and suburbs are transformed to make it easy to get around by foot, bicycle, transit and light rail using unified ticketing, as they do in Switzerland, Frankfurt and Maryland. Carsharing, ridesharing, teleworking and videoconferencing are part of everyday life, and high-speed electric trains zip across the country. Electric and plug-in hybrid electric vehicles are powered by green electricity and sustainable biofuels, while road-pricing and car-free zones discourage cars from dominating our cities.

Light rail transit in Eskisehir, Turkey.

Every generation faces a challenge. In the 1930s, it was the creation of Social Security. In the 1960s, it was putting a man on the moon. In the 1980s, it was ending the Cold War. Our generation's challenge will be addressing global climate change while sustaining a growing global economy.
— Eileen Clausen, president of the Pew Center on Global Climate Change

City Transportation

Governments need to invest heavily in transit, bus rapid transit, trams and light rail, creating a seamless network in which travelers can connect smoothly, as they do in many European cities. When electrified with green power, the emissions are zero. The goal might be that 50% of all local trips are by public transport by 2030. A family that switches to commuting by transit will save $1,800 a year, in 2008 prices.

Governments must also invest in safe, off-road bicycle paths, helping city mayors to create at least 300 miles of segregated cycle route in every city, and in safe, long-distance bike routes that connect the regions, such as Britain's 12,000-mile National Cycle Network and the US dream of a National Bicycle Greenway. The goal might be that 15% of all local trips are by bicycle by 2030, up from 0.4% in 2008.

- Invest heavily in transit, bus rapid transit, light rail transit, bicycle paths and long-distance bike routes.
- Give tax breaks for the purchase of bicycles and travel passes.
- Place a 20-year freeze on all new highway construction.

Rail Transportation

Governments must bite the bullet and invest in a national high-speed rail network powered by 100% green electricity, with trains traveling at 320 km/hr (200 mph) and regional commuter trains at

200 km/hr (125 mph), connected to local transit and light rail. California is proposing a high-speed line from Sacramento to San Diego, costing about $40 billion, for an increased electrical demand of only 1%. Existing rail lines need to be upgraded and electrified, and services such as Amtrak and Canada's VIA Rail need to be expanded while the high-speed network is being built.

- Invest in a high-speed rail network.
- Electrify and upgrade all existing rail lines.

Freight Transportation

Heavy and light commercial trucks produce 20% of US CO_2 emissions from transportation. A variety of initiatives could reduce this considerably. (See #47)

- Develop a green freight strategy to shift 50% of road freight to electrified rail by 2030.[2]
- Work with city mayors to build more self-reliant local economies.
- Use electric trucks for local delivery from freight transfer stations, as CityCargo is planning in Amsterdam.
- Eliminate black-carbon pollution from inefficient trucks.
- Develop a goods movement action plan, as California has done.
- Fund these investments by road tolls and carbon taxes.

Shipping and Flying

When oil prices are high, airlines and shipping companies are desperate to reduce their use of fuel, but there are few solutions on the horizon. (See #49 and #50.) Since 2001 the US government has given

- California's Goods Movement Action Plan: arb.ca.gov/gmp/gmp.htm
- CityCargo: citycargo.nl
- National Bicycle Greenway: nationalbicyclegreenway.com
- Public Transportation: publictransportation.org
- Rails-to-Trails Conservancy: railtrails.org
- Sustrans (UK): sustrans.org.uk

$32 billion to the airlines; from 1972–1992 Boeing received $29 billion in grants and subsidies.[3]

- Require all airlines, ports and shipping companies to report their GHG and black-carbon emissions.
- Require all ports to become electrified so that ships can switch off their diesel generators.
- Take steps to eliminate black-carbon emissions from ocean-going ships.
- Require all ships to polish their propellers once a year when they are in port.
- Remove all tax-breaks and subsidies from airlines, airports and shipping.
- Tax aircraft kerosene at the same rate as vehicle fuel.
- Require airlines to include the cost of offsets in their ticket prices.
- When applying carbon taxes or cap and trade, use a multiplier of 2.7 to account for aviation's additional radiative forcing.[4]
- Place a freeze on all airport expansions.
- Package all these measures into a Sustainable Transport Act.

76

Develop a Sustainable Vehicles Strategy

> The time has come for a public-private community partnership to fix this country and put it back to work.
>
> — Van Jones

One hundred percent zero-carbon vehicles by 2030: that must be the goal.

The global measure for fuel efficiency is grams of CO_2 per kilometer, not miles per gallon. In the European Union, where the average 2008 new car produced 158 g/km, the requirement is 120 g/km by 2015. California requires 127 g/km by 2016. The Obama administration's requirement of 35 mpg by 2020 represents 156 g/km.

Will this be sufficient? Absolutely not. The potential exists today for electric and plug-in hybrid vehicles (PHEVs) that use green electricity for 80% of their travel and carbon-neutral biofuel for the rest. Battery recharging and replacement

By 2010, over 100 electric Mercedes-Benz Smart cars will be in Berlin, Germany, with RWE recharge stations throughout the city. e-mobility Berlin is a joint project of RWE and Daimler AG.

are advancing rapidly, and most automakers are promising to produce EVs and PHEVs by 2012.

Automakers say they need five to six years to bring a new model to market, and the average car is on the road for 10 years. If all new vehicles produced 50 g/km of CO_2 by 2020, backed with sustainable biofuels, most vehicles would produce zero emissions by 2030. In 1942, facing a similar crisis, the US auto industry retooled in nine months to produce tanks and planes, reassured by a guarantee of continued profits.

Andy Grove, former CEO and chair of Intel, has proposed that instead of waiting for the auto-industry to deliver a new fleet, we should develop a huge retrofit plan to convert 10 million pickups, SUVs and vans to PHEVs by 2012, and 60 million by 2019.[1]

- Require all new cars and light trucks to produce 50 g/km of CO_2 by 2020.
- Starting in 2015, ban all new vehicles that produce more than 200 g/km of CO_2.
- Give grants of $7,500 per vehicle for the first 200,000 plug-in vehicles from each of the world's major auto-makers, as the US government is doing.
- Give similar grants for the mass conversion of older vehicles to electric or plug-in electric, as Andy Grove and Cal Cars (calcars.org) are proposing.
- Introduce a "scrap-and-replace" incentive for cars that produce more than 240 g/km leaving the body for EV conversion, tied to the purchase of a car that produces less than 120 g/km of CO_2.

- Give rebates up to $4,500 for efficient vehicles, as France does, financed by a fee on inefficient vehicles ("feebates").
- Place EV and PHEV orders for government fleets and postal delivery trucks.
- Establish a federal battery guarantee corporation to address concerns about battery life.
- Require all new vehicles to be flex-fueled ($50–$100 per vehicle).
- If need be, buy a majority stake in an auto company, as the government of Saxony has done in Germany.

Create a Better Place

In Israel, Denmark, Australia, Hawaii, California and many other states and countries, Shai Agassi's company Better Place is working with governments to create networks that integrate green power, electric vehicles, EV charging spots and battery exchange stations. (See betterplace.com)

- Encourage and assist the widespread development of integrated EV recharging networks.

Sustainable Freight Trucks

There are many solutions available (see #47) that governments can encourage:

- Limit all new trucks to 466 grams/km by 2015 (12 mpg).
- Ban engine idling.
- Support the development of electric tram cargo transport systems in urban areas.
- Work with trucking companies to develop multi-fuel trucks, combining electric and liquid biofuel capacity.

More Sustainable Vehicle Actions:

- Require carbon labeling on all vehicles for sale.
- Adopt regulations to reduce black-carbon (soot) emissions from diesel to almost zero.
- Require car advertising to devote 20% of the space/time to the vehicle's carbon rating.
- Allow Low Emission Vehicles (LEVs) to use High Occupancy Vehicle lanes.
- Make LEVs exempt from emissions testing.
- Allow Neighborhood Electric Vehicles on roads posted up to 50 kph (35 mph).
- Require all new vehicles to be 95% recyclable, as Europe does.
- Require all new cars to carry fuel-efficiency indicators and automatic tire pressure monitors, to encourage eco-driving habits.
- Legislate the use of fuel-efficient tires, as California is doing.
- Add ecodriving skills to the driving test, including for truckers.
- Require all new cars to use CO_2-based, not HFC-based air conditioning systems.
- Reduce the speed limit to 90 kph (55 mph), which is the fuel-efficient "sweet spot."
- Require insurance companies to offer distance-based insurance to encourage less driving.
- Introduce road pricing on all busy roads to reduce congestion.
- Support carsharing and ridesharing schemes.

77

Develop a Sustainable Bioenergy Strategy

There are a host of ways in which the energy that plants gather can be used to generate heat, electricity, gas, fuel and plastics. The challenge for governments is to support the right, not the wrong, kinds of bioenergy. The small Austrian town of Güssing, which reduced its carbon emissions by 93% between 1995 and 2007 did so, among other things, by generating energy from nine types of plant material in 30 bioenergy projects. (See #26.)

WORLD BIOENERGY ASSOCIATION

Jönköping, Sweden, produces biogas and biofertilizer from the sewage sludge of 60,000 people at a wastewater treatment plant in the heart of the city. The biogas is used to operate the municipality's buses and other vehicles.

This is not some slow, controlled change we're talking about. It's fast, it's unpredictable, and it's unprecedented during human civilization.

— Adam Markham, WWF

How much biofuel might we need in a climate-friendly world? As a back-of-the-envelope calculation, we could start with the 15 million barrels of oil that are used every day for transportation in the US. If we reduce the miles traveled by 50% by using rail, transit, cycling, ridesharing and other means, and if half the vehicles are electric, the biofuel needed falls to 3.75 million barrels a day. If 80% of the non-electric vehicles are plug-in hybrids that use electricity for 80% of their travel, it falls to 1.35 million barrels. And if all new vehicles are made from lightweight materials weighing half as much, it falls to 675,000 barrels a day, for a 95% overall reduction.[1]

This is highly theoretical, and we will also need fuel for ships and trucks — but there are many potential sources of bioenergy. In Britain 6.7 million tonnes of annual food waste could produce biogas for 420,000 cars, representing 1.25% of Britain's 33 million vehicles — or 75% if we use the fuel-reduction math above.

What about airplanes? Globally, they use five million barrels of jet fuel a day. In 2008 *New Scientist* magazine calculated that if biofuel for air transport came from algae. If it came from willow or miscanthus, producing 2 tonnes per hectare, we'd need 1.2 million km². However, if it came from algae grown in ponds of seawater, producing 36 tonnes per hectare, we'd need 66,000 km² — about the size of Ireland.[2]

That's a lot of land — but it is only 0.15% of the world's 43 million square kilometers of farm and pastureland.[3] On this basis, the global shipping industry (7.5 million barrels a day) would need 100,000 square kilometers (0.22%). Producing the

Actions

- Develop a national sustainable bioenergy strategy to meet all sustainable fuel and heat needs by 2030, emphasizing local and regional planning.

- Work with other nations to develop sustainability rules, certification and compliance mechanisms to govern the global biofuels market, using the *Principles for Sustainable Biofuels* developed by the Roundtable on Sustainable Biofuels.[4] cgse.epfl.ch/page65660-en.html

- Require all bioenergy GHG calculations to use full lifecycle analysis, including land-use changes, and allow only those that achieve a large GHG reduction.

- Support second-generation biofuels from cellulosic wastes, algae, seaweed, sewage and crops grown sustainably on non-agricultural land.

- Encourage carbon-negative biofuels which store their carbon as biochar or CO_2.

- Require all plastics to be made from 20% bioplastic by 2020, 50% by 2030.

- Double the incentives for biofuels from sustainable feedstocks and eliminate incentives for all others.

- Encourage all sewage treatment plants to maximize their bioenergy production.

- Algae Biofuels Challenge: carbontrust.co.uk/technology/directedresearch/algae.htm
- Bioenergy Wiki: bioenergywiki.net
- Biofuel Watch: biofuelwatch.org
- Canadian Renewable Fuels Association: greenfuels.org
- Carbohydrate Economy Clearing house: carbohydrateeconomy.org
- *Green Fuel for the Airline Industry:* inyurl.com/cn459n
- Renewable Fuels Association: ethanolrfa.org
- Renewable petroleum: ls9.com
- Sustainable Jatropha: jatrophabook.com
- US DoE Biofuels: genomicsgtl.energy.gov/biofuels

Biosource →	Method →	Result
Algae	Anaerobic digestion	
Animal wastes	Biogas collection	Biobutanol
Crops	Combustion	Biochar
Crop wastes	Crush, pulp and purify	Biocrude
Food wastes	Enzymatic digestion	Biodiesel
Garbage	Fermentation	Biodimethyl-ether
Landfill gas	Photobioreaction	Bioelectricity
Sawdust	Liquid pyrolysis	Bioethanol
Seaweed	Thermal gasification	Biogas
Sewage sludge	Transesterification	Bioheat
Switchgrass		Biohydrogen
Waste oils		Bioplastics
Wood wastes		
Woody plants		

entire world's oil supply (85 million barrels a day) from algae biofuel would need 1.1 million square kilometers — 2.6% of the world's farm and pasture land. 70% of the world's farm and pastureland is used to raise livestock, so if everyone in the world were to eat a vegetarian or vegan diet just one day a week, it would free up 4 million square km of land. In 2008 Britain's Carbon Trust launched the Algae Biofuels Challenge to commercialize the use of algae biofuel by 2020.

The key to realizing bioenergy's potential is the word "local" — every region will need a network of biorefining plants, as Güssing does. But it does appear doable, as long as we start planning now. Germany's Green Party has estimated that Germany's farms could produce enough biogas to replace the country's entire supply of natural gas, which is currently imported from Russia.

78

Put a Price on Carbon

We are living proof the world should not fear a tax on carbon. Sweden has the highest carbon taxation in the world but we are not living in the Stone Age.

— Per Rosenqvist,
Swedish Environment Authority

"Planet Earth: RIP. Died from an excess of economic negative externalities." Might this be our awful epitaph?

An externality is a cost that is not captured in the price of something. The use of fossil fuels carries an externality in the damage that climate change will cause — a cost that Sir Nicholas Stern put at $85 per tonne of CO_2 in 2006.[1]

Because the prices of coal, oil, gas, cement, timber, and food fail to capture this cost, today's carbon users are enjoying a free ride at the expense of our children and grandchildren, who will have to pay for the damage.

Levy a Carbon Tax

The solution is to put a price on carbon and other greenhouse gases by introducing a carbon tax — marketed as a climate-solutions tax. Sweden's carbon tax, starting in 1991, now stands at $150 a tonne, increasing the price of gas by 35 cents a liter ($1.10 a gallon) — and yet from 1990–2006

Sweden's economy grew by 44%, its carbon emissions fell by 9%, and Sweden made enormous progress in developing district heating systems, bioenergy from forest wastes and biogas from sewage and animal wastes. Sweden uses the income for general revenues and to support climate-friendly technologies.

In the same period Denmark, with a carbon tax of $14/tonne, had a 4% rise in emissions and Norway, at $65/tonne, had a 15% rise, but this is less than would have happened without carbon taxes and populations rising because of immigration. All three nations also introduced many other policies that helped reduce their emissions.

In North America, British Columbia introduced a $10 carbon tax in 2008, rising to $30/tonne by 2012. To make the tax 100% revenue-neutral, the government reduced personal, small-business and corporate taxes, sent every citizen a $100 carbon rebate and gave additional carbon rebates to low-income families. In spite of

Country	$/tonne*	Notes
Finland 1990	$27	Combined energy/carbon tax.
Norway 1991	$65	Fishing industry gets 100% exemption. F-gases included.
Sweden 1991	$150	No tax by industry on fuels used to generate electricity; industries pay 50%. Citizens pay in full.
Denmark 1992	$14	Businesses pay 50%. Cap for exporting businesses.
UK 2001	Variable	Climate change levy. $58/tonne on oil. Revenue neutral.
Holland 2004	$18	Applies to coal only.
Quebec 2007	$3.50	Applies to 50 energy companies.
British Columbia 2008	$10	Rising to $30 by 2012. Revenue neutral.
France 290	$25	To be phased in gradually.

Enjoying the trail around the reservoir in Central Park, Manhattan, New York. A Manhattanite's carbon footprint is 30% smaller than the average American's, because citizens rejected a big freeway proposal in the 1960s in favor of old-fashioned urbanism.

Actions

- Levy a carbon tax, rising to $300/tonne by 2020.
- Include aviation (as Britain is doing) and shipping.
- Alternatively, levy the tax at the source where the carbon leaves the ground and/or the port of entry.
- Make the tax revenue-neutral, returning 100% as a dividend to all citizens.
- Alternatively, make it revenue-neutral for low-income people only, using tax and welfare adjustments. Place the income in a Climate Solutions Fund, governed by a team of advisors whom people trust.
- Be flexible with industries competing with foreign companies that don't pay the tax.
- Allow certain industries to place their tax in a special fund dedicated to developing industry-specific solutions, such as for the cement industry.
- Create exemptions for local governments, schools and public health agencies on condition that they sign a Climate Action Charter and develop plans to reduce their emissions, as British Columbia is doing for local governments.
- Extend the tax to the F-gases, repayable on delivery of the gases to a collection center, as Norway does.[3]
- When prices are rising rapidly, publish a monthly fuel surcharge figure, and require anyone burning fossil fuels for a living to pass it on in their prices.
- Promote the tax widely, with endorsements from well-known people.
- Publish clear information on ways in which people can reduce their emissions and pay a lower tax.
- Prepare a financial exit strategy as success reduces income from the tax, replacing it with other taxes such as road-pricing and windfall profits taxes on oil companies.

- Carbon Tax Center: carbontax.org
- Ecological Fiscal Reform: fiscallygreen.ca
- Price Carbon: pricecarbon.org
- Sustainable Nordic Gateway: sum.uio.no/susnordic
- The Carbon Tax Song: earthfuture.com/carbontaxsong

this, it ran into considerable opposition when the left-leaning opposition party ran a campaign to "Axe the Gas Tax," fueling public misunderstanding and encouraging the belief that climate change was not really a problem.

Because we have become used to cheap fossil fuels, $150 a tonne (36 cents a liter, $1.14 a gallon) is needed to change people's behavior. This was convincingly demonstrated in 2008 when oil rose to $146 a barrel (the equivalent of $150 a tonne), causing people to look urgently for ways to cut costs. A British study found that $100 to $200 a tonne is needed to stimulate innovation in low carbon technologies.[2]

79

Introduce Cap and Trade for Industry

Cap and trade is a complementary approach to carbon taxes that may offer more flexibility for industry. It is also known as ".emissions trading," "carbon allowances" or "carbon trading."

The system was developed in the 1980s to reduce sulfur dioxide pollution (acid rain) in the US. Instead of each polluter being required to reduce its SOx emissions, a limit was set for how much SOx each company could produce, and a market certificate (or pollution permit) was created for certified reductions. A cap was imposed, and a company could either reduce its emissions to the cap or buy certificates from a company that had gone below the cap. The cap is reduced each year, and the system has reduced SOx by 50% since 1990 at only $150 a ton, compared to the projected price of $600. In 2002, the *Economist* magazine called it "probably the greatest green success story of the past decade." [1]

Can the same method work for carbon emissions? That's the hope. A company's emissions are capped below the current level, and it must either

Cisco's TelePresence™ enables people to hold meetings on a face-to-face basis, reducing costs, travel-time and carbon footprints.

CISCO SYSTEMS

Having been a lobbyist myself you can't overstate how easy it is to get things done your way. It's a complicated policy and many lobbyists are trying to wreck amendments as we speak.

— Bryony Worthington, British emissions trading expert, founder of Sandbag.org

hit the target or buy permits from a company that has gone below the target. This creates a market for carbon permits and a price that fluctuates according to the cost of reducing emissions.

It sounds fine in theory — but creating a market for 2,000 sources of one pollutant (SOx) in one country is a far cry from creating a market for a million sources of six different greenhouse gases all over the world.

The European Union's Emissions Trading Scheme (EUETS) has been operating since 2003, learning from its mistakes as it goes along. Norway, New Zealand, Japan and Australia have all set up systems.

American Clean Energy and Security Act

In North America, the US government looks set to pass the American Clean Energy and Security Act of 2009, establishing a nation-wide cap and trade scheme which Canada will probably join.

The Act's weaknesses are that it is not fully phased in until 2016; aims at a very low 17% reduction in emissions below the 2005 level; allows two billion tonnes a year of reductions to come through eligible offsets; only auctions 15% of its carbon permits at first; gives too many permits away to energy-intensive companies; and places no restrictions on coal-fired power plants being built without carbon capture technologies up to 2020.

The Act's strengths are that it is very comprehensive, and will cover 86% of US emissions by 2020. It sets an "80% below 1990" goal for 2050; avoids the problem of a million sources of CO_2 by targeting upstream emissions, focused on roughly 7,400 companies; auctions 15% of the permits

Actions

- Require all companies and organizations to start reporting their emissions. -
- Join or set up a cap-and-trade scheme and drive the cap down to achieve 90–100% reduction by 2030.
- Invite NGO climate specialists into the negotiating process, with a supporting budget.
- Sell the emissions certificates by auction, and use the income to finance the shift to renewables, reduce the impact on low-income people, or distribute them as a citizens' dividend ("cap and dividend").
- Require it to cover all GHGs from all sectors, including aviation, shipping, cement, agriculture, forestry, waste and buildings.
- Place the income from the cement industry in a dedicated cement solutions fund to help the industry develop new kinds of cement.
- Set up a scheme for smaller organizations, as Britain has done.

- American Clean Energy and Security Act: pewclimate.org/acesa
- Cap and Dividend: capanddividend.org
- Carbon Finance: carbon-financeonline.com
- Chicago Climate Exchange: chicagoclimatex.com
- Europe's Emissions Trading Scheme: ec.europa.eu/environment/climat/emission
- European Climate Exchange: europeanclimateexchange.com
- International Carbon Action Partnership: icapcarbonaction.com
- Kyoto2: kyoto2.org
- Sandbag: sandbag.org.uk

with the income going to low-income families; distributes the income from unallocated permits to all legal US citizens, creating a "cap and dividend" (starting in 2026); includes many protections against market manipulation; and allows states to enact more stringent climate regulations.

The Act contains many other features:

- 20% of US electricity must come from renewables by 2020, excluding nuclear and large hydro (up to 5% from efficiency gains);
- States may use feed-in tariffs to meet their renewables goals;
- New energy efficiency standards;
- New buildings 50% more efficient by 2016;
- It provides grants for students building careers in renewable energy and energy efficiency;
- It establishes a bank for clean energy loans;

- The development of a smart supergrid;
- Financial support for the development of EVs and PHEVs;
- Regulations to limit black carbon (see p. 14).

In Britain the government is requiring all energy-intensive organizations, including supermarkets, banks, hotel chains, schools and government departments to participate in a carbon reduction commitment trading scheme.

Cap and trade may work if its problems are ironed out. If they are not, we will have spent ten crucial years tinkering with a failed system, instead of developing regulations and carbon taxes that can achieve real reductions.

80

Launch a Green New Deal

Urgency and vision are the twin
pillars on which humanity's hope
now hangs.

— Christopher Flavin,[2]
Worldwatch Institute

There is a belief among some that acting to reduce carbon emissions will harm the economy, destroying their jobs and businesses.

They could not be more wrong. It is not acting on climate change that poses by far the greater danger. The economic and financial consequences of a two-meter sea level rise, plus the other disastrous impacts of a 6°C warmer world, would be far greater that the 20% loss of GDP estimated in Sir Nicholas Stern's report, which did not include extreme events.[1] In his devastating book *Six Degrees*, Mark Lynas spells out what such a world would look like — and there are not many humans left alive. (See p. 24.)

The impact of rising oil and gas prices also needs to be considered, as supplies peak and then become so expensive that no one can afford them. What are the economic consequences of a world in which many people can no longer drive or heat their homes and farmers can no longer grow food?

- Architecture 2030: architecture2030.org
- Clean Energy 2030: tinyurl.com/3qfmnw
- Global Warming Solution: globalwarmingsolution.org
- Green Bonds: greenbonds.ca
- Green New Deal: greennewdealgroup.org
- New Apollo Program: apolloalliance.org
- Redefining Progress: rprogress.org
- Repower America: repoweramerica.org
- Worldwatch Institute: worldwatch.org

These two alarming outlooks share the same solution, which is a civilizational shift to a whole new world, powered by renewable energy. Smog vanishes, coal-mining and oil-drilling ends, and without oil to fight over, so does most war.

Civilization does not stand still. A hundred years ago, we were all riding horses. Every time the world makes a civilizational shift it has been associated with progress, not decline. No economist who retains an oil-defined mindset should be trusted to analyze the benefits of a world that operates without oil.

What are the economics of a climate-friendly world? Fuel for a conventional car costs $50 a week, and rising. Power for an electric car costs $5 a week, and falling. Heating a home with oil costs $4,000 a year. Heating a zero-energy home costs $0. The difference can be spent on restaurants, theatres, or marriage counseling. Also, while most money spent on oil disappears to Saudi Arabia, most money spent on renewables remains at home, stimulating more activity.

The urgency of the crisis calls for a response similar to the New Deal, which pulled the US out of the Great Depression of the 1930s. We need a Green New Deal, as a group of British writers and economists has proposed, with massive investments in efficiency and renewable energy that would create millions of green-collar jobs, while reining in the reckless financial games that caused the near-collapse of the global banking system.

In 2008, many such calls were made, all singing from the same songbook:

- Al Gore issued his challenge to achieve 100% clean electricity within ten years.

Advocates, policymakers, and practitioners of green jobs meet in San Francisco.

GREEN JOBS NOW

- The New Apollo Alliance presented a comprehensive strategy to invest $500 billion over ten years, creating more than five million green-collar jobs.

- Google launched its Clean Energy 2030 plan to wean the US off coal for power and cut oil for cars by 40% over 22 years to produce $5.4 trillion in savings, for a net saving of $1 trillion, reducing US greenhouse gas emissions by 40%.

- The Worldwatch Institute published *Low-Carbon Energy: A Roadmap*, which showed — among other things — how many jobs could be created by designing a new energy system.

- Global Warming Solutions, a non-profit society in Missoula, Montana, published *Rosie Revisited: A US-Led Solution to Global Warming*, calling for a 15-year, wartime-speed approach to reduce US carbon emissions to 80% below the 1990 level by 2025.

- Architecture 2030 presented its 2030 Challenge Stimulus Plan. (See #67).

Actions

- Analyze the financial costs and benefits of unmitigated climate change with temperatures rising by 6°C by 2100, continually rising oil and gas prices, and switching to 100% renewable energy by 2030.
- Develop a GHG Abatement Curve to cost the full array of solutions in an orderly manner, and lay down a pathway to a climate-friendly future. (See p. 70.)
- Launch a Green New Deal, combining the best of the proposals.
- Issue Green Bonds, encouraging the public to invest in climate solutions.

In 2009, the Obama Administration acted on many of the suggestions in these proposals.

With financial confidence shaken by Wall Street's debacle, this "Think Big" approach, is by far the best medicine to put the global economy back on its feet. All other solutions simply reinvigorate the world's lurch toward disaster, instead of changing course onto a new, green path.

81

Invest in a Climate - Friendly Future

A rise of 5°C would be a temperature the world has not seen for 30 to 50 million years. We've been around only 100,000 years as human beings. We don't know what that's like.
— Sir Nicholas Stern

The turning of a global era is a huge event, carrying enormous significance. It happened at the end of the Neolithic Age when we changed from hunting and gathering to settled agriculture. It happened at the birth of the Industrial Age when we used the newly discovered power of fossilized fuels to build the 20th century civilization.

And it is happening again today as we step into a new Ecological Age powered by renewable energy, inspired by the discovery that we can manage our forests, farms, cities, industries, homes, and transportation, at peace with nature and in harmony with our own hearts.

Poetic sentiments to one side, the practical work of these transitions was done by working toolmakers, engineers and scientists. The Watts, Edisons and Pasteurs spent years bent over their experiments and calculations, supported by investors or praying desperately that they would be.

From Beijing to Bogotá, inventors are hard at work in basements and laboratories, piecing together the breakthroughs we need. There is an enormous ferment of creativity underway — but it needs help from governments for research and development. Here are ten green technologies that need urgent funding:

1. Geothermal Energy

The energy that sits 6–10 km under our feet is enormous. In 2007 a team at the Massachusetts Institute of Technology estimated that with technical advances, geothermal could supply 20,000 times more energy than the US needs. (See #55.)

2. Solar PV Energy

The faster solar PV advances, the sooner we will reach price parity, making it acceptable for building

Using a 10 kW solar PV system, the catamaran "sun21" arrives in New York in May 2007, after sailing across the Atlantic in 30 days at 5–6 knots, equivalent to an average sailing yacht.

TRANSATLANTIC21.ORG

codes to require solar roofs on all new buildings and as a retrofit on all existing buildings.

3. Solar Thermal Energy

Every week the solar energy that reaches the world's deserts is equivalent to the world's entire remaining supply of fossil fuels — and we know how to capture it. The faster we can accelerate its development, the sooner solar thermal can become the steam engine of the green future. (See #55.)

4. Sustainable Bioenergy

There are many ways in which bioenergy can be produced from plants and wastes in a sustainable manner. The chief reason why Sweden and Germany are so far ahead is because their governments supported research and development, accelerating the day when every sewer bioheat, every farm biogas and every waste-stream will produce biofuel. (See #76.)

5. Ocean Energy

The oceans hold an enormous store of energy that could be available to us through tidal, wave and ocean-thermal-exchange technologies. Government support is essential so that the industry can mature. (See p. 48.)

6. Supergrid, Smart Grid, Batteries

The first electrical grid was constructed in Germany in 1891. We need a new grid for the 21st century to connect the new sources of renewable power. We also need to accelerate the development of battery technologies, both for the grid and for electric vehicles. (See #58 and #70.)

7. Climate-Friendly Cement

We depend on cement so much, yet its manufacture produces so much CO_2. Stimulated by the crisis, new formulations are being developed, but the industry is extremely conservative, and government intervention is needed to ensure that trials take place. (See #36.)

8. Soil Carbon

Earth's soil is the sleeping savior that could capture enormous quantities of carbon. Ecosystem-based forestry, farming, grasslands management and biochar production could lock away enormous quantities of carbon if we became seriously engaged. We need to help every forester, farmer and rancher to adopt the new approaches.

9. Sustainable Cities

There is much innovation happening in some of the world's cities, from electric vehicle networks to urban farming initiatives and zero-carbon neighborhoods. To grow further, cities need a thousand small grants that can stimulate initiative and help people experiment with their dreams.

10. Sustainable Buildings

There are apartments that recycle compost into biogas, and buildings that use no oil or gas for heat. There are projects underway in Cambridge, MA, and Berlin, Germany to retrofit half their cities' buildings. If we are to achieve carbon neutrality for all buildings by 2030, we must give initiatives like these the funding they need.

82

Build a Zero-Waste Economy

We are living on this planet as if we had another one to go to. As far as raw materials are concerned, we simply can't run a throwaway society on a finite planet.

— Paul Connett, professor of chemistry at St. Lawrence University, New York

Zero waste. What a concept. We are the most wasteful generation that has ever lived. As a world, we throw away more than two billion tonnes of municipal solid waste a year, a total that has doubled since 1990.[1] 260 million tons are generated in the US[2] and 35 million tonnes in Canada.

Every day, Americans and Canadians fill 50,000 garbage trucks with their wastes, and all of it, from discarded popcorn bags to demolished buildings, carries a carbon footprint — and worse, because the garbage in a landfill produces methane, which over 10 years is 125 times more powerful than CO_2.

In Japan, where 120 million people discard 20 million home appliances a year, the Japanese Mining Industry Association has estimated that if all home appliances, cars and office automation equipment was recycled, 200,000 tonnes of copper could be recovered each year, representing 25% of Japan's copper output.[3] The same goes for many other valuable resources.

In 2008 the US recycled 32% of its municipal solid waste. In doing so, it stopped nearly 200 million tonnes of greenhouse gases (3% of the US's carbon footprint) from entering the atmosphere. If 100% of the waste was composted or recycled, a further 400 million tonnes of GHGs could be eliminated. Every tonne of food waste that goes to the landfill creates almost a tonne of greenhouse gases.

Energy also goes down the drain in sewage, which contains ten times the energy needed to treat it. The Swedish city of Stockholm heats

Recycling containers in Llinars del Vallès, Vallès Oriental, Catalunya, Spain

WIKIMEDIA

- Californians Against Waste: cawrecycles.org
- National Recycling Coalition: nrc-recycle.org
- Product Stewardship Institute: productstewardship.us
- Recycling Facts: recycling-revolution.com/recycling-facts.html
- Recycling Market Development Zones: ciwmb.ca.gov/RMDZ
- The Story of Stuff: storyofstuff.com
- Zero Waste: A Key Move Towards a Sustainable Society: americanhealthstudies.org/zerowaste.pdf

Actions

- Adopt binding legislation requiring zero waste by 2030, taxing landfills after that at $1,000 per tonne.
- Ban incineration and encourage facilities that separate materials into their various streams for successful recycling.
- Require the capture of methane emissions from all landfills and phase in a methane tax for non-compliers.
- Phase in landfill disposal bans for organics, plastics, metals, paper, cardboard, electronics and concrete.
- Phase out the use of toxic and non-recyclable materials, such as PVC packaging.
- Extend curbside recycling programs to all multi-family dwelling units and businesses.
- Adopt "pay as you throw" for garbage, creating a financial incentive to recycle.
- Introduce a 33-cent user fee on ultra-thin plastic bags, as Ireland did, achieving a 94% reduction.
- Require an increasing level of recycled content in manufactured goods and supplies purchased by government.
- Require all take-out food packaging to be made from recyclable or compostable material.
- Adopt beverage container recycling laws that require producers to take back their containers, as California has done since 1988.
- Support state-wide composting programs.
- Support deconstruction by increasing demolition fees, distributing the income as deconstruction rebates.
- Support Recycling Market Development Zones where businesses that use recycled materials receive incentives to help them grow, as they do in California.
- Phase in product stewardship or laws, requiring manufacturers to take back their goods at the end of their useful life, as they do in Germany.

80,000 apartments from its sewage; and other cities extract methane and biodiesel from their sewage.

What will it take to stop wasting so much and move to zero waste? The technologies and policies we need already exist, as San Francisco demonstrates (see #28), but in North America, higher levels of government have typically washed their hands of the problem, leaving it to local communities to sort out.

California is the exception, achieving 50% recycling thanks to detailed legislation to keep things out of landfills, ban the use of toxic and non-recyclables, support the use of recycled materials and build a recycling economy. Compared to landfilling, each ton of recycled waste in California pays $101 more in wages, produces $275 more in goods and services, and generates $135 more in sales, and creates 5 jobs, compared to 2.5 jobs for landfilling, making it an essential part of the green-collar economy.[4]

Incineration may seem an intelligent option, but in reality it converts three or four tonnes of trash into a tonne of toxic ash that nobody wants, produces carcinogenic emissions from burned plastics, and undermines the drive to zero waste because investors need to burn 25 years' worth of garbage to get their money back. Recycling PET plastic saves 26 times more energy than burning it delivers.[5]

83

Build a Green - Collar Economy

We stand at a moment in history without precedent. Decisions that are ours to make over the next ten years will have a sweeping impact on the future direction of life on the planet.
— Eban Goodstein

For years, conservative-minded commentators have tried to kill climate solutions legislation by saying "It's a job killer" or "It will ruin the economy" on the simple and unproven assumption that increased energy costs from putting a price on carbon will destroy jobs.

They could not be more wrong.

Building a climate-friendly economy will *create* millions of new jobs, not kill them. It is the very thing we need at a time of economic turmoil, when investment is needed to build confidence and get the money flowing. In his stirring book *The Green Collar Economy: How One Solution Can Fix Our Two Biggest Problems*, Van Jones, founder of Green For All, shows how a major investment to tackle global warming could also tackle poverty in the cities and growing unemployment:

- Retrofitting buildings for greater efficiency
- Manufacturing electric cars and bikes
- Manufacturing transit, rail and light rail systems
- Manufacturing wind turbines and solar panels
- Installing new transmission grids

- Apollo Alliance: apolloalliance.org
- Blue Green Alliance: bluegreenalliance.org
- Ella Baker Center: ellabakercenter.org
- Five Million Green Jobs: 1sky.org
- Green for All: greenforall.org
- *The Green Collar Economy,* by Van Jones: greenforall.org/resources/the-green-collar-economy

- Harvesting bioenergy
- Installing solar PV and solar hot water systems
- Recycling and reclaiming wastes
- Developing urban agriculture, planting trees
- Restoring farmlands and ranchlands
- Deconstructing old buildings

In Germany, weatherizing 200,000 apartments created 25,000 new jobs and helped retain 116,000 existing jobs. In Bangladesh, where Grameen Shakti installed 100,000 solar home systems by 2007, installing a million systems by 2015 will create 100,000 jobs for local youth and women. In Brazil the ethanol industry employs some 300,000 people — though the misery and near-slavery of cutting cane is not a good advertisement for the new green economy. Globally, as wind power advances, the industry will create at least 2.1 million new jobs by 2030.[1]

1Sky's report *Five Million* Green Jobs describes how the US could generate two million green-collar jobs within two years by investing $100 billion in a green economic recovery plan, and five million by 2030 in rural areas alone, as long as bold measures are taken to incentivize the next generation of low-carbon fuels and renewable electricity.

This raises four major challenges:

1. To put policies in place that stimulate rapid investment in climate solutions;

2. To make sure local colleges are training people with the skills they need;

3. To include the people and regions most affected by the decline of traditional manufacturing and people stuck in poverty;

Trainees of green construction programs in Richmond, California, install a solar panel system

SOLAR RICHMOND, RICHMOND BUILD, GRID ALTERNATIVES, AND THE ELLA BAKER CENTER FOR HUMAN RIGHTS

4. To make sure that workers in the declining coal, oil and gas industries are able to make a just transition into the new green-collar economy.

In Oakland, CA, the Green Jobs Corps, funded by the city, provides job-training for at-risk youth and people who are underemployed, low-income or ex-prisoners. The Oakland Apollo Alliance, a coalition of labor unions, environmentalists and community organizations, wants Oakland to become a shining national example of a blue-collar becoming a green-collar powerhouse.

In California, legislation will hopefully be signed in 2009 to invest $2.25 billion in green careers education and training, especially for California's disadvantaged communities, and create a Green Collar Jobs Council that will fund California's green workforce needs.

The nonprofit organization Green For All, founded by Van Jones, has a goal to secure $1 billion in funding for green-collar job training, lifting 250,000 people out of poverty across the country. As Green For All says, "Now is the time for an inclusive green economy. Now is the time for action."

Actions

- Pass a Green Jobs Act to fund green-collar training.
- Establish a Green Corps similar to the 1933 Civilian Conservation Corps, to combine service, training and employment focused on building retrofits, urban ecological restoration and renewable energy production, as California has done with its Green Corps, a statewide effort to train 1,000 young adults for the new green tech economy.
- Provide 50:50 funding community green-collar jobs initiatives.
- Establish a green-jobs training partnership to provide the training, certification and apprenticeships needed, with strategies for each industrial sector.
- Require all colleges and universities to undertake a green upgrade of all building, plumbing, electrical trades, business, engineering and similar courses as a condition of funding.
- Write community benefit agreements into all green investment contracts, requiring contractors to train and hire residents in the communities where they work.

84

Ditch Neo - Classical Economics

> How is it that we have created
> an economic system that tells
> us it is cheaper to destroy the
> Earth and exhaust its people
> than to nurture them?
>
> — Amory Lovins, Hunter Lovins
> and Paul Hawken

Can we solve the problems of climate change and peak oil if we retrofit our economy with green energy, green buildings and organic farming? The answer might appear to be "yes", but as soon as we scratch deeper, a host of factors make such a retrofit seem like an attempted invasion against a coalition of hostile forces which are determined to resist us.

There are people all over the world who have embraced the belief that it is good to consume more, whatever the larger cost.

There are corporations that have profited very well from the existing arrangements and have no wish to lose their power.

There are politicians, obligated by campaign donations, who use their power to win subsidies, loopholes and lucrative policies for these same corporations.

There are bankers and financiers who make loans and investments to these corporations on the principle of profit maximization alone, regardless of the consequences.

- *Agenda for a New Economy,* by David Korten, Berrett-Koehler, 2009.
- Climate Change Economics: climatechangeecon.net
- Fiscally Green: greeneconomics.ca
- Natural Capital Project: naturalcapitalproject.org
- Redefining Progress: rprogress.org
- *The Economics of Happiness: Building Genuine Wealth,* by Mark Anielski, New Society Publishers, 2007
- Ethical Markets: ethicalmarkets.com

And there are economists who provide intellectual justification for this behavior, using complex equations to argue that free-market capitalism, following the precepts of neo-classical economics, is the only game in town.

Neo-classical economics uses 19th-century physics to postulate the existence of a perfect world in which all humans behave rationally, defined as seeking economic gain in all decisions, in which free market competition produces the best outcome for all, and bankers always make sound, wise decisions. In this world, there are no limits to economic growth. When a resource becomes scarce, its price rises, stimulating investment in a substitute. For this world to work as it should, government regulation must be kept to a minimum. Some economists go so far as to suggest that the free market is all we need to solve global warming.

This is the paradigm on which our global economy operates — or collapses, as we saw so alarmingly in 2008 when the banks and financiers lost, by some estimates, more money than they had made in their entire history, and governments had to bail them out with trillions of dollars of public money, because of gambling in speculative derivatives using computerized risk models that almost nobody understood, and the absence of government regulation, supported by this neo-classical philosophy.[1]

While this was going on, the very economy that everyone was so desperate to save was causing a meltdown of far greater significance in Earth's forests, oceans, farmlands and atmosphere. None of this existed in the world the bankers

To neo-classical economics, the extinction of the polar bear is an unfortunate "negative externality". We can either keep neo-classical economics, or we can keep polar bear. We cannot keep them both.

DAN GURAVICH, POLAR BEARS ALIVE, POLARBEARSINTERNATIONAL.ORG, WITH APPRECIATION

inhabited; it was an "externality" that had nothing to do with their world.

This is the real crisis we need to address — for our world does have limits, and the "externalities" that neo-classical economics fails to address are the foundation for all life on Earth. If we continue to follow these beliefs, most life will cease to exist.

So how should we replace them — and what should governments do?

The fundamental new reality we need to accept is that Earth's economy is not a world unto itself but a subset of Earth's ecology. Where Earth's ecology has rules, the disregard of which causes civilizations and species to collapse, these rules, must be reflected in Earth's economy's operating principles.

The 2008–2009 financial crisis showed very clearly that the era of economic deregulation that President Reagan started in 1981 has failed. When Reagan said "government is not the solution to our problem; government is the problem," he could not have been more wrong.

The causes of global warming are full of instances where deregulation and the absence of carbon pricing encourage deforestation, coal burning, soil loss, cheap meat, big highways and gas guzzling cars, sending the signal that encourages us to continue down the wrong path.

It is time to change direction. We need to build a new economy on new principles that recognize our total dependence on Earth's ecosystems, and harmonize our human activities with them.

- Internalize nature's externalities, using fiscal reform to make markets include ecological costs. Require every government department to produce a full list of taxes, fees, rebates, credits and saleable permits that could be used to recover the present and future cost of ecologically harmful activities. Implement green fiscal reform across the board.

- Replace Gross Domestic Product as the primary measure of progress with Genuine Progress Indicators, or Gross National Happiness, as Bhutan has done.

- Launch a public debate about "progress," to build support for a new direction. (See also #99.)

85

Become a Global Player

Our problems are manmade —
therefore, they can be solved by
man. And man can be as big as he
wants. No problem of human destiny is beyond human beings.
— President John F. Kennedy[3]

There is not a single nation that can solve this crisis on its own. We have to work together to craft a new civilization as a matter of absolute urgency. A government that is truly committed can make an enormous difference if it steps outside the business-as-usual box and empowers its citizens to work globally as well as locally.

In every embassy, staff could build climate-solutions partnerships with their host countries, sharing knowledge, funding projects, putting schools in touch with each other and organizing visits to cities, farms and businesses that are demonstrating solutions.

Such trips may involve flying, but the benefits of carbon-neutral flights to climate solutions centers such as Ashton Hayes, Berlin, Chicago, Copenhagen, Curitiba, Findhorn, Freiburg, Güssing, Rizhao, Samsø, Stockholm and Växjö could be enormous if they were linked to a commitment to write about the trip and give public talks.

- Brief all embassy staff on the science of climate change and the likely impacts on the host country. Ask them to build databases of local solutions, and encourage exchanges.

Small partnerships can yield very significant results. Norway's government is giving $1 billion over seven years to Brazil's Amazon funding research into water conservation and sustainable agriculture. Britain's Department for International Development and the Canadian International Research Development Centre are helping Africans deal with climate-related floods and disease outbreaks and funding research water conservation, agriculture.

- Screen all development aid for climate sensitivity. Prioritize projects that include climate solutions, and cancel projects that support fossil fuels and deforestation.

There is also a great need to change the way the world financial order operates, unhooking its agencies from the assumption that fossil fuels are good for development:

- Withhold funding from the World Bank until it stops financing fossil fuel projects and rainforest destruction. (See #98.)

- Prohibit export agencies such as the US Overseas Private Investment Corporation and the Export-Import Bank from financing any projects that increase the risk of climate change, as the End Oil Aid Act[1] proposed and the European Parliament called for in 2007.[2]

- Use your membership in the International Energy Agency to push for less fossil-fuelled thinking and greater urgency around climate change and peak oil.

- Work with other nations to establish a Global Climate Fund and implement a global carbon tax on all fossil fuels. (See #94.)

Promote Climate Solutions Treaties

The most important action that a government can pursue globally is to support and promote a strong post-Kyoto treaty and work with other nations to promote Climate Solutions Treaties.

The Kyoto Protocol was designed to reduce the problem, a challenge at which most countries have failed miserably. Kyoto's successor will probably use the same approach, calling perhaps for a 20% reduction below the 1990 level by 2020.

Equally important is the need for climate solutions treaties in which nations cooperate to build the solutions. Because the solutions will help build climate-friendly economies with many new jobs, participation by developing nations may be easier to achieve. We need treaties to increase the use of renewable electricity, to establish global efficiency standards for all new appliances, and much more. (See #93.)

The clearest sign of progress in this direction is the birth of the International Renewable Energy Agency in 2009, led by Germany and 60 other nations, with the goal of becoming the driving force in promoting the widespread and sustainable use of renewable energy on a global scale. Similar initiatives are needed to promote the uptake of zero- and low-carbon vehicles, zero-carbon buildings and and much more.

- Work with other governments and global leaders to introduce international climate solutions treaties.
- Join the International Renewable Energy Agency, and work to bring other nations on board.
- Establish initiatives to engage young people in international efforts: for it is they who will live in the future we create.

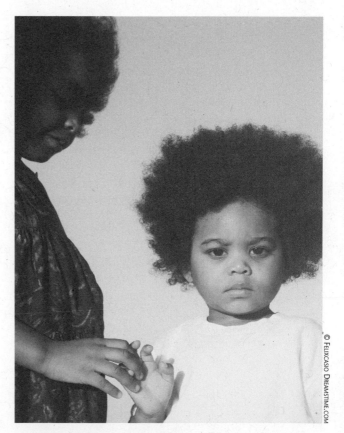

© FELIXCASIO DREAMSTIME.COM

Act responsibly. Please think about my future, as well as your own.

86

Scramble! This Is Serious

Once the snow is gone, I don't know what we'll do. Only God knows our fate, once the water and snow are gone.

— Felipe, Andean alpaca herder whose flocks depend on Peru's melting Ausangate glacier.

M any people in the developing world resent being told by people in the polluting nations that they must reduce their carbon emissions.

"We have enough to think about," they reply. "Our cities are bursting at the seams. We have terrible poverty. Our fresh water supplies are disappearing, and we've a growing problem with AIDS. We have ethnic and tribal conflicts, and we're being devastated by natural disasters. We are losing our forests and natural habitat. Our economic growth is just getting going — and now you are demanding that we stop burning fossil fuels?"

These things are all true, but it is also true that if we don't work together as a planet to reduce our emissions as rapidly as possible, the results will be appalling — and the developing nations will get the worst of it.

In Bangladesh the rising sea level will force 20 million people to abandon their ancestral fields and homelands[1] — and India is building a 3,000-km steel, concrete and razor-wire fence to keep them out. In 2007 the combination of flooding and

a cyclone claimed 4,000 lives, affected 8.7 million people and cost Bangladesh almost $3 billion.

The Shanghai region of China, home to 14.6 million people, is totally vulnerable to the rising sea level. A two-meter rise would inundate the heart of China's economy. Six thousand kilometers away, the glaciers on the Qinghai-Tibet plateau are melting.[2] 17% of the glaciers that feed China's Yellow River — which feeds 130 million people on the agricultural lands it supports — have already gone. In China's arid western region, 300 million farmers will see a decline in the waters that flow from the glaciers. Likewise the Yangtze River, whose waters enable the production of 70% of China's rice, 50% of its grain and 70% of its fisheries. The same Himalayan glaciers feed the Indus, Ganges and Brahmaputra Rivers, supporting the rich agricultural plains of Pakistan, India and Bangladesh. When they are gone, these rivers will carry only the seasonal monsoon downpours.

The most devastating impacts will come from storms, downpours and floods. In 2008 torrential rain lasting 20 days ruined more than 900,000 hectares of farmland in China, caused a million people to flee their homes, and destroyed 45,000 homes. Similar floods are occurring all over the world, setting back hard-won development, often by decades. The World Meteorological Organization reported in 2007 that the number of damaging weather incidents is increasing by 6% a year, including heavy precipitation and more intense hurricane and cyclones.[3]

In South America the glaciers in the Andes are also melting, threatening the drinking water for 30 million people and the hydro power that supplies

- Assessment of Impacts and Adaptation: aiaccproject.org
- Group of 77: g77.org
- Institute for Environment and Development: iied.org
- Millennium Goals: un.org/millenniumgoals
- PANOS: panos.org
- Peru — The Retreating Glacier: tinyurl.com/6cjvjg
- Red Cross Red Crescent Climate Guide: climatecentre.org
- START: start.org

Gangotri Glacier, at the source of the Ganges River, in Uttaranchal, India, is receding by 10-25 metres a year. Note the dirt, caused by black carbon from south Asian air pollution, which accelerates the melting.

© TOTOREAN DREAMSTIME.COM

most of the electricity for Peru, Colombia, Ecuador and Bolivia.[4] Since 1970, the glaciers have lost 20% of their volume.[5]

Elsewhere, especially in Africa, climate change will bring more droughts, shorter rainy seasons and huge impact on agriculture. In 2008, Greenpeace found that China's food production could fall by 23% by 2050 because of climate change and other factors.[6] Fisheries will also be severely affected by the warming oceans and the loss of coral reefs.

This is only half the picture. The other half is that the world's oil supply may peak as early as 2010, depending how soon the world recovers from financial collapse. For most developing nations, oil above $150 a barrel will bring financial chaos, unemployment and civil unrest.

Even though the developed world is responsible for the vast majority of the planet's cumulative carbon emissions, everything points in one direction, which is the need to join the global effort to abandon fossil fuels and build instead a clean, healthy, sustainable, climate-friendly economy. The same strategies that lead to the development of ecological cities, solar villages, renewable energy, protected forests and organic farms will produce better and more lasting results as a means to overcome poverty.

- Work within the Group of 77, representing 132 developing nations and China, to persuade them to accept the need for greenhouse gases reductions.

- Join the European Union's Global Climate Change Alliance, accepting their help to integrate climate change into poverty reduction strategies.

- Work with START to build capacity for the best climate adaptation policies.

- Participate in the global carbon market, using the Clean Development Mechanism and Joint Implementation to protect and restore forests, restore grasslands, capture methane and close down coal-fired power plants.

- Invest heavily in energy efficiency, which could reduce GHGs from energy use by 25%.[7]

- Shift to renewable energy, which could save countries in East Asia as much as $2 trillion in fuel costs by 2030.[8]

87

Build a Politics of Change

We have a responsibility to protect the rights of generations, of all species, that cannot speak for themselves today. The global challenge of climate change requires that we ask no less of our leaders, or ourselves.

— Wangari Maathai,
founder of Kenya's Green Belt Movement

The challenge that developing nations face in making the transition to a climate-friendly future is enormous — but the prospect of failure is appalling. A new generation of younger leaders is emerging, and it is essential that they have the courage and the ability to tackle corrupt and incompetent rulers.

Developing nations are investing billions of dollars in new power capacity that will last for 30–40 years, so it is essential that it be in safe, 21st-century renewable technologies, not dangerous, polluting 20th-century technologies.

Many nations are showing rapid progress. Brazil is producing 45% of its transportation fuel from sugar-cane ethanol grown on 1% of its agricultural land.[1] The Philippines and Indonesia are ramping up their geothermal capacity.[2] Vietnam is

Wangari Maathai, founder of the Green Belt Movement.

MARTIN ROWE

promoting energy-efficient lighting in its streets, schools and hospitals. India is fourth in the world in wind power is making a huge commitment to solar energy. Algeria is preparing a solar thermal revolution that will send power to Europe, while supporting a more sustainable future at home. China is active in every sector. (See #90.)

To build a climate-friendly society, governments must remove the fossil-fuel subsidies people have grown used to, and overcome the special interests that get rich from fossil fuels and deforestation. This can sometimes mean enormous struggle. In southern Thailand, villagers have spent years trying to stop the construction of two new large coal-fired power plants near Prachuap Khiri Khan on the Gulf of Thailand. Sixteen environmentalists and human rights organizers have been killed or "disappeared" since 2000, including Charoen Wat-aksorn, who was shot dead by gunmen on June 21, 2004 while riding his motorbike.

Overcoming this kind of resistance needs leaders who have the audacity to hope and to articulate the vision of a green, sustainable future that is strong enough to convince people and build the political support it needs.

The sustainable future will create millions of new jobs, enabling people to escape poverty and build a future for their families. Governments must argue it aggressively, and remind people how catastrophic global warming will be if we don't change our ways. This requires widespread public support, and the best way to build it is to support citizens' organizations that are actively promoting and building sustainable future.

Muhammad Yunus, founder of the Grameen Bank.

GRAMEEN AMERICA

- Centre for Science and Environment, India: cseindia.org
- Global Greengrants Fund: greengrants.org
- Grameen Bank: grameen-info.org
- Grameen Shakti: gshakti.org
- Greenbelt Movement: greenbeltmovement.org
- Right Livelihood Award: rightlivelihood.org
- Sekem: sekem.com and earthfuture.com/economy/sekemegypt.asp
- "We're sheep, not citizens": tinyurl.com/ngvc74
- *Unbowed: A Memoir,* by Wangari Maathai, Knopf, 2006
- *Banker to the Poor: Micro-Lending and the Battle Against World Poverty,* by Muhammad Yunus, Public Affairs, 2003

In the southern Indian state of Kerala, members of the Kerala Sastra Sahitya Parishad have helped half a million homes install high-efficiency smokeless wood-burning ovens, led programs to replace the state's 20 million 60-watt light bulbs with compact fluorescent bulbs, helped local governments install micro-hydro and encouraged farmers to adopt agro-forestry, conserving soil and trees.

In Kenya the Green Belt Movement, founded by Wangari Maathai, has encouraged women to plant more than 40 million trees on community lands, protecting against soil erosion, stopping deforestation and creating self-sufficiency in firewood for the women involved.

In Egypt, Sekem, founded by Dr. Ibrahim Abouleish in 1977, has shown how a community of 2,000 people can thrive sustainably on near-desert farmland in a new kind of social, economic and environmental harmony. As a result, the use of pesticides on Egyptian cotton fields had fallen by more than 90%, 80% of Egypt's cotton is grown without chemical insecticides, and yields have risen by nearly 30%.

In India, the Centre for Science and the Environment, inspired by Anil Agarwal, caused the city of Delhi to convert its entire public transport fleet of 80,000 vehicles to natural gas, ban taxis, buses and rickshaws that are more than 15 years old, bring in tough penalties for vehicles that emit visible pollution and launch the first underground trains.

A sustainable green economy educates and empowers those who are often the poorest. The Grameen Bank, founded in Bangladesh in 1976 by Muhammad Yunus, developed a highly successful way to lift the poorest families out of poverty, using women's ability to help each other to issue micro-loans to 7.6 million people. Some of these women now work in Grameen Shakti, spreading renewable technologies such as solar, biogas, wind and improved cooking stoves, and creating 100,000 green jobs through the Grameen Technology Centers, which train women in renewable energy technologies and help them to set up their own businesses, benefiting more than two million people.

That's a political base. These are the organizations that are building the foundations for a healthy, sustainable world. Support them, and you support the changes we all need.

88

Build Ecological Cities

> There is no endeavor more noble than the attempt to achieve a collective dream.
>
> — Jaime Lerner, past Mayor of Curitiba

We often call our cities lively, colorful or bustling, but we rarely call them "ecological." All over the world, cities suck in resources and spew out wastes. Most cities in the developing world are a chaotic jumble of noisy vehicles, polluted rivers and crowded shantytowns, with a minimum of green space.

Life changes, however. Two hundred years ago most of Europe's cities were a fetid mass of pollution, with no drains or clean water. A hundred years from now, if we play our cards right, most cities in the developing world will be well-managed urban paradises with clean-flowing rivers, parks, bikeways, pedestrian areas, efficient bus

A bus shelter being built in Curitiba, Brazil.

transportation, electric bikes and tuk-tuks, thriving green economies and millions of trees.

Brazil's Ecological Capital

The residents of Curitiba, in southeast Brazil, think they live in the best city in the world. With a metropolitan population of more than 3.5 million people, Curitiba is known as Brazil's ecological capital because of the many initiatives it has taken.

The story started in 1965, when the city planned two major arterial routes to take pressure off the downtown. When Jaime Lerner was appointed Mayor in 1971, he turned the arterials into high-density transit corridors, adding three more later as the city grew. A single fare was introduced with shorter trips subsidizing the longer ones, and raised plexiglass tube stations were built for easy loading. Today, even though Curitiba has a high rate of car ownership, 2.4 million people ride the city's 2,100 buses every day. The city also has 115 km of bicycle paths and 26 car-free downtown blocks, including the famous Rua das Flores. One and a half million tree seedlings were given away to neighborhoods, and now there are trees everywhere; builders get a tax break if their projects include green space. In 2005, 99% of Curitibans told pollsters they were happy with their city.[1]

Colombia's Ecological Capital

Similar initiatives are happening elsewhere, and if you piece them together it is possible to see the emergence of the ecological city. In Bogotá, Colombia, Mayor Enrique Peñalosa transformed urban life in just three years by redesigning his

- Afribike: afribike.org
- City Farmer News: cityfarmer.info
- City of Curitiba: curitiba.pr.gov.br
- EcoCity Trust, South Africa: ecocity.org.za
- Electric Solar Tuk-Tuks: jcwinnie.biz/wordpress/?p=1625
- Electric Tuk Tuk: elektrischetuktuk.nl
- Envirofit 2-stroke retrofit: envirofit.org
- Institute for Transportation and Development Policy: itdp.org
- Interface for Cycling Expertise: cycling.nl
- International Bicycle Fund: ibike.org
- Song of Sustainable Cities: youtube.com/watch?v=haKh9mCk3xk

city for people and bicycles, not cars. He banned cars from parking on the sidewalks; created the 17 km Alameda, the longest pedestrian street in Latin America; created or renovated 1,200 parks; introduced the highly popular Transmilenio bus rapid transit system with 800,000 daily riders, modeled on Curitiba's example; created 300 km of bicycle paths that now have 350,000 trips a day; planted 100,000 trees; and involved citizens directly in improving their neighbourhoods. The result has been a drop in CO_2 emissions and a 70% decline in the murder rate. Ten years before the changes, people hated their city, thinking it a disaster — *un enfierno*, a living hell — with no hope and no direction. By changing the public space and reclaiming people's delight in walking and cycling, Bogotá has regained its self-esteem.

Sustainable Tuk-Tuks

The developing world is home to an estimated 100 million two-stroke tuk-tuks and tricycles, which pour their pollution into the air, losing up to 70% of their energy in the exhaust stream. There are two solutions. The first is electric conversions, being developed in Nepal and Holland, and solar electric tuk-tuks, being built on a small scale in India.

The second is to convert the two-strokes to four-strokes, which the US-based charity Envirofit is doing in the Philippines and elsewhere with a specially designed direct in-cylinder fuel injection retrofit kit, reducing carbon monoxide emissions by 76%, CO_2 by 35% and other pollutants by 89%, producing a cleaner and more fuel-efficient option than replacing their engines with four-strokes.

Urban Farming

Food is an essential part of any ecological city. In Caracas, Venezuela, 8,000 tiny one-square-meter microgardens have been created in the barrios, each able to produce either 330 heads of lettuce, 18 kilos of tomatoes, or 16 kilos of cabbages a year in multiple harvests.[2] The country's goal is to create 100,000 such microgardens, along with 1,000 hectares of urban compost-based gardens. Dar es Salaam, Tanzania, has 650 hectares of land in or near the city where 4,000 farmers produce fresh produce for the city. In Kolkata, India, the city's wastewater treatment ponds, covering nearly 4,000 hectares, are used to raise 18,000 tonnes of fish a year that feed off the algae.[3]

If there is a key, it is building a participatory politics that empowers people to take responsibility for their communities. Without local organizations, people feel passive and disempowered, and when they are angry they become vulnerable to negative political forces. Community organizations enable people to participate, and join the journey to sustainability.

89

Build Solar Villages

> The excitement has rubbed off on the learners.... You will never understand how much difference SELF has made in the education of an African child.
>
> — Melusi Zwane,
> Principal of Myeka High School, South Africa

Karma Dorjee lives in a small village in the Phobjikha Valley of Bhutan with his mother and three sisters. Candles and a smoking kerosene lamp provide light — along with fumes and the risk of fire. Power for the radio comes from a dry battery, and cooking fuel comes from firewood that his mother gathers in the forest. When Karma grows up, he wants to move to Thimphu, the capital of Bhutan, where there are bright lights and a chance to get ahead. He may not know it but he shares his life with two billion people who live in rural villages in the developing world.

One day, a young man arrives in the village carrying a large black case. He meets with the elders and, as Karma and the other children cluster around, he opens his case and takes out a solar panel, assembling it to show them how it works. When the sun shines on the panel, the light comes on. When he shades it, the light goes off. He takes out a laptop computer and a small satellite receiver, plugs it into the solar panel, logs onto the Internet, and invites the kids to explore. Karma is enthralled. He has heard about this "Internet," but he never imagined he would try it for himself in his own village. His thoughts about moving to Thimphu fade a little.

Karma's mother is persuaded, and two months later they buy a 50W solar home system that powers a fluorescent light, a radio and a television. The system costs $450 and is financed through a revolving loan fund, thanks to the Solar Electric Light Fund (SELF), a non-profit group which has helped villagers install solar systems in 20 countries around the world. It costs only slightly more than they were paying for their kerosene, candles and dry-cell batteries.

Sustainable Villages

Around the world, 400 million households use kerosene for lighting and firewood for cooking. In Bangladesh, Grameen Shakti is helping women to become solar agents in 83,000 villages where the Grameen Bank has established a micro-lending program, enabling the villagers to hook up to light, telephones and the Internet.

SELF and Grameen Shakti have demonstrated that villagers will buy a solar system when there's a good partnership with a local organization. Globally, 5% of the 400 million households could pay in cash; 25% could pay with short-term

The Many Advantages of Solar Light

1. Reduced air pollution. Inhaling kerosene fumes is equivalent to smoking two packs of cigarettes a day.
2. Reduced CO_2. A 50W PV system saves 6 tonnes of CO_2 over its 20-year life.
3. Reduced ground pollution from discarded lead-acid batteries.
4. Improved literacy through more evening schoolwork.
5. Increased evening work-time for home craft industries.
6. Reduced need for urban migration, because villagers no longer feel isolated.
7. Reduced birth-rate, as there's more to do in the evenings.
8. Reduced risk of fire and burns from kerosene.
9. With credit, monthly solar payments are the same as for kerosene and batteries.

credit; 25% could pay with long-term credit; and 45% would need subsidies.[1] A 50-watt solar system costing $300 will prevent 6 tonnes of CO_2 from being released by kerosene over 20 years, so solar offsets at $25 a tonne provides $150 toward the cost. In India, the Barefoot College in Tilonia, Rajasthan, led and inspired by Sanjit Bunker Roy, trains women from India and twelve other countries to become barefoot solar engineers, even if they have no formal education, and to take their skills back to their villages, from Mauritania to Afghanistan.

Solar power also reduces the use of firewood and the loss of trees for cooking: at Mount Abu, India, the world's largest solar cooker can cook up to 38,500 meals a day. These and other alternative cooking stoves such as the cardboard Kyoto Box cooker are urgently needed to reduce the emissions of black carbon. (See p. 14.)

Solar can also help with agriculture. In the Kalalé district of northern Benin, Africa, where the average income is $1 a day, SELF is helping villagers install solar-powered drip irrigation systems in Solar Market Gardens, enabling them to grow food during the six-month dry season, eat more food, earn $7–$8 a week by selling produce and use the time previously spent carrying water to engage in activities such as gathering seeds and (for the girls) going to school. SELF's goal is to deliver similar systems to every village in the region, along with solar electrification of the schools, community centers and water pumps. The same approach could work in most of sub-Saharan Africa, south and south-east Asia, and Central and South America.

- Barefoot College: barefootcollege.org
- Grameen Bank: grameen-info.org
- Grameen Shakti: gshakti.org
- Greenstar Solar Community Center: greenstar.org
- International Solar Energy Society: ises.org
- Kyoto Box Solar cooker: kyoto-energy.com/kyoto-box.html
- One Million Lights: onemillionlights.org
- Sarvodaya Movement of Sri Lanka: sarvodaya.org
- Solar Cooking International: solarcooking.org
- Solar Electric Light Fund: self.org

SOLAR ELECTRIC LIGHT FUND

Tending the crops in the Kalalé District of Benin, West Africa, where the Solar Electric Light Fund (SELF) has installed solar water pumping and drip irrigation.

With determination, public support and political will, most of the world's 400 million poorest households could be lifted into a solar future.

90

Solutions for China

> Do politicians understand just how difficult it could be, just how devastating, rises of 4°C, 5°C, or 6°C could be? I think, not yet.
>
> — Sir Nicholas Stern

It is common to hear people say, "But... what about China? What's the point when China's burning so much coal?"

China is responsible for 8% of the world's cumulative carbon emissions since 1750 and 19% of the current emissions, 33% of which come from Chinese industries making products for export, mostly to North America and Europe.[1] China also pours 50% of the world's concrete.

China gets 70% of its energy from burning coal, which is responsible for 80% of the country's CO_2 emissions. In 2008 it burned 2.7 billion tonnes of raw coal which produced 6 billion tonnes of CO_2, representing 15% of the world's CO_2 emissions. In 2007 China added 90 GW of new coal-fired electrical power plants, at two new plants per week. The environmental damage caused in part by burning coal is reducing China's

GDP by 8–13% a year — almost as much as it is growing.

In 2009 China ranked 49th out of 60 countries in the Climate Change Performance Index, reflecting emissions, emissions trends and climate policy — well ahead of Canada (59th) and the USA (60th). China is making a huge effort to reduce its emissions, while still needing to satisfy the need of its people to escape the poverty, famine and hardships that previous generations have suffered.

China's Climate Action Initiatives[2]

China's one-child population policy has avoided 300 million births, reducing its CO_2 emissions in 2005 by 1.3 billion tonnes. China's goal is to reduce the CO_2 intensity (not the absolute emissions) of its economy by 20% by 2010, and 80% by 2050. If successful, this would cause a 1.5-billion-tonne fall in emissions by 2010.

This is supported by a drive to increase energy efficiency by 20% by 2010, using mandatory energy efficiency appliance labeling, the distribution of 150 million half-price efficient light bulbs and a Top 1,000 Enterprises Program that targets efficiency in China's most energy-intensive enterprises, representing a third of China's energy use. China's building codes, if enforced, will cut energy consumption in new buildings by 65%.

On the supply side, China has adopted Europe's feed-in tariff for renewable energy (see #69) and set a goal to provide 10% of all electricity from renewables by 2010, and 15% by 2020, totaling 137 GW. For wind energy China's goal is 3% (5 GW) by 2010 and 30 GW by 2020, though experts believe wind could provide 40 GW by

- Asia Alternative Energy Program: worldbank.org/astae
- Beijing Energy Efficiency Center: beconchina.org
- China Climate Change Info-Net: ccchina.gov.cn/en
- China Energy Group: eande.lbl.gov/EAP/China
- China Goes Green (CNN): tinyurl.com/5us9e8
- China Sustainable Energy Program: efchina.org
- China's Copenhagen Commitments — a Workable Solution (Climate Progress blog): tinyurl.com/dmpsq3
- China's Green Beat: chinasgreenbeat.com
- China's Green Buildings: greendragonfilm.com
- Electric Bikes in China: tinyurl.com/57xcpd
- Solar hot water in China: tinyurl.com/6mjjgc

Solar evacuated tubes heat the water on a roof in Rizhao, China.

2020. China has 253 GW of wind energy potential, mainly in Inner Mongolia and along the coast.

China is the world's largest exporter of solar panels, and the largest consumer of solar hot water systems with over 40 million heaters — 60% of the world's total. In Kunming, capital of Yunnan Province, half of the city's 3.2 million inhabitants use solar heaters. In the coastal resort city of Rizhao (population 3 million), which is aiming to become fully carbon neutral, almost all buildings in the urban area use solar hot water. The city also has 60,000 solar-heated greenhouses. Between 2000 and 2008 Rizhao cut its CO_2 emissions by almost 50%, partly by closing small inefficient coal-fired operations.

In 2008 China's fuel-efficiency standard for new vehicles was 36 mpg, compared to the US goal of 35 mpg by 2020: most new American cars would not be allowed on China's roads. There is also a goal that China's vehicle fuels should be 10% renewable content by 2010, 15% by 2015. Chinese citizens are buying 25–30 million electric bikes a year, making China the world leader in the field, and China is likely to emerge as a leader in the production of electric cars.

Finally, China is making an enormous effort to plant trees in an initiative known as the Green Wall of China, to stop the Gobi Desert from advancing, end the appalling dust storms that are happening because of the ecologically destructive management of Inner Mongolia's grasslands, and increase the forest carbon sink by 50 million tonnes by 2010, compared to 2005. Between 1982 and 2007, volunteers and workers planted 52 billion trees, increasing China's forest cover from 12% in the early 1980s to 18% in 2008. It may not be working,[3] however, in part because proper restoration needs the return of the Mongolian wolf to restore natural rotational grazing patterns.[4] The loss of carbon from China's grasslands has not been tallied.

What more can China do? Double every effort; impose a carbon tax; make China's citizens aware of the terrible danger they face from global warming; and aim to make China a fully climate-friendly country by 2040.

91

Restore a Climate - Friendly World

> You should never even ask if a campaign is winnable, because the question is not answerable. No one can predict the course of the future. Time and time again I have seen completely unforeseeable shifts in the tide of events that result in campaign victory.
>
> — Elizabeth May

Finally, we come to the global solutions — and this is, above all, a global crisis. There is nothing that anyone can do to stop Greenland thawing or the ocean warming unless we do it all together.

The greatest problem is that we don't have a consensus as to what's needed. The discussion thus far has been phrased as the acceptable level of atmosphere CO_2 we think "safe" (450ppm, 400ppm, 350ppm) and the reduction needed to get there (e.g., 80% by 2050).

Every time there is a scientific update, the projections are at the bad news end of the spectrum. Even the supposedly "safe" goal of 450 ppm creates a 33% chance that we will not avoid a collapse of Earth's ecosystems and civilizations. There's not a parent on Earth who would stand their child in the middle of a dangerous road to face a one-in-three chance of death.

There are only two rational goals that we should pursue if we value the future of our planet.

Goal 1: Zero Emissions by 2030–2040

The ideal goal would be "zero emissions yesterday" — for that's what is needed to minimize the risk. Since this can't be achieved, we are left with the second best option, which is "as quickly as humanly possible."

Goal 2: Atmospheric Restitution to 280 ppm of CO_2

This addresses the need to draw the excess carbon out of the atmosphere. Before the industrial age, the atmosphere contained 560 gigatonnes (GT) of carbon, at 280 parts of CO_2 per million. Today,

thanks to our consumption of Earth's fossil fuels, it contains 800 GT of carbon at 390 ppm CO_2 (2009), increasing by 2 ppm (3 GT of carbon) a year.

The only truly safe goal is to get back down to 280 ppm. Stabilizing at 450 is far too dangerous, and so is 350, though it is an important intermediate goal. If we could cease producing anthropogenic CO_2 and restore Earth's forests, farmlands and grasslands to holistic ecosystems-based management, they could capture up to 6.5 GT of carbon a year from the atmosphere, enabling us — theoretically — to return to 560 GT (280 ppm) by 2100, for full atmospheric restitution to a climate friendly world.

To achieve these, we need to engage in two more goals:

Goal 3: Awaken the world to the danger we are in.

The public has absorbed the general message, but almost no one has grasped what climate change will cost us if we fail to act. In addition to three-yearly regional and national reports that lay out the financial and other consequences of not acting (see #61), we need a three-yearly report from the United Nations that spells out the cost for the whole world.

Goal 4: Inspire people with a vision of what victory looks like.

During World War II, as well as desiring to defeat the Nazis, the Allies were motivated by their hopes for the world that would follow victory. When Vera Lynn sang, "There'll be blue birds over the white cliffs of Dover…" she conjured up a powerful longing for the peace that would follow the war.

Without vision, people perish. We must be able to visualize a positive vision of a sustainable, climate-friendly world, backed with modeling that examines the impact of such a rapid reduction in emissions, followed by a drawdown. None of the existing models addresses such rapid reductions, leaving a major hole in our understanding of what is possible.

To win this climate peace, we need also to embrace a fifth goal:

Goal 5: Organize as if our lives depended on it — for they do.

We need to act as if we have been attacked in war: pull out every stop, invoke every emergency power and expect everyone to pull their weight. But this time, for the first time in human evolution, we need to do it not as a nation but as the entire planet.

This is an incredible moment. It is our generation that has the responsibility to guide our global civilization through the most critical years of its existence. If we fail, we lose everything. If we succeed, we reap enormous benefits:

• We remove the global warming Sword of Damocles that hangs over our planet, and the fear that we're on the path to extinction;

• We create a climate-friendly world that operates on a permanent supply of safe, renewable energy, and we eliminate the looming worries about peak oil and gas;

She is depending on us to do what is right…

• We remove a primary cause of warfare in the 20th and 21st centuries — conflict over oil and gas — and enjoy an enormous peace dividend;

• We accelerate the willingness of the world's nations to cooperate together for global goals;

• We restore hope to our children and grandchildren, who can resume the journey of human civilization toward whatever goals they choose.

92

Forge a New Global Treaty

No generation before us has faced a decade of choices that will so profoundly impact the course of life on this planet as those we now face. And no generation before us has had the opportunity to enrich the future so vastly.

— Eban Goodstein

There can be no doubt that we need a new global treaty. Our first attempt — the Kyoto Protocol, signed and ratified by 183 nations — has been a good beginning, even though few will meet their goals.

In the years since Kyoto was created in 1997, the climate deniers have mounted an enormous campaign of disinformation through the organized efforts of oil-industry-funded think tanks and voluntary websites that have poured vitriol and abuse on the science of climate change and on people like Al Gore, who have made a heroic

effort to alert the world to the dangers we face. Most of the delay in reducing our emissions can be put down to this disinformation, which has reinforced people's natural conservatism and their desire to resist changes they don't understand.

The sooner we dispel the disinformation and get people excited about a climate-friendly world, the sooner we will get traction on reducing our emissions. Once we begin to see success, however, the danger is that people will go back to sleep and the sense of urgency will evaporate. This is why we need a strong global treaty, with mechanisms to normalize, legislate and reward the process of carbon reduction.

Three options have attracted serious discussion:

1. The mainstream post-Kyoto Treaty, finalized in Copenhagen in 2009;
2. Oliver Tickell's Kyoto2, proposing a global system of cap and trade;
3. Aubrey Meyer's Contraction and Convergence solution, proposing a rights- and equity-based approach.

Post-Kyoto: The Mainstream Option

At the time of writing, the Copenhagen Agreement was still in the future. The European Union had agreed to a 20% reduction below 1990 by 2020, combined with a 20% increase in energy efficiency and 20% of its energy to come from renewables.

The weakness of this option is that it does not involve the whole world or go far enough in the urgency of its reductions. Its strength is that it is mainstream, and that it commands the loyalty of most of the world's nations.

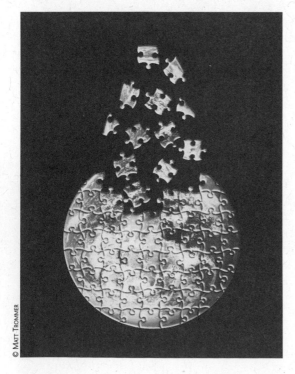

© MATT TROMMER

- Contraction & Convergence: gci.org.uk
- Kyoto2: kyoto2.org
- UNFCCC: unfccc.int

Ideally, the new treaty should:

- impose a legally binding 30% reduction in emissions below 1990 by 2020,
- limit developing world emissions to 15–35% below business-as-usual by 2020,
- include black carbon (soot) in the basket of greenhouse gases,
- reduce atmospheric carbon to 350 ppm, en route to 280 ppm,
- include emissions from international aviation and shipping,
- include a mechanism to protect forests from deforestation and degradation,
- include a mechanism to restore carbon in farmlands and grasslands, and
- establish a framework for global carbon trading that is measurable, reportable and verifiable.

Kyoto 2

In his book *Kyoto2: How to Manage the Global Greenhouse,* Oliver Tickell proposed a unified global system of cap and trade for carbon emissions, under which all greenhouse gases in every country would be capped as close as possible to their point of production, and the annual emission rights would be sold at a global auction, with the income — which could reach $1 trillion a year — being used to help finance the solutions and climate adaptation costs.

Many notable climate activists have expressed support for Kyoto2, including Jonathon Porritt, Bill McKibben, George Monbiot and Mark Lynas, and even though it is not being adopted at Copenhagen, there is every opportunity for the world's carbon-trading schemes to unite and create Kyoto2 in parallel to the new Copenhagen mechanisms. The weakness of the proposal is its faith in carbon trading.

Contraction and Convergence

Aubrey Meyer, a former viola player and composer, has been promoting a third solution, Contraction and Convergence (C&C) since 1998, soon after he became aware of the dangers that climate change presented, and how little Kyoto would do to solve the problem.

C&C would establish an annual global cap on emissions based on the best science; create an entitlement to these emissions that would be shared by all humans, reflecting the principles of justice and equity; allocate each country a share of the emissions based on population; trade the allowable emissions internationally, creating a considerable flow of money from the richer to the poorer nations; and shrink the availability of emissions certificates as the annual allowable emissions level was lowered.

C&C has won considerable support, including from some national leaders and many climate leaders. The consensus among its supporters is that C&C should provide the foundation for a future global treaty. The weakness of the proposal is the enormous transfer of money that it would create from the rich to the poorer nations without any linkage to investments in climate solutions.

The weakness of all three approaches is that they focus on reducing the problem. To address this, we need climate solutions treaties (see #93).

93

Adopt Climate Solutions Treaties

The point is to provide the next generation with the tools to restore Earth and themselves.... The security of humanity will not require armies of soldiers but battalions of conscious environmentalists.

— Hanne Strong

When climate change first became a public concern, most people assumed we had time to fix the problem, so the focus was on shaving bits off the problem by reducing our emissions.

This negative mindset is one reason why we have not made better progress: we have been fixated on removing the problem, rather than generating the solution. This does not build a vision of a sustainable, climate-friendly world that can make people excited and eager to act. We also need climate solutions treaties.

A Global Renewable Energy Treaty

The price of all renewable energy is falling as mass production increases. If nations agree to increase their use of renewables by a set amount each year, with specific targets for solar, wind and geothermal, investors would have the security they need to finance new projects. The formation of the International Renewable Energy Agency in 2009 is a huge step forward.

A Global Energy Efficiency Treaty

We need global energy efficiency standards that maximize the efficiency of all appliances, following Japan's Top Runner Program (see #66) — and we need an International Energy Conservation Agency to help make it happen.

A Global Low-Carbon Vehicles Treaty

We need a global agreement in which nations agree to purchase large volumes of zero- and low-carbon-emission vehicles by set dates, as California has done. Volume production is critical to drive down the price — we can't wait until 2015 or 2020

for cheap EVs and PHEVs. If vehicle manufacturers knew there was a treaty among nations to purchase 5 million PHEVs in 2012 and 50 million in 2014, production lines would be converted overnight.

A Global Zero-Carbon Buildings Treaty

Architects are already designing zero-carbon buildings, and some jurisdictions are writing green building codes, such as Britain's requirement that all new buildings must be zero carbon by 2016. We need a treaty in which nations agree to make all new buildings zero carbon by 2020, and help each other with resources.

A Global Zero-Waste Treaty

We need to eliminate all garbage, both for its methane emissions and to stop energy-intensive materials from being landfilled and incinerated, not just in a few scattered cities, like San Francisco. We need a global zero-waste treaty in which nations agree to achieve zero waste by 2030, and help each other get there.

A Global Farmlands Carbon-Recovery Treaty

We need to reverse decades of carbon loss from the world's farmlands through chemical use and poor farming practices. A carbon-recovery treaty would commit nations to increase the percentage of farmlands being monitored and rewarded for carbon recovery, aiming at 100% global compliance by 2030, and give farmers the support they need.

A Global Grasslands Carbon-Recovery Treaty

We need a similar treaty among grasslands nations that commits them to holistic grazing management and other practices that restore the grasslands' carbon, rewarded by inclusion in a world carbon-trading market, with the goal of 100% global compliance by 2030.

A Global Forests Carbon-Recovery Treaty

We need a treaty by which the world's forest nations agree to restore the world's forests and maximize their ability to store carbon. (See #95.)

A Global Carbon-Tax Treaty

A global carbon tax, levied by as many nations as possible, would internalize the price of pollution into the global economy and level the economic playing field among nations. A percentage of the income could go into a global climate-solutions fund (see #94) and the rest be kept by participating nations.

A Global Carbon-Subsidies-Removal Treaty

Subsidizing the use of fossil fuels is one of our worst global habits: we are paying to accelerate our own destruction. We need a treaty among nations to eliminate all such subsidies by 2015.

A Global Methane Treaty

In the critical years leading up to 2020, and thereafter, the impact of methane must be measured by its global warming potential over its 8-year lifetime, not the artificial 100-year lifetime set by the IPCC. A global treaty would bind nations to eliminate their methane emissions from landfills, animal manure, natural gas leakage and other sources.

BETTER PLACE

A Better Place electric charging post in Sacramento, California. A global treaty to accelerate the production of EVs and plug-in hybrid EVs would drive prices down.

A Global Black-Carbon Treaty

Black carbon, or soot, has only recently begun to ring the alarm bells as a cause of global warming. (See p. 14.) A global treaty is urgently needed to replace smoky cook-stoves and dirty diesel vehicles and accelerate other known solutions.

A Global F-Gas Treaty

The F-Gases are usually forgotten and ignored, but the air-conditioning and coolant gas HFC-134a has a staggeringly high GWP — 5,000 times more powerful than CO_2 over its short, 14-year lifetime. A global treaty would bind all nations to phase out all F-Gases by 2020. (See #97.)

94

Establish a Global Climate Fund

We, the human species, are confronting a planetary emergency. But there is hopeful news as well: we have the ability to solve this crisis and avoid the worst — though not all — of its consequences, if we act boldly, decisively and quickly.

— Al Gore

A global climate fund large enough to help the world's poorer nations protect themselves against the impacts of climate change and make the transition to a climate-friendly world is essential. Everyone knows that it has to happen — the question is how.

In 2008 Britain's prime minister proposed that such a fund be established by the World Bank. China has proposed that all developed nations should contribute 1% of their GDP to help the poorer nations reduce their greenhouse gas emissions ($130 billion a year for the US; $160 billion for the European Union). They also floated the idea that if a nation failed to meet its climate targets, it would be obliged to increase its payment into the Fund.[1] The Group of 77 developing nations, has proposed that the fund be substantial, obligatory

and automatic, and that it be overseen by the UN, not the World Bank, which they distrust.[2]

Oxfam International has proposed that for climate adaptation alone at least $50 billion a year is needed; more if nations fail to reduce their emissions enough to stop the temperature from rising by 2°C. The full fund might need $150 billion a year to cover adaptation ($50 bn), accelerating the climate solutions ($50 bn) and restoring forests and grasslands ($50 bn).

That seems like a lot of money, but it is only 0.23% of the world's 2008 GDP of $65 trillion. The post-war Marshall Plan, when the US invested 1.7% of its GDP ($13 billion) between 1948 and 1951 to help Western Europe rebuild its shattered economy after World War I, was seven times greater than this.[3]

The world is not short of money:

- In 2008–2009 the US spent $4 trillion trying to bail itself out of its financial crisis. Globally, the total may have been close to $8 trillion.

- In 2007 the world's nations spent $1.3 trillion on their militaries ($700 billion for the US military), many of which are focused on energy conflicts that would disappear in a sustainable, climate-friendly world.

- The world's nations spend $240 to $310 billion a year subsidizing fossil fuels, using public money to pour gasoline on the fire of global warming.[4]

- In 2007 the top 50 US hedge and private equity fund managers earned an average of $588 million each, totaling $26 billion. It would take the average US worker 19,000

- Association for the Taxation of Financial Transactions for the Aid of Citizens: attac.org
- Campaign for a Global Climate Fund: choike.org
- Currency Transaction Tax: currencytax.org
- Financing Climate Change: cseindia.org/equitywatch/financing_climate.html
- Global Green Economy Initiative: unep.org/greeneconomy
- Global Taxes: igc.org/globalpolicy/socecon/glotax
- Sovereign Wealth Fund Institute: swfinstitute.org
- Tobin Tax Initiative: ceedweb.org/iirp
- UNEP Green Economy Initiative: unep.org/greeneconomy
- Climate Funds Update: climatefundsupdate.org

years to earn this much money. US taxpayers subsidize executive compensation by $20 billion a year through a variety of tax and accounting loopholes.[5]

- In 2008 Sovereign Wealth Funds totaled $3.3 trillion before the financial crisis; 64% came from the earnings of oil-and gas exporting countries, socked away for a rainy day (Abu Dhabi $875 billion; Norway $350 billion; Kuwait $250 billion).[6] Norway is contributing a small amount to tackle the problem, spending $300 million a year on forest protection in the Congo Basin, and $1 billion to support the Amazon Fund, but no others have stepped up to the plate.

The next question is 'how could we raise the money'?

A Global Carbon Tax

In 2007 the world produced 31 billion tonnes of CO_2 by burning fossil fuels. A global tax at $25 per tonne would produce $775 billion a year. If 80% remained in the country of origin, 20% ($145 billion) could finance the global climate fund.

A Global Tax on Currency Dealing

In 2008 international currency dealers traded $4 trillion a day, of which only 5% financed any actual trade in goods and services — the rest was speculative activity. In 1972 economist James Tobin proposed levying a 1% tax on foreign exchange transactions to mitigate the predominance of speculation over enterprise.[7] Assuming 240 trading days per year, and that 20% of the transactions were tax-exempt, 20% evaded the tax, and the

Algae Ponds at HR BioPetroleum Pilot Facility located in Kona, Hawaii.

HR BioPetroleum

volume of trading fell by 50% because of the tax, a 0.25% tax would generate $720 billion a year — enough to pay for the climate solutions fund, the United Nations, the worldwide elimination of AIDS, leprosy and diphtheria, and the dismantling of all nuclear warheads.

A Global Levy on Cap-and-Trade Auctions

The world's emerging carbon market, worth $60 billion in 2007, could allocate 10% of the income from the auction of carbon permits to the global climate fund, providing $6 billion a year.

These are all major proposals that some financiers and politicians may dismiss out of hand. The Marshall Fund was only created *after* the shock and chaos of World War II. With climate change, there is no "after," when we can pick up the pieces. We have to do it *now*.

95

Protect the World's Forests and Grasslands

The Costa Rica experience [shows that] with dedicated resources, creative institutions, and a sound legal framework, deforestation can be reversed and forest cover expanded.

— Bruno Stagno-Ugarte,
Costa Rica's Foreign Minister

The world's tropical rainforests and their soils store 400 GT of carbon. If this were lost, it would raise the atmospheric CO_2 by 100 ppm and the global temperature by a further 0.6°C.

And losing it we are at 13 million hectares a year, because of illegal logging and forest clearance for farming and biofuels, especially in the Amazon, Indonesia, Congo Basin and Nigeria. Governments are constantly getting proposals from investors who want to convert their forests into land for agriculture or biofuels.

The annual carbon loss from the tropics comes to 5.8 GT of CO_2, representing 16% of global CO_2 emissions. The loss to nature, and to the species that live there, the forest dwellers and the 60 million people who depend on the forests for their livelihood, is immeasurable. Untouched natural forests store 60% more carbon than plantation

The buttress roots of a rainforest tree in Leticia-Amazon, Colombia.

forests,[1] and research has discovered that intact tropical forests are absorbing 18% of the annual CO_2 from fossil fuels.[2] If we don't protect them, the emissions from forest loss will increase atmospheric CO_2 by an additional 33 ppm by 2100.[3]

We know that progress is possible. Costa Rica is moving toward carbon neutrality by 2021 by creating national parks, banning deforestation, planting trees and negotiating a debt-for-nature swap with the US government. Guyana has expressed a willingness to place almost its entire rainforest — larger than England — under internationally verified supervision, if the right incentives can be created. In Paraguay, a moratorium on deforestation cut illegal logging by 83% in just one year.[4] In the utopian community of Las Gaviotas, Colombia, 8,000 hectares of Amazon rainforest have been regrown on the previously treeless savannah, and plans are underway to reforest 6.3 million hectares.

We know that when forest-dwelling communities are involved, they do an excellent job of protecting the forest. One of the reasons rainforests are being destroyed is that the forest dwellers do not have secure land tenure and can't use legal means to protect their rights.

There are lots of good intentions. In the Amazon, Brazil has committed to reduce deforestation by 70% by 2018, and Norway is giving $1 billion over seven years to the Amazon Fund. Peru is seeking $20 million a year to help it save 54 million hectares of the Amazon over ten years — that's only 37 cents per hectare per year.

The Eliasch Review, commissioned by the British government, urged the world to achieve a

RHETT A. BUTLER MONGABAY.COM

50% cut in carbon emissions from deforestation by 2020, and net carbon neutrality by 2030. If we succeed, the long-term financial benefits from the ecosystem services that tropical rainforests provide would come to $3.7 trillion. If we fail, the additional cost of climate change damages could amount to $1 trillion a year by 2100.[5]

The boreal forests in Canada and Russia must also be protected, because they store 22% of the Earth's terrestrial carbon, almost twice as much per acre as the tropical forests. Canada's boreal forest stores some 186 GT of carbon.

There is a global consensus that the means must be found to compensate for carbon preserved in protected forests. It's called REDD — Reducing Emissions from Deforestation and Forest Degradation. Skating over a wealth of complexities, this is the outcome we need to achieve:

- Protection of land tenure rights for forest-dwelling people and tribes
- Full participation by forest communities, including women and elders
- An end to corruption, and increased policing
- Satellite and on-the-ground monitoring
- Strengthened local institutional capacity to provide good governance
- Local economic opportunities, enabling forest communities to benefit by protecting the forest
- Integrated forest-management policies that protect the whole ecosystem
- A sufficient price on forest carbon to attract investors in the global carbon market

- Amazon Watch: amazonwatch.org
- Canopy Capital: canopycapital.co.uk
- Coalition for Rainforest Nations: rainforestcoalition.org
- Congo Basin Forest Fund: cbf-fund.org
- Forest People's Programme: forestpeoples.org
- Forest Stewardship Council: fsc.org
- Forests Now: forestsnow.org
- Friends of Gaviotas: friendsofgaviotas.org
- Global Canopy Programme: globalcanopy.org
- Global Forest Alliance: worldwildlife.org/alliance
- Global Forest Watch: globalforestwatch.org
- International Boreal Conservation Campaign: interboreal.org/globalwarming
- Little REDD Book: littleREDDbook.org
- Mongabay: mongabay.com
- Prince's Rainforest Project: princesrainforestsproject.org
- REDD Monitor: redd-monitor.org
- Rights and Resources Initiative: rightsandresources.org
- The Borneo Project: borneoproject.org
- World Rainforest Movement: wrm.org.uy

- Elimination of the perverse incentives that entice investors into clearing forests
- Limitation of the purchase of forest carbon credits by developed nations to a small portion of their required carbon reductions.

Develop a Global Grasslands Mechanism

The same is needed for the world's grasslands, which have an even larger capacity to store carbon. This discussion has hardly begun, and negotiations are urgently needed to create a mechanism that will encourage pastoralists and grasslands managers to adopt new practices.

96

Slow Global Population Growth

> If global warming is not contained, the West will face a choice of a refugee crisis of unimaginable proportions, or direct complicity in crimes against humanity.
>
> — George Monbiot

If we are struggling to reduce our global carbon footprint with 6.8 billion people, how will we cope when Earth's population in 2050 is 9.3 billion, just 40 short years away?[1]

It took us six million years to reach a population of one billion, in 1804. The Industrial Revolution brought huge improvements in public health, causing the numbers of deaths in childbirth and childhood to fall like a stone. Go back a few generations in any developed country and you'll find families with six to ten children.

Our sex drive is ancient, and when you take away nature's means of controlling our numbers by maternal and infant death without adding birth control, our population explodes. By 1927 it had reached two billion. By 1999 it had reached six billion — and while the rate of growth is slowing,

A Korean girl pleads with the world to reduce the atmospheric level of CO_2 to 350 ppm.

our total population is still increasing by 80 million people a year.

The pressure that we create on Earth's soils, oceans, forests, atmosphere and ecosystems is enormous, for each of us wants to enjoy life to the fullest, whether we live among the millionaires in Hollywood Hills or the slum-dwellers in Dharavi, Mumbai, India, where 44,000 people crowd into every hectare, sharing one toilet for every 5,000 people.

It is true that most of the world's carbon emissions have been produced by the developed nations, and it is also true that the carbon footprint of a Dharavian is a hundred times smaller than that of a Hollywood film star. To conclude that population growth is not therefore a problem, however, is a disastrous mistake.

All of the additional 2.5 billion people by 2050 will want to live a better life. They will want to eat — pushing up food prices and increasing the pressure to clear the rainforests. They will want space to live in — increasing the pressure on natural ecosystems. And they will want to consume — increasing the amount of energy we need to produce.

As soon as a nation educates its women and achieves a certain level of economic development, its women choose to have fewer children. We need this economic growth — and we must have confidence that nations can achieve it without burning fossil fuels. A life lived sustainably in an ecological city or solar village can satisfy most people's desire for a better life.

As part of our effort to reduce our carbon emissions, we *must* continue to reduce our global

- International Planned Parenthood Federation: ippf.org
- Population Reference Bureau: prb.org
- United Nations Population Fund: unfpa.org

birthrate. A generation ago, most women in the developing world had six children. Today, they have 2.6 children, or 2.3 if you include girls who don't live to adulthood, which is getting close to the level needed to end population growth.

We know what works — ready access to family planning techniques, including abortion, and a strong focus on the health, education and empowerment of women. As a world, however, we are failing to deliver.

Birth control programs must be broadened and made more effective. In sub-Saharan Africa, only 10% of the poorest women and 31% of the wealthiest women use family planning. At least 200 million women around the world do not have access to effective and affordable family planning services — and demand for these services is expected to increase by 40% by 2020, while funding for family planning has fallen.[2]

In 1965, Egypt and Ethiopia had similar populations (21.8 million and 18.4 million), and the women in both countries had an average of 6.6 children. Egypt started family planning in the 1960s, and its population in 2050 is expected to be 121 million. Ethiopia did not start until 2000, and its population in 2050 is expected to be 183 million. Only 6% of women in Ethiopia use modern contraceptive methods — yet Ethiopian clinics have reduced services, laid off staff, cut community health programs, and there is a shortage of contraceptive supplies.[3]

This failure is in part because of the decision by the US administration under George Bush to allow fundamentalist religious interests to control

Population in millions			
	1950	2008	2050
India	357	1150	1755
Pakistan	37	170	292
Afghanistan	6	28	79
Bangladesh	42	147	215
Indonesia	80	240	343
China	554	1325	1437
Brazil	54	195	260
Congo, DR	0.8	67	189
Nigeria	33	148	282
Africa	242	922	2000
USA	158	305	438
World	2500	6700	9300

federal policy, which led the government to impose a "Global Gag Rule," prohibiting any organization that received US funds from using its own money to provide abortion information, services and care. The UN Population Fund lost $34 million for every year of the Bush Administration because of the dispute over abortion. This policy was scrapped as soon as the Obama administration entered office in 2009.

Nations must fund population programs adequately. If we met women's current unmet needs for contraception, we could prevent 23 million unplanned births, 22 million induced abortions and 1.4 million infant deaths every year.[4] We must integrate population control into our portfolio of climate solutions.

97

Phase Out the F-Gases

> In almost all areas of application, it is possible to replace fluorinated greenhouse gases by halogen-free alternatives.
>
> — Germany's EPA[10]

Among the greenhouse gases that are warming Earth's atmosphere, the F-gases (CFCs, HCFCs, HFCs, PFCs, and SF6) are often forgotten. They are called F-gases because they contain fluorine, but among them, the HFCs (hydrofluorocarbons) are 95% of the problem.

HFCs were introduced as a replacement for CFCs in fridges, air conditioning and insulation, because the CFCs were destroying Earth's ozone layer. Unfortunately, the ozone people didn't talk to the climate people, and the HFCs that are being phased in are themselves a powerful greenhouse gas. This puts us the ridiculous situation where we are being asked to phase HFCs *in* to save the ozone layer and out to save the climate at the same time. We should laugh, if the problem were not so serious.

The most common HFC, which carries the glamorous name HFC-134a, is used mostly in air conditioning. It accounts for two-thirds of all HFCs in use, and two thirds of the HFC-134a is used in mobile air conditioning — mostly in vehicles. In 2007, HFC-134a global production was about 245,000 tonnes.[1] It survives in the atmosphere for only 14 years, but its global warming potential is such that over 100 years it traps 1,430 times more heat than CO_2. Over its natural life of 14 years, it traps up to 5,000 times more heat,[2] so during those 14 years — the critical period for global action on climate change — each kilogram of escaped HFC-134a acts like 5 tonnes of CO_2 equivalent.[3]

The critical question is: how much escapes? The industry claims a leakage rate of less than 5%, but independent research backed by fluorocarbon industry data indicates that 25–30% a year is more likely,[4] in which case the entire annual production is escaping within five years. Evidence for high leakage also comes from the Norwegian Institute for Air Research, which found that concentrations of HFC-134a above Mount Zeppelin had doubled between 2001 and 2004, and from China, where industry data assumes twice as much HFC-134a sales for servicing as for new uses.[5,6]

If a whole year's production of 245,000 tonnes escapes over five years, it will trap as much heat as 350 million tonnes of CO_2 — representing 1% of the total CO_2 emissions. That's comparing it to CO_2 over 100 years, however. Compared over the critical 14 years of its actual life in the atmosphere, it will act like 1.25 billion tonnes of CO_2, representing 3.3% of global CO_2 emissions.[7] Production is also growing — and hence the urgency of eliminating it. If the current growth rate were to continue to 2040, an additional 90 billion tonnes of CO_2e would be added to the atmosphere.[8]

- Alliance for CO_2 Solutions: alliance-co2-solutions.org
- Avoiding HFCs: mipiggs.org/library/pdfs/climatefriendly3.pdf
- Chilling Facts: chillingfacts.org.uk
- CO_2 substitutes for HFC-134a: r744.com
- Environmental Investigation Agency: eia-international.org
- Government Action: mipiggs.org/govaction
- Greenfreeze Technology: archive.greenpeace.org/climate/greenfreeze
- Multisectorial Initiative on Potent Greenhouse Gases: mipiggs.org

A molecule of HFC-134a traps 5,000 times more heat than CO_2 while it is in Earth's atmosphere. Each molecule of PFC-14, used in semi-conductor manufacturing, will trap heat for up to 50,000 years.

Seizure of illegal CFC refrigerant canisters in Indonesia, 2004.

The obvious solution is the rapid global phase-out of HFCs through the coordinated integration of the Montreal and Kyoto Protocols, before the fluorine industry builds too many new factories, as it is currently doing in China. There are safe alternatives for almost all HFCs and other F-gases, including CO_2-based systems that are more energy efficient. Unilever has already installed more than 200,000 non-HFC chiller systems worldwide, and Coca-Cola is planning to install 100,000 CO_2-based bottle coolers by 2010.

In its efforts to phase out HFC-134a in car air-conditioning, the European Union had a huge battle against organized lobbying by the fluorine industry. They eventually legislated a phase-out in new cars designed by the automakers by 2011 and in all new cars by 2017, while allowing F-gases with a GWP below 150 (such as HFC-152a). They have no phase-out for the other uses of HFC-134a, only a requirement for containment. Globally there is no legislation at all.

The other F-gases are similarly powerful which is why Denmark and Austria are phasing them out by 2011, and why Norway has imposed an F-gas tax of €30 ($42 US) per tonne of CO_2-equivalent, and Sweden similarly.[9] Leadership for a phase-out, thankfully, is being provided by the non-profit Environmental Investigation Agency, based in London, which has good background materials.

We must negotiate the full global phase-out of all F-Gases and not leave them to simmer away, cooking Earth's atmosphere for millennia.

F-Gases	Lifetime (years)	GWP over 20 years	GWP over 100 years	Developed world phase-out	Developing world phase-out	Status
CFCs	45–1700	Up to 11,000	Up to 14,400	1996	2010	Falling
HCFCs	1.3–18	Up to 2,300	273–5490	2020	2030	Rising
HFC-134a	14	3,830	1,430	EU Partial	None	Rising
HFC-152a	1.4	437	124	None	None	Rising
PFC-14	50,000	5210	7390	None	None	Rising
SF6	3,200	16,300	22,800	None	None	Rising
NF3	550	n/a	17,000	None	None	Rising

98

Decarbonize Global Financing

> Our kids are going to be so angry with us one day. We've charged their future on our Visa cards.
>
> — Thomas Friedman

Around the world, open-cast coal mines, coal-fired power plants and oil pipelines are still being built, even as floodwaters and tropical storms sweep away people's homes. What gives? Everyone knows we have to switch fuels, but down in Earth's boiler-room, nothing is happening.

The leaders of the G-8 nations — USA, Japan, Germany, France, Britain, Italy, Russia and Canada — who control 65% of the world's GDP, could be in control if they wanted to, but instead of sending a clear signal to decarbonize global financing, they use their power in the World Bank, the International Finance Corporation, the Inter-American Development Bank, the European Investment Bank, the European Bank for Reconstruction and Development and their national Export Credit Agencies to finance the fossil fuel projects.[1] It's business as usual out there, and they are doing nothing to stop it.

Coincidentally, many of these projects direct their contracts back to corporations from the G-8 nations. In 1995 Larry Summers, undersecretary in the US Treasury, told Congress that for every dollar the US government contributes to the World Bank's coffers, it gets $1.30 back in procurement contracts for US corporations.[2] If you doubt any of this, read *The Confessions of an Economic Hitman*, by John Perkins.

Loving the Oil and Gas Industries, 2000-2007[8]	
US Export-Import Bank	$12.3 billion
European Investment Bank	$7.3 billion
European Bank for Reconstruction and Development	$5.6 billion
World Bank	$8 billion

The World Bank's Carbon Funding

The World Bank says it's committed to tackling climate change, but it continues lending for fossil-fueled projects "to tackle poverty." It has ignored its own internal recommendations to phase out support for fossil fuels by 2008.[3] In that year, its lending to coal, oil and gas projects increased to more than $3 billion, and coal-fired project lending increased by 256%.[4] From 1997–2007 the projects it financed produced 26 Gt of CO_2 emissions.[5] In 2005, by contrast, financing for renewable energy and energy efficiency made up less than 5% of its financing, and only 2% of the financing by the International Finance Corporation (IFC), its private sector arm.[6]

The IFC is financing oil palm plantations in Indonesia, which release up to 28 times more carbon than doing nothing, and stimulating hundreds of millions of dollars' worth of cattle ranching in Amazonia, accelerating the destruction of the rainforest.[7] The Bank has refused to act on its commitment to analyze the carbon footprint of every project it considers.

Funding The Oil Industry

More money pours out to the oil industry. From 2000–2007 global lending agencies assigned at least $61.3 billion to the oil and gas industries, in addition to the domestic subsidies they were receiving of up to $210 billion a year.

Destroying Orissa[9]

The Indian state of Orissa is being devastated by coal mining and related projects that have unleashed a torrent of destruction. Rivers carry

- Bank Information Center: bicusa.org
- *Confessions of an Economic Hitman* by John Perkins: economichitman.com
- End Oil Aid: endoilaid.org
- Export Credit Agencies Watch: eca-watch.org
- Oil Aid Database: oilaid.priceofoil.org
- SEFI Public Finance Alliance: sefalliance.org
- Sustainable Energy and Economy Network: seen.org
- Sustainable Energy Finance Initiative: sefi.unep.org
- World Bank Climate Change: worldbank.org/climatechange
- World Bank Energy Projects Spreadsheets: bicusa.org/ei_spreadsheets

The Wairakei Geothermal Power Station, near Lake Taupo, North Island, New Zealand, produces 1550 GWh of electricity a year (4.3% of NZ's electricity).

toxic effluent through villages where people rely on the blackened waters for bathing, drinking and washing their clothes; the black waters of the Nandira, once a life-sustaining river, are slowly poisoning and killing people, animals, fish and plants as far as 50 miles downstream. Agricultural productivity has dropped for farmers who depend on its water; fishing communities have been wiped out; and the incidence of cancer, bronchitis and skin diseases is soaring. All for the sake of coal, financed by the World Bank. If Orissa were to invest the money in solar and biomass energy, it could trigger a statewide shift toward sustainability and the elimination of poverty.

Switch off the Dirty Gusher

The leaders of the G-8 nations must require the World Bank and the other international financial institutions to sign a sustainability pledge not to use public resources for projects that promote climate change, destroy rainforests, pollute rivers, destroy people's livelihoods or undermine human rights. In Chad, instead of financing Exxon's oil pipeline, the World Bank could finance microlending for village-based solar systems throughout the country. In China, instead of financing coal-fired power plants, it could finance efficiency and wind turbines.

Open the Green Energy Gusher

Several international NGOs are working to achieve such a goal, and the United Nations Environment Program has established a Sustainable Energy Finance Initiative to stimulate the use of public finance mechanisms to mobilize investment in climate solutions, including debt and equity financing, carbon financing and innovative grants. UNEP's analysis suggests that the careful use of public money can achieve a leverage ratio to 3 to 15, so that $10 billion in public money would leverage $50–$150 billion in total investment in climate solutions.[10]

We are past the time for more studies. Global leaders must just do it.

99

Adopt Natural Capitalism

What a great time to be born, what a great time to be alive, because this generation gets to completely change this world.

— Paul Hawken

We live at a time of multiple global crises — but as Albert Einstein so wisely said, "The problems that exist in the world today cannot be solved by the level of thinking that created them."

The climate crisis is clearly the most dangerous, as it has the ability to end most life on Earth if it is allowed to run out of control. It is accompanied by eleven other global crises, however, each of which poses a growing danger to humanity:

- The freshwater crisis
- The fisheries and oceans crisis
- The food and hunger crisis
- The poverty crisis
- The population crisis
- The species extinction crisis
- The deforestation crisis
- The chemical pollution crisis
- The looming peak oil and gas energy crisis
- The warfare and terrorism crisis
- The current financial crisis

One crisis is bad enough. How can anyone think about twelve, and stay sane? A smart person might pause to ask if they share a common cause

that might lend itself to a common solution. They share two common characteristics:

1. The lack of a coherent system of global governance and regulation, allowing freelancing and piratical nations, corporations and individuals to run rampant, endangering the wellbeing of all while stashing their ill-gotten gains in offshore bank accounts. (Shared by all 12 crises.)

2. The lack of any understanding that the global economy is a wholly owned subsidiary of nature, and that nature needs to be represented in every cabinet and board room meeting, every school and college, and every set of accounts. (Shared by all but the last two crises.)

Thanks to the urgency of the financial crisis, there is finally an agreement that we need a fundamental reform of the "Washington Consensus" neo-liberal ideology of deregulation, and a total rethink of the role of the state in the economy. The free market "don't regulate or intervene" attitude that got us into this mess has failed miserably, causing western governments to have to bail out their own economies. Hopefully, 2009 was to unregulated free market capitalism what 1989 was to communism.

This is *not* just another economic depression, needing deficit spending to reboot consumer confidence and get the engines of the economy rolling again. Because the fundamental character of the failed economy is a prime cause of all the other crises, if we reboot the same economy, we're going to make all the other crises worse. This is not just a financial crisis that needs the infusion

- Bretton Woods Project: brettonwoodsproject.org
- Choike — a portal on Southern civil societies: choike.org
- Natural Capitalism: natcap.org
- New Economics Foundation: neweconomics.org

Capitalism 1.0 Mercantile Pre-Capitalism (1550–1776) launched the foundations of modern trade and commerce.

Capitalism 2.0 Classic Capitalism (1776–1933) facilitated the Industrial Revolution and achieved many wonders, at the cost of great poverty and hardship.

Capitalism 3.0 Welfare Capitalism (1933–1980) added labor union organizing and social and healthcare protection in most developed nations.

Capitalism 4.0 Hyper Capitalism (1980–2008) removed government constraints, accelerated financial flows and invented digital derivative trading, but financially and ecologically it failed disastrously.

Capitalism 5.0 Natural Capitalism (2008–) restores government oversight, adds global governance and integrates nature into capitalism's software.

Clean Current's tidal turbine being installed at Race Rocks Ecological Reserve at Pearson College, near Victoria, replacing two diesel generators.

of ten trillion dollars and a new global regulatory regime to get it back on track — it goes far deeper than that.

This is a truly critical moment. As a world, we are about to adopt a new architecture for the governance of the global economy. If we change course now, we'll be able to address all the other crises too. Return to the same track, and it will carry us to a human and ecological train wreck of the worst possible kind.

Yes, we need a new global authority that can regulate the world's financial markets, but that is just the beginning. We also need a whole new paradigm of thought, reflecting new assumptions:

- Capitalism is a subset of natural capitalism, which integrates human activities into nature's activities.
- Since nature's resources and their services are the foundation of all wealth and wellbeing, their sustainability must be protected.
- Human economic activity must be constrained by the ability of local and global ecosystems to support it.

- Since the "invisible hand" of the market is blind and amoral, governments must to steer the economy toward society's chosen goals.
- In additional to personal gain, most people are also motivated by the desire to participate in nature and community.
- Community economies are an essential focus for local initiative, employment and sustainability.
- There is an income threshold beyond which more wealth no longer brings more personal happiness.
- Fabulous personal wealth indicates that greed is out of control. It is a sign that governance is failing and that offshore tax shelters need to be closed.

Capitalism is not something fixed — that is its greatest strength. It has evolved through at least four versions. Since Capitalism 4.0 has just crashed so miserably, the time has come to upgrade to Capitalism 5.0 — Natural Capitalism.

100

Build A Global Movement

> Our most basic common link is that we all inhabit this small planet. We all breathe the same air. We all cherish our children's future.
> — President John F. Kennedy[1]

Every journey to success starts with a visualization. If we can visualize something clearly, we can achieve it. If we can't, it means we need to dream up something more.

Many of us can visualize a peaceful, sustainable climate-friendly world, but it's not happening as fast as it needs to. Something more is needed.

We need a movement that unites people all over the world. We need children, students, worshippers, scientists, economists, politicians, writers, musicians, astronauts, artists, bird-watchers, Nobel Prize winners, doctors, nurses, parents, grandparents, farmers, philosophers, labor unions, business organizations, lawyers, social investors, parliamentary groups and senior citizens to join with environmentalists in demanding action.

We need coordinated protests with simultaneous actions, as the climate action group 350.org is achieving. We need to learn, so that we are well prepared for radio interviews and letters to the editor. Most climate deniers are not well informed — they take their opinions from the blogosphere, where anger and slander hook up with myth and nonsense to produce balderdash and baloney.

When I imagine a peaceful, climate-friendly 2040, I ask myself, "What made the difference?" Five images come to mind.

The Railroad Tracks

Imagine a Global Day to Stop Coal, when people all around the world sit down on the railroad tracks to stop the coal-fired juggernauts, and chain themselves to the doors of the banks that finance them. Some may defend themselves successfully, arguing in court as the Kingsnorth Six did in England in 2008 that breaking a small law is legally justifiable if it helps prevent a greater harm. Imagine such actions being accompanied by a torrent of letters and phone calls to the power-plant owners and the banks that finance them, declaring their refusal to buy their electricity or use their banks until they close the coal plants.

The Schoolchildren

Imagine young people around the world refusing to go to school for a week unless they are able to spend their time learning about climate change

In the pre-dawn hours of April 27, 2009, seven Greenpeace activists climbed a 140-foot construction crane to greet the sunrise and leaders of the world's 17 largest global warming polluters, who were assembled for climate talks at the State Department in Washington D.C.

and what they can do, sharing learning resources over the Internet. Imagine them linking together in shared actions that shake the adult world out of its sleep.

The Young People in the Courts

Imagine young people grouping together to bring class action lawsuits against their governments, arguing that their failure to act will cause them to suffer huge material and emotional damage and to pay far higher taxes in the future. Imagine judges and juries ruling in favor, requiring governments to act with sufficient urgency to prevent the predicted outcomes. Imagine them requiring governments to undertake a rigorous study every three years to quantify the cost of inaction to future generations, and requiring them to create enormous funds that the next generation could use to cover those costs.

The Business Case

Imagine the financial arguments for investing in climate solutions being so strong, and the arguments for inaction so disastrous, that everyone could see what must be done. Imagine leaders in the business community splitting ranks to sideline the delayers, using their collective weight to lobby their governments for action.

The Singing

Imagine need a global movement that is motivated not by the fear of catastrophe but by a love for this Earth — a love that is expressed in songs that become our anthems of hope, our inspiration to persist.

- 350: 350.org
- Arab Climate Campaign: indyact.org/environmental.php
- Rising Tide (Australia): risingtide.org.au
- Australia's Climate Action Summit: climatemovement.org.au
- Austria's Climate Movement: sos-klima.at
- Avaaz: avaaz.org
- Campaign Against Climate Change (UK): campaigncc.org
- Climate Action Network: climatenetwork.org
- Network for Climate Action (UK): networkforclimateaction.org.uk
- Rising Tide (UK): risingtide.org.uk
- Canada Climate Action Network: climateactionnetwork.ca
- Climate Action Network Europe: climnet.org
- Earth Hour: earthhour.org
- Fossil Fools Day: fossilfoolsdayofaction.org
- Friends of the Earth International: foei.org
- Germany's Climate Alliance: die-klima-allianz.de
- Global Climate Campaign: globalclimatecampaign.org
- Greenpeace: greenpeace.org
- India's Climate Movement: whatswiththeclimate.org
- One Blue Sky — the music: tinyurl.com/98a7wu
- One Sky: 1sky.org
- Rainforest Action Network: ran.org
- Réseau Action Climat France: rac-f.org
- US Climate Action Network: usclimatenetwork.org
- WWF: panda.org
- Youth Climate Movement: itsgettinghotinhere.org

101

Don't Sit This One Out

Anything else you're interested in is not going to happen if you can't breathe the air and drink the water. Don't sit this one out. Do something. You are by accident of fate alive at an absolutely critical moment in the history of our planet.

— Carl Sagan

And now, my friends, this is your moment.

By whatever comparisons you choose, we are living through the most astonishing period in humanity's evolution on this Earth. The Universe from which we evolved created itself 13.5 billion years ago. The Earth was formed 4.5 billion years ago. The first living cells emerged 3.5 billion years ago. Our modern ancestors left Africa 100,000 years ago. We embarked on the Industrial Revolution 250 years ago.

Within the last 100 years, we have waged the most destructive and painful wars the world has ever seen. We have rid the world of most dictatorships and tyrannies. We have removed many of the barriers that prevented women from participating in the affairs of the world. We have launched ourselves into space and looked back on our planet for the first time. We are moving beyond narrow nationalism and tribalism to create a truly global society. We have created the first global communications systems, forming a global brain.

During these past 100 years we have tripled our human population; multiplied our use of the Earth's resources twenty-fold; and poured a vast burden of pollution into Earth's oceans, soils and air, and into our own bodies. We have eliminated more species than at any other period of evolution.

By burning fossil fuels and destroying Earth rainforests and grasslands, we have put ourselves on a path that will lead to the collapse of Earth's ecosystems and human civilization. A rapid course adjustment is needed before it is too late.

We have done these things before, however. As humans, we are *good* at change, especially if we allow the world's youth to play their part. Maybe the problem with democracy is that too many old people vote and not enough young people.

As a planet, these are like our teenage years - we have a great sense of power, but not the wisdom to accompany it. Can we build a sustainable future? That is the great unknown. If we succeed, an incredible future awaits us. If we fail, it may be curtains for this particular planet. All of the sacrifices made by our ancestors, whether on the beaches of Normandy, in the struggle to abolish slavery, or through a million other heroic achievements, will have been in vain. Will we succeed? That is up to you, me, and all of us.

Within every human, a hero awaits. If you are a dancer, dance your concern. If you are a teacher, teach it. If you are just an ordinary person, join up with other ordinary people and change the world.

I have done my best to lay the solutions before you. At this point, I stand aside. The stage is yours.

- Joanna Macy: joannamacy.net
- The Great Turning: thegreatturning.net
- YES! Magazine: yesmagazine.org
- Positive News: PositiveNews.org.uk

© ANTONIA. DREAMSTIME.COM

Driven by the forces of love, the fragments of the world seek each other, so that the world may come into being.

— Pierre Teilhard de Chardin (1881–1955)

Our task must be to widen our circle of compassion to embrace all living creatures and the whole of nature in its beauty.

— Albert Einstein (1879–1955)

If the world is to be healed through human efforts, I am convinced it will be by ordinary people, whose love for this life is even greater than their fear.

— Joanna Macy (1929–)

Notes

PART ONE: THE CHALLENGE

A Gift from the Past

1 See *The Medieval Machine: The Industrial Revolution of the Middle Ages,* by Jean Gimpel. Pimlico, 1992
2 "Optimist Predicts World Oil Demand Will Outstrip Production In 2020," University of Colorado at Boulder Science Blog, 1998.

Our Story

1 Based on *The Cosmic Walk,* developed by Sister Miriam Therese McGillis based on the works of Thomas Berry and Brian Swimme and adapted by Ruth Rosenhek using the works and words of Elisabet Sahtouris, John Fowler and Lynn Margulis.
2 The Eemian period may have begun anywhere between 132,000 and 140,000 years ago. See *Sudden climate transitions during the Quaternary,* by Jonathan Adams, Mark Maslin and Ellen Thomas, esd.ornl.gov/projects/ qen/transit.html.
3 Maybe as early as 20,000 years ago.
4 The Lunar Orbiter 1 carried an Eastman Kodak camera that took the first ever photo of the Earth from a spacecraft in the vicinity of the Moon on August 23 1966, showing a cloud-covered, crescent Earth above a lunar landscape. For the actual photo, see lpi.usra.edu/resources/lunarorbiter/ frame/?1102.

Earth's Miraculous Atmosphere

1 "Atmospheric carbon dioxide concentrations over the past 60 million years," by Paul N. Pearson and Martin R. Palmer, *Nature,* Aug. 17, 2000.
2 "Oceanography: Stirring times in the Southern Ocean," by Sallie Chisholm (MIT), *Nature,* Oct. 12, 2000.
3 "Oceans weakening as warming tool," MSNBC, May 17, 2007, reporting on study by the British Antarctic Survey and the Max-Planck Institute for Biogeochemistry.
4 "Effects on carbon storage of conversion of old-growth forests to younger forests," by M.E. Harmon, W.K. Ferrell and J.F. Franklin, *Science* 247, 1990, pp. 699–702.
5 "Acceleration of global warming due to carbon-cycle feedbacks in a coupled climate model," by Peter M. Cox et al., *Nature,* Nov. 9, 2000.

The Greenhouse Gases

1 "Climate Change and Trace Gases," by James Hansen et al., *Philosophical Transactions of the Royal Society,* 365, May 2007, pp. 1925–1954.
2 See *Global Warming in the 21st Century: an Alternative Scenario,* by James Hansen et al (NASA Goddard Institute for Space Studies), 2000. See also "Strong radiative heating due to the mixing state of black carbon in atmospheric aerosols," by Mark Jacobson, *Nature* 409, 2001, pp. 695–697.

Methane — The Forgotten Gas

1 "Methane bugs warmed the young planet," Myles McLeod, *New Scientist,* March 24, 2006.
2 *IPCC Fourth Annual Report, The Physical Science,* Chapter 2, 2007.
3 In the IPCC Second Assessment Report (1995), methane's GWP was listed as 21. In the 3rd Report (2001) it was increased to 23; in the 4th report (2007) it was increased to 25, to account for the amplifying impact of increased water in the atmosphere.
4 Methane emissions soar as China booms. Steve Connor, *The Independent,* Sept. 28, 2006.
5 "Methane's impacts on climate change may be twice that of previous estimates," by Drew Shindell, NASA Goddard Institute for Space Studies, July 18, 2005.
6 "Methane emission controls can save thousands of lives," Princeton University News Release, March 29, 2006.
7 "Methane prime suspect for greatest mass extinction," Jeff Hecht, *New Scientist,* March 26, 2002.

Black Carbon

1 The 2007 IPCC Report estimates +0.5. In his testimony to the Congressional hearings on black carbon on October 18, 2007, *Role of Black Carbon on Global and Regional Climate Change,* Professor V. Ramanathan estimated +0.9.
2 0.8 watts per square meter is 21% of the human-caused radiative forcing of 3.79 watts per square meter.
3 Dr. Tami Bond, University of Illinois, Urbana Champaign, quoted in the "Black Carbon Briefing Paper."

Greenhouse Gases Chart

1 *IPCC Fourth Assessment Report: The Physical Science Basis, Summary for Policy Makers,* 2007

2 About 50% of the increased CO_2 will be removed within 30 years, and 30% of the remaining CO_2 within a few centuries. The remaining 20% may stay in the atmosphere for many thousands of years. IPCC 2007 FAR. See "Carbon is forever," *Nature Reports Climate Change,* Vol. 2, Dec. 2008.

3 GWP represents the cumulative radiative forcing between the present and a chosen time horizon (100 years) caused by a unit mass of gas emitted now, expressed relative to the reference gas CO_2, including any indirect effects of the emitted gases.

4 "Changes in Atmospheric Constituents and in Radiative Forcing," *IPCC Fourth Assessment Report: The Physical Science Basis, Summary for Policy Makers,* 2007. The figures represent increased forcing since 1750.

5 It is likely that total atmospheric water vapor has increased by several percent per decade over many regions of the northern hemisphere (IPCC).

6 Increased water vapor is caused by increased ocean precipitation, caused indirectly by human activities. There is uncertainty about the extent to which it causes increased radiative forcing due to an increase in high level cirrus cloud cover (which traps heat) or decreased forcing due to increased albedo on lower stratus clouds (which reflect heat back into space).

7 Water vapor feedback approximately doubles the warming for fixed water vapor (IPCC).

8 The numbers add up to 104% because the CO_2 from cement is counted twice, under both cement and coal.

9 "Global CO_2 Emissions from Fossil-Fuel Burning, Cement Manufacture and Gas Flaring: 1751–2004," Carbon Dioxide Information Analysis Center, Oak Ridge National Laboratory, May 29, 2007, cdiac.ornl.gov/ftp/ndp030/global.1751_2005.ems.

10 *Reducing Black Carbon May Be the Fastest Strategy for Slowing Climate Change,* Black Carbon Briefing Note, Institute for Governance and Sustainable Development, 2008.

11 Estimates range from 0.5 (IPCC, 2007) to 1.2 (Ramanathan, 2007), making 0.8 a mid-range number (Hansen, 2001 and 2005). For a comprehensive summary, see *Reducing Black Carbon* (supra).

12 Global methane sources and sinks are not well understood. Methane emissions in the atmosphere continue to increase, but the rate has slowed since 1990.

13 These are very approximate numbers, derived from IPCC Special Report on Emissions Scenario, 2001, 3.6.5 Ozone Precursors, which lists the sources of O_3's precursors — nitrogen oxides (50% fossil fuels, 16% biomass burning); carbon monoxide (26% fossil fuels, 30% biomass burning, 43% methane); and volatile organic compounds (72% fossil fuels, 14% biomass burning, 14% solvents.)

14 James Hansen suggests that the radiative forcing of HCFCs and HFCs will be similar to that of the CFCs they replace. If HFC-134a were phased out and stockpiles of CFC-21 destroyed, the overall forcing might fall to 0.25 Wm2. Range of uncertainty for radiative forcing: +/–0.5 Wm2.

15 Yoshio Makide and a team of researchers at Tokyo University reported in 1998 that HFCs are accumulating rapidly as they replace CFCs, though they are still at a very low level. 1995: 2 ppt. The figure of 8 for 2007 is based on an estimated increase of 0.55 ppt per year.

16 SF5CF3, chem.hawaii.edu/Bil301/SF5CF3.html (viewed July 2, 2007).

17 Data on various aerosols is difficult, partly because they are so short-lived.

18 In its 2007 Fourth Assessment Report, the IPCC put the radiative forcing of aerosols at –0.5 Wm2 for sulfate, organic carbon, black carbon, nitrate and dust.

19 –0.2 Wm2 for land cover changes largely due to net deforestation; + 0.1 Wm2 for the effect of black carbon on snow, but with a low to medium level of scientific understanding.

The First and Second Alarm Bells

1 "Ice-capped roof of the world turns to desert," Geoffrey Lean, *The Independent,* May 7, 2006.

2   "Rising seas, shrinking streams may leave state dying of thirst," Mike Taugher, *Contra Costa Times,* Jan. 22, 2007.
3   "The Century of Drought," Michael McCarthy, *The Independent,* October 4, 2006.
4   "World's most important crops hit by global warming effects," Steve Connor, *The Independent,* March 19, 2007.
5   "Rice yields decline with higher night temperature from global warming," by Shaobing Peng et al, National Academy of Sciences, June 29, 2004.
6   *Outgrowing the Earth: The Food Security Challenge in an Age of Falling Water Tables and Rising Temperatures,* by Lester Brown, W.W. Norton & Co., 2005.
7   *The Last Revolution,* by Fred Pearce, Key Porter, 2007.
8   Ibid, p 95.

The Third and Fourth Alarm Bells
1   "Earth faces 'catastrophic loss of species,'" Steve Connor, *The Independent,* July 20, 2006.
2   "Global warming and terrestrial biodiversity decline," WWF, August 1, 2000.
3   "Long-term decline in krill stock and increase in salps within the Southern Ocean," by Angus Atkinson et al., *Nature* 432, Nov. 4, 2004, pp. 100–103.
4   "Climate blamed for mass extinctions, Zeeya Merali, *New Scientist,* April 1, 2006.
5   See Graham Hancock's website for photos and interpretation: grahamhancock.com.
6   The IPCC sea level numbers. Posting by Stefan Rahmstorf at realclimate.org, March 27, 2007.
7   "Forecast for big sea level rise," BBC, April 15, 2008.
8   "How much will sea level rise?" Real Climate, Sept 4, 2008. realclimate.org/index.php?p=598.
9   "Huge sea level rises are coming — unless we act now," James Hansen, *New Scientist,* July 25, 2007.
10 "Bangladesh faces climate change refugee nightmare," Reuters, April 14, 2008.

The Fifth Alarm Bell: Global Economic Disaster
1   *The Stern Review on The Economics of Climate Change,* by Nicholas Herbert Stern, UK HM Treasury, 2006, p. 1.

2   "Climate change to cost Germany 800 billion Euros," German Institute for Economic Research, March 13, 2007.
3   See ecy.wa.gov/climatechange.
4   "Green Jobs: Towards Decent Work in a Sustainable, Low Carbon World," UNEP/ILO, 2008.

How to Talk to a Climate Denier
1   "The Scientific Consensus on Climate Change" by Naomi Oreskes, *Science,* Dec. 3, 2004.
2   "No Sun link to climate change," BBC News, July 11, 2007.
3   "Climate myths: It was warmer during the Medieval period, with vineyards in England," by Michael Le Page, *New Scientist,* May 16, 2007.

The Crucial Data
1   Fossil fuels produce 53.48% of the radiative forcing: 81% of CO_2's radiative forcing (36.64%)
    50% of the tropospheric ozone (4.5%)
    40% of the black carbon (8.4%)
    26% of the CH_4 (3.38%)
    14% of the N_2O (0.56%)
2   Deforestation produces 17% of the radiative forcing:
    40% of the black carbon (8.4%)
    15% of the CO_2 (6.6%)
    6% of the CH_4 (0.78%)
    12.5% of the tropospheric ozone (1.125%)
3   The data is extrapolated from the 2005 CDIAC data, increased by 6.3% for two years' growth. The total comes to more than 36.6 Gt because the emissions from cement include the use of coal, causing a double counting.
4   Tropospheric ozone (O_3) is created by atmospheric chemicals known as ozone precursors. Establishing their anthropogenic origins is complicated; these numbers are from the IPCC's 2001 *Special Report on Emission Scenarios,* 3.6.5 Ozone Precursors:
    • Nitrogen Oxides (NOx): fossil fuels 50%, biomass burning 16%, natural and anthropogenic soil release 24%, lightning 10%.
    • Carbon Monoxide (CO): fossil fuels 26%, biomass burning 30%, methane 43% (26% fossil fuel origins, 50% farming).

- Volatile Organic Compounds (VOCs): fossil fuels 72%, biomass burning 14%, solvents 14%. Looking at the data, it seems reasonable to assign 50% of O_3's origins to fossil fuels and 25% to biomass burning, 12.5% from tropical rainforest burning and 12.5% from other sources.
5 The total is greater than 100% because farming's emissions include deforestation and the use of fossil fuels, and cement's emissions include the use of coal.
6 The Greenpeace Report *Cool Farming* estimates that farming is responsible for 17–32% of all greenhouse gas emissions. The UN report *Livestock's Long Shadow* found that the livestock industry alone is responsible for up to 18% of global GHGs.
7 "Worst case global warming scenario revealed," Fred Pearce, *New Scientist,* Feb. 18, 2006.
8 *BP Statistical Review of World Energy,* June 2008. The oil reserves data is an average from the *BP Statistical Review, the Oil and Gas Journal,* and *World Oil,* which may be higher than reality — see p. 36.
9 Three tonnes of CO_2 per tonne of coal. Natural Resources Canada CO_2 Emissions Factors.
10 Each liter of oil on average releases 2.5 kg of CO_2 when burned. A barrel of oil (158 liters) releases 397 kg of CO_2. Shell Oil estimates an additional 60 kg per barrel for processing, for a total of 457 kg of CO_2 per barrel (see #60, Footnote 2). Burning 1,390 billion barrels of oil would release 635 billion tonnes of CO_2, or 173 Gt of carbon.
11 *Navigating the Numbers: Greenhouse Gas Data and International Climate Policy,* by Kevin A. Baumert, Tim Herzog and Jonathan Pershing, Chapter 6, World Resources Institute, 2006.
12 *Larry King Live,* Dec. 25, 1999.

What Targets Should We Adopt?

1 *Tipping elements in the Earth's climate system.* Proceedings of the National Academy of Sciences, by Timothy M. Lenton et al. Feb. 4, 2008. The other tipping points include:
- melting of Arctic sea ice
- decay of the Greenland ice sheet
- collapse of the West Antarctic ice sheet
- collapse of the Atlantic thermohaline circulation
- increase in the El Nino Southern Oscillation
- collapse of the Indian summer monsoon
- greening of the Sahara/Sahel and disruption of the West African monsoon
- dieback of the Amazon rainforest
- dieback of the Boreal Forest
2 "Warming hits tipping point" *Guardian,* Aug. 11, 2005.
3 Oil produces 33% of the CO_2, plus a share of the methane (fugitive emissions) and 40% of the black carbon (dirty diesel); its use is also a precursor for tropospheric ozone and nitrous oxide.
4 For a detailed discussion of the risks around 2°C, see *The 2° Target: How Far Should Carbon Emissions be Cut?* carbonequity.info/PDFs/2degree.pdf.
5 *Climate Change: On the Edge,* by James Hansen, UK *Independent,* Feb. 17, 2006.

Is the Oil Running Out?

1 "When will the oil run out?" George Monbiot interview with Fatih Birol, chief economist of the IEA, *Guardian,* Dec. 15, 2008.
2 "World will struggle to meet oil demand," *Financial Times,* October 28, 2008.
3 "When will we reach peak oil?" The Leading Edge, Signals, October 2008, tinyurl.com/dbr64c
4 *Study of World Oil Resources with a Comparison to IPCC Emissions Scenarios,* by Anders Sivertsson, Department of Radiation Sciences, Uppsala University, 2004.
5 "Too little oil for global warming," *New Scientist,* Oct. 5, 2003.

Energy Security

1 This assumes oil at $100/barrel. It rises to $56 trillion with oil at $200 a barrel.
2 *The Hidden Cost of Oil,* 2003, updated January 8, 2007. National Defense Council Foundation. Peer review comment from *The Hidden Cost of Our Oil Dependence.* Milton Copulos interview with Bill Moore, *EV World,* April 23, 2006.
3 20.7 million barrels of oil a day "cost" $825 billion a year, or $2.26 billion a day. 1 barrel (42 gallons) costs $109 a day. 1 gallon costs $2.60.

Earth's Future Buildings

1 *The 2030 Blueprint,* Architecture 2030, April 7, 2008.

Earth's Future Transport

1 20% of CO_2 (Radiative Forcing RF 44%) = 8.8%
+ 20% of black carbon (RF 21%) = 4.2%
+ 0% of methane
+ 25% of tropospheric ozone (RF 9%) = 2.25%
+ 25% of F gases (HFC-134a) (RF 9%) = 2.25%
+ 14% of N_2O (RF 4%) = 0.56%
Total = 18%, plus the other two factors.

Those Pesky Energy Data

1 US Energy Information Agency. World per capita CO_2 from the combustion and flaring of fossil fuels, 2006. This data does not include CO_2 from deforestation or other greenhouse gases.

Earth's Future Electricity

1 US Energy Information Agency.
2 This assumes a 15% capacity factor, averaging 3.6 hours of sunshine a day.
3 Algeria 2,381,740 sq. km. Texas 691,030 sq. km.
4 The Energy Watch Group's 2008 report *Wind Power In Context: A Clean Revolution in the Energy Sector* estimated 16,400 TWh by 2030, assuming a sustained 30% growth rate.
5 *Contribution of Geothermal Energy to Sustainable Development,* International Geothermal Association Submission to UN CSD 9, 2001.
6 *The Future of Geothermal Energy,* an assessment by an MIT-led interdisciplinary panel, 2007.

Earth's Future Water Power

1 See *China, Land of Discover and Invention,* by Robert Temple, p.54 (Patrick Stevens, 1986).
2 *World Energy Assessment 2000,* Chapter 5, p 154.
3 For a more detailed discussion see *Comparing CH_4 Emissions from Hydropower to CO_2 from Fossil Fuel Plants,* by Stuart Gaffin, Ph.D., (Environmental Defense Fund), submission to the World Commission on Dams. edf.org/documents/723_reservoirgases.pdf
4 Project aims to extract dam methane. BBC News, May 10, 2007.

Renewable Energy — The Small Print

1 Except where endnoted otherwise, all data comes from "Review of solutions to global warming, air pollution, and energy security," by Mark Jacobson, *Energy and Environmental Science,* Dec 1, 2008. stanford.edu/group/efmh/jacobson/revsolglobwarmairpol.htm
2 For updated data, see the US Energy Information Administration's website eia.doe.gov/emeu/international/electricitygeneration.html
3 Assumes 33% possible saving of global primary energy use of 145,000 TWh.
4 Assumes displacement of average North America electricity, at 600 grams CO_2/kWh.
5 Same as 3, since many efficiency projects can begin immediately.
6 *Nuclear Power: Climate Fix or Folly?* by Amory Lovins et al., RMI Solutions, April 2008, updated Dec. 31, 2008.
7 The price increases as the easiest efficiency projects are completed.
8 There is no known measurement for the growth of efficiency retrofits.
9 Lovins.
10 "Low-Carbon Energy: A Roadmap," by Christopher Flavin, *Worldwatch Report* 178, 2008. p. 18.
11 General market data.
12 Flavin, p.18.
13 The more desert you use, the more power you get. "Concentrating solar plants on less than 0.3% of the desert areas of North Africa and the Middle East could generate enough electricity to meet the needs of these two regions plus the European Union." *Solar Thermal Power Coming to a Boil,* Earth Policy Institute, July 2008.
14 *Solar Thermal Power Coming to a Boil,* Earth Policy Institute, July 2008.
15 *Assessment of Parabolic Trough and Power Tower Solar Technology Cost and Performance Forecasts,* NREL, October 2003.
16 "Concentrating solar thermal power capacity expected to double every 16 months over the next

five years," *Solar Thermal Power Coming to a Boil,* Earth Policy Institute, July 2008.

17 The MIT study of enhanced geothermal power found that the US has enough capacity for 2,000 times its current energy needs, or 20,000 times its needs with technological advances. *The Future of Geothermal Energy,* MIT, January 2007.

18 Geothermal Energy Association.

19 1990: 5831 MW; 2005: 9064 MW. International Geothermal Association.

20 World Energy Council, 450,000 MW, 27% capacity factor = 1064 TWh a year

21 Data from Renewable Northwest Project. rnp.org/RenewTech/tech_wave.html.

22 World Energy Council, European Ocean Energy Association.

23 Data from Renewable Northwest Project. rnp.org/RenewTech/tech_wave.html.

24 Flavin, p.18.

25 The low number is for current thermal reactors and known uranium reserves. The high number is for light water and fast-spectrum reactors and possible future uranium reserves (Jacobson).

26 *Nuclear Power: Climate Fix or Folly?* by Amory Lovins et al, RMI Solutions April 2008, updated 31 Dec 2008.

27 Flavin, p.18.

28 This is based on the MIT report *The Future of Coal,* which found that CCS coal would not be technically practical at a utility scale before 2030.

29 This is Jacobson's number, based on a 5- to 8-year construction delay. The 20-year delay that is realistic will add to this number enormously.

30 As for coal, with 75% increased cost for CCS.

Earth's Future Bioenergy

1 *Biofuel less sustainable than realized,* Wetlands International, Dec. 8, 2006. Quoted in "A Lethal Solution," by George Monbiot, *Guardian,* March 27, 2007.

2 There's a big debate as to whether ethanol produces a positive or negative net energy return. A 2004 US Department of Agriculture study found that using modern agricultural methods, it has 67% net energy gain. A 2006 study by David Pimentel, an ecologist from Cornell University, however, found that when you account for all the energy inputs, biofuel from corn has a 29% net energy loss. ("Cornell ecologist's study finds that producing ethanol and biodiesel from corn and other crops is not worth the energy," Cornell University News Service, July 5, 2005.) Researchers from the prestigious Rocky Mountain Institute, on the other hand, say that, "most scientific studies, especially those in recent years reflecting modern techniques, do not support this concern." (*Setting the Record Straight on Ethanol* by Nathan Glasgow and Lena Hansen, Rocky Mountain Institute, Oct 31, 2005.)

3 "Exploding US Grain Demand for Automotive Fuel Threatens World Food Security and Political Stability," by Lester Brown. Earth Policy Institute, Nov. 3, 2006.

4 "Review of solutions to global warming, air pollution, and energy security," by Mark Jacobson. *Energy and Environmental Science,* Dec. 1, 2008.

5 Ibid.

6 "The High Price of Clean, Cheap Ethanol," Clemens Höges, *Der Spiegel,* Jan. 22, 2009.

7 "New Technology Foresees Trees, Not Grain, in the Tank," Christian Wüst, *Der Spiegel,* April 15 2008.

8 "Carbon-negative biofuels from low-input high-diversity grassland biomass," by David Tilman et al., *Science,* Dec. 8, 2006.

9 "Tree fungus could provide green transport fuel," *Guardian,* Nov. 4, 2008.

10 The lower yields come from "Resetting global expectations from agricultural biofuels," by Matt Johnston et al., *Environmental Research Letters,* Jan. 13, 2009. The higher yields come from the Worldwatch Institute and other quoted sources.

11 Assuming a new, efficient car that burns 5 liters per 100 km (20 km per liter).

Earth's Future Wastes

1 EPA, 2005, epa.gov/epaoswer/non-hw/muncpl/ pubs/ex-sum05.pdf.

2 "Experimental Determination of Energy Content of Unknown Organics in Municipal Wastewater Streams," by Ioannis Shizas and David M. Bagley, *Journal of Energy Engineering,* Vol. 130, No. 2, August 1, 2004.

3 "Sewage turned into hydrogen fuel," Will Knight, *New Scientist,* April 29, 2002.

4 "Treatment Through Resource Recovery: Options for Core Area Sewage," by Stephen Salter, Peng, April 6, 2006, georgiastrait.org/files/Resource-Recovery-Submission-SETAC.pdf.

5 The Leadership in Energy and Environmental Design (LEED) Green Building Rating System™ certifies buildings at four levels: certified, silver, gold and platinum. See cagbc.org/leed/what/index.php.

Nuclear — Hope or Hype?

1 "High Cost Energy: The Economics of Nuclear Power," Ontario Clean Air Alliance, *Air Quality Issues Fact Sheet #20,* March 2006.

2 "Final Study: Choosing a Way Forward," Nuclear Waste Management Organization, nwmo.ca.

3 *Decommissioning Nuclear Facilities,* Australian Uranium Association Nuclear Issues Briefing Paper, June 13, 2007

4 *"Why expanding nuclear power would reduce and retard climate protection and energy security... but can't survive free-market capitalism,"* invited testimony to the Select Committee on Energy Independence and Global Warming, United States House of Representatives, Washington, DC. Hearing on "Nuclear Power in a Warming World: Solution or Illusion?" March 12, 2008, by Amory Lovins, Chief Scientist, Rocky Mountain Institute.

5 "The 2030 Blueprint," Architecture 2030, April 7, 2008.

6 Lovins, as above.

7 The calculation assumes the need for 184.5 tonnes of uranium per GW of capacity.

8 For more evidence, see *Cancer: 101 Solutions to a Preventable Epidemic,* by Liz Armstrong, Guy Dauncey and Anne Wordsworth, New Society, 2007.

9 *Nuclear Power — the Energy Balance.* See www.stormsmith.nl. In "Review of solutions to global warming, air pollution, and energy security"

(*Energy & Environmental Science,* December 1, 2008), Stanford's Mark Jacobson finds that when you combine a life-cycle analysis and opportunity cost CO_2 emissions because of delays in nuclear construction compared to wind energy, nuclear power produces 68–180 grams of CO_2 per kWh. (See p. 51 for the full chart.)

Hydrogen — Hope or Hype?

1 "The Hydrogen Economy," UIC Nuclear Issues Briefing Paper #73, August 2005.

2 "The Hydrogen Energy Economy: Its Long Term Role in Greenhouse Gas Reduction," by Geoff Dutton et al., *Tyndall Centre Technical Report, No. 18,* Jan. 2005, p. 12.

Clean Coal — Hope or Hype?

1 World Coal Institute, 2006 data.

2 "'Clean' Coal? Don't Try to Shovel That," *Washington Post,* March 2, 2008.

3 "Top Ten Reasons Coal is Dirty," from *Coal is Dirty,* a joint project managed by The DeSmog Project, Rainforest Action Network, and Greenpeace USA.

4 Dr. Anupma Prakash, Geophysical Institute, University of Alaska Fairbanks.

5 "Carbon Dioxide Capture and Storage," IPCC Summary for Policymakers, 2006.

6 "Can coal live up to its clean promise?" *New Scientist,* March 27, 2008, and "Review of solutions to global warming, air pollution, and energy security," by Mark Jacobson, *Energy and Environmental Science,* Dec. 1, 2008.

7 Jacobson study, see p. 51.

8 *The Future of Coal,* MIT, 2007.

9 See "Carbon Capture and Storage," by Peter Viebahn et al., in *State of the World 2009: Into a Warming World,* Worldwatch Institute.

Earth's Future Farms

1 *Livestock's Long Shadow,* by H. Steinfeld et al., UN FAO, Nov. 2006.

2 *Cool Farming* (Greenpeace 2008) says 17%–32%. In her paper "Organic Agriculture and Localized Food and Energy Systems for Mitigating Climate Change," Dr. Mae-Wan Ho, from the Institute of Science in

Society, suggests that 34% may be a more accurate number. Invited lecture at the East and Southeast Asian Conference on Sustainable Agriculture, Food Security, and Climate Change, Philippines, October 2008. Institute for Science in Society.

3 Data from *Cool Farming* (Greenpeace, 2008) except where noted.

4 1.7 MT of N_2O x 298 = 506 MT CO_2e.

5 86 MT of methane a year x 25 = 2150 MT of CO_2e.

6 18 MT x 25 = 450 MT CO_2e.

7 The Consumer Society, "State of the World Report," Worldwatch Institute, 2004.

8 "Global methane emissions from landfills: New methodology and annual estimates 1980–1996," by J. Bogner et al., *Global Biogeochemical Cycles*, Vol. 17, No. 2, 2003.

9 Estimated total global soil carbon losses range from 44–537 Gt, with a common range of 55–78 Gt. ("Sustainable Food System for Sustainable Development," by Mae-Wan Ho, Sustainable World International Conference, 2008). If we assume a 60 Gt loss spread over 200 years, this suggests an annual incremental loss of 300 Mt carbon loss, or 1100 Mt of CO_2.

10 "Organic agriculture and the global food supply," by C. Badgley et al., *Renewable Agriculture and Food Systems 2007*, 22 (2), 86–108.

Earth's Future Forests

1 Global Forestry Resource Assessment, FAO, 2005.

2 "World faces megafire threat," Reuters, Jan. 19, 2007.

3 UN FAO says 25%; Nature Conservancy says 20%; CIFOR says 20–25% from land-use changes in general, most of which involve deforestation.

4 "Parks effectively protect rainforest in Peru," Mongabay, Aug. 9, 2007.

Earth's Future Atmosphere

1 For a good layperson's summary, see "A Safe Landing for the Climate," by Bill Hare, scientist in Earth System Analysis at the Potsdam Institute for Climate Impact Research, Germany, published in *State of the World 2009*, Worldwatch Institute, worldwatch.org/node/5984.

2 Some believe such an experiment is already underway, using airplanes to create contrails containing materials such as aluminum that reflect the sun's heat back into space, known as "chemtrails." Others think this nonsense, but US Patent 5003186 was taken out by the Hughes Aircraft Company in 1991 for the spray nozzles (tinyurl.com/an375s); the approach is described in Chapter 28 of *Policy Implications of Greenhouse Warming* (1992) (tinyurl.com/cwe9ho); and one of the concept's pioneers, the physicist Edward Teller (founder of the A-Bomb) describes the details in a 1997 paper *Global Warming and Ice Ages: Prospects for Physics-Based Modulation of Global Change*, prepared for the 22nd International Seminar on Planetary Emergencies in Sicily, August 20–23, 1997 (tinyurl.com/69nj2b). It is unfortunate that the subject has been tarnished with conspiracy theories when it needs full and open discussion.

3 "The radiative forcing potential of different climate geoengineering options," by T. M. Lenton and N. E. Vaughan, *Atmospheric Chemistry and Physics*, January 28, 2009, tinyurl.com/cg24wa.

4 Lenton and Vaughan.

5 Lenton and Vaughan.

6 "Bio-Char Sequestration in Terrestrial Ecosystems — a Review," by J. Lehmann et al., *Mitigation and Adaptation Strategies for Global Change*, March 2006, pp 395–419.

7 "Victoria University of Wellington, Institute of Policy Studies Working Paper 07/01: Holistic greenhouse gas management: mitigating the threat of abrupt climate change in the next few decades," by P. Read and A. Parshotam, tinyurl.com/agbjwk.

The Climate Solutions Dividend

1 *The Illness Costs of Air Pollution in Ontario*, Ontario Medical Association, 2005.

2 *Lives per Gallon: The True Cost of Our Oil Addiction*, by Terry Taminen, Island Press, 2006. Chapter 3 summarizes the various studies and references their sources.

3 *The Trillion-Dollar Defense Budget Is Already Here*, by Robert Higgs, The Independent Institute, March 15, 2007.

Getting Serious

1 "Cost of tackling global climate change has doubled, warns Stern," *Guardian*, June 26 2008.

2 "Nicholas Stern: Spend billion on green investments now to reverse economic downturn and halt climate change," *Guardian*, Feb. 11, 2009.

3 CIA World Factbook.

4 The inflation adjustment data is in US dollars, from thepeoplehistory.com/1940s.html, multiplying by 14.33.

5 "16th Asian Parliamentarians' Meeting on Population and Development," by Lester Brown, March 2000.

TEN SOLUTIONS FOR INDIVIDUALS

2. Change the Way You Eat

1 *Livestock's Long Shadow*, Food And Agriculture Organization of the United Nations, 2006. fao.org/docrep/010/a0701e/a0701e00.HTM.

2 In "Meat is Murder on the Environment" (July 21, 2007), *New Scientist* reported on a Japanese study that concluded 36.4 kg of CO_2e is produced from one kg of beef. See chart below for data for the 96kg of CO_2e per kg of beef figure. The *UN Report Livestock's Long Shadow* reported that deforestation is responsible for 34% of the overall impact of livestock, which is not covered in this data. If included, it adds approximately another 2 tonnes of CO_2, for a total of 6 tonnes per meat eater (excluding pork, chicken and other types of meat).

3 An average meat eater consumes 31 kg of beef and 28 kg of pork a year. One kg of beef produces 96 kg of CO_2e (others say far less, but see my numbers above), so 31 kg of beef produces 3 tonnes of CO_2e. Pork produces 4.25 kg of CO_2 per kilogram, so 28 kg produce 119 kg of CO_2 a year. When you add dairy, the total is probably around 3.5 tonnes of CO_2e a year from meat and dairy. An average small car that travels 20,000 kilometers a year at 7 liters per 100 km (33 mpg) produces 3.27 tonnes of CO_2e (2.34 kg of CO_2 per kilometer). If the vegan's car is less fuel efficient than this, it will produce more CO_2e than the meat eater.

4 For detailed evidence, see *Cancer: 101 Solutions to a Preventable Epidemic*, by Liz Armstrong, Guy Dauncey and Anne Wordsworth, New Society Publishers, 2007.

5 *Organic farming combats global warming ... big time.* Rodale Institute, New Farm field trials, March 2004. newfarm.org/depts/NFfield_trials/1003/carbonsequest.shtml.

6 "Dreaming of a Green Christmas," *Toronto Star*, December 18, 2005.

7 *Estimating and Addressing America's Food Losses*, Economic Research Service, US Dept. Agriculture, 1997.

3. Wake Up to Green Electricity

1 Using the GHG Chart on page 17: CO_2 causes 44% of global warming, so 19% of CO_2 is 9% of the cause. Add methane from coal mining and gas wells

	Weight of greenhouse gases produced	GWP	Kg CO_2e per year
CO_2	208 gallons of oil per feedlot cow = 787 liters @2.34kg CO_2 per liter = 1841 kg of lifetime CO_2. Distributed over the 5-year life of a cow = 368 kg of CO_2 per year	1	368
CH_4	375 grams a day = 137 kg a year	25	3425
N_2O	9 kg a year	298	2682
Total			6475
1 cow living for 5 years, producing 6.475 tonnes of CO_2e a year, produces 32 tonnes of CO_2e, and yields 331 kg of beef. 1 kg of beef produced = 96 kg of CO_2e			

— maybe 13% of 13% = 1.7%. Add tropospheric ozone — maybe 25% of 9% = 2.25%. Total = 9+1.7+2.25 = 13%.

4. Keep It Warm, Keep It Cool

1 *Cool Citizens: Everyday Solutions to Climate Change,* by Richard Heede, Rocky Mountain Institute, 2002, rmi.org/images/PDFs/Climate/C02-12_CoolCitizensBrief.pdf.

2 "Tree Ordinances Protect Canopy, Lower A/C Bills, UF Study Shows," July 2000, news.ufl.edu/2000/07/06/tree.

5. Heat Your Home Without Carbon

1 *Cool Citizens.* See Solution #4.

6. Change the Way You Travel

1 Assumptions: Electricity: 600 grams of CO_2 per kWh. Car: 9.4 l/100km (25 mpg). Electric bike: 10 watt-hours per km. Electric scooter: 28 watt-hours per km.

2 Edinburgh Centre for Carbon Management, based on an average coach load. nationalexpress.com/utilities/press36.cfm.

7. Drive a Greener Car

1 Assumptions: Electricity: 600 grams of CO_2 per kWh. Car: 9.4 l/100km (25 mpg). Electric bike: 10 watt-hours per km. Electric scooter: 28 watt-hours per km.

2 The PHEV Prius Test Program by Sacramento Municipal Utility District 9/27/06 showed these results, with my translation into CO_2. PHEV fuel: 98 mpg = 157.7 km/hr (1 gallon = 19.56 pounds CO_2 = 8.87 kg. 157.7 km = 8870 grams = 56 g/km. PHEV electricity: 154 watt-hours /mile/per 1.6 km. 1 km = 96.26 wh @ 600 grams CO_2/kwh = 57.75 grams. Fue: 56 grams + electricity 57.75 grams = 113.75 grams per kilometer). See arb.ca.gov/msprog/zevprog/symposium/presentations/maccurdy.pdf.

3 250 watt-hours/mile = 156 wh/km @ 600 grams CO_2 per kWh = 94 grams CO_2 per kilometer.

4 Automobiles: Manufacture vs. Use, Institute for Lifetime Environmental Assessment, 2003, ilea.org/lcas/macleanlave1998.html.

5 Renault estimates 6 tonnes of CO_2 per car, including the assembly plants, suppliers and supply chain

distributors. "Emissions — Developments in the Pipeline," Automotive Engineer on the Web, June 2007, ae-plus.com/Key%20topics/kt-emissions-news4.htm.

8. Take a Climate-Friendly Vacation

1 The report *Calculating the Environmental Impacts of Aviation Emissions,* by Dr Christian Jardine, Environmental Change Institute, Oxford (Climate Care) concludes that 2.0 is a sound factor. See jpmorganclimatecare.com/media/documents/pdf/aviation_emissions__offsets.pdf. The report *GHG Emissions Resulting from Aircraft Travel,* by Dr. Davide Ross, June 1 2007 (Carbon Planet) concludes 2.7. See carbonplanet.com/downloads/ ghg_emission_factors_for_flights.pdf.

2 Data from Princess Cruises: 585 tons of fuel for a 7-day cruise carrying 1500 passengers. Residual fuel, producing 3.12 kg of CO_2 per liter. Quoted in "Top ten ways to cool the planet while you're traveling", *San Diego Earth Times,* August 1996.

3 A well-tuned two-stroke outboard motor burns 0.6 to 0.8 pounds of fuel for each unit of horsepower per hour; a four-stroke motor burns 0.4 pounds. The oil produces 2.36 kg of CO_2 per liter (19.56 lbs per gallon).

9. Change Your Consumer Habits

1 EPA data.

2 41pounds.org.

3 "Sunday Times clarifies figures in Google carbon emissions debate," *Guardian,* Jan. 16. 2009.

4 "Plastic or paper?" *Washington Post,* 2007, tinyurl.com/38t24r.

5 Carbon Trust labeling program, UK.

6 In 2006, Americans bought 31.2 billion liters of bottled water, the plastic manufacturing of which produced 2.5 million tonnes of CO_2 = 80 grams per liter. Bottled Water and Energy — Pacific Institute Fact Sheet.

7 Carbon Rally, carbonrally.com/challenges/5.

8 "AskPablo: Exotic Bottled Water," triplepundit.com/pages/askpablo-exotic-bottled-water-002401.php.

9 At the sawmill, 50% of the tree becomes lumber, 50% becomes chips and sawdust. At the pulp mill, 50% becomes pulp, 50% becomes fuel. Energy

needed at mills for the virgin fibers: 0.3 kWh = 109.5 kWh/yr = 175 lbs CO_2. Energy needed for the 50% recycled fibers: 0.15 kWh = 54.75 kWh/yr = 88 lbs CO_2. Energy needed to make the ink, print, deliver and recycle: unknown. CO_2 emissions: 263 lbs per year = 0.72 lbs per day = 326 grams. From *Stuff: The Secret Lives of Everyday Things,* by Alan Thein Durning and John C. Ryan, The Futurist, 1998.

10 "Pepsi tests the carbon footprint of orange juice," *International Herald Tribune,* Jan. 23, 2009. Research by The Carbon Trust.

11 "Hamburgers are the Hummers of Food in Global Warming," *Agence France Presse,* Feb 16, 2009. Nathan Pelletier (Dalhousie University) presentation to AAAS, February 2009.

12 "By reducing your junk mail for five years, you'll conserve 1.7 trees and 700 gallons of water, and prevent 460 pounds of carbon dioxide from being released into the atmosphere." 460 lbs = 92 lbs per year = 42 kg of CO_2. Junk mail impact: 41pounds.org/impact.

13 Green Progress Calculator, greenprogress.com.

14 Carbon Rally carbonrally.com/challenges/5.

15 Ibid.

16 Each recycled bottle saves 1lb/453 gms of CO_2 when used to make a new one.

17 Each recycled newspaper saves 0.25 lb/113 gms of CO_2 when used to make a new one.

18 Environmental Defence. Virgin paper = 3.24 kg CO_2 per kg paper. 100% post-consumer recycled paper = 1.76 kg CO_2 per kg paper, environmentaldefense.org/article.cfm?contentid=1689.

TEN SOLUTIONS FOR CHAMPIONS

11. Solutions for Friends, Family and Neighbors

1 Inspired by Patricia Lane of Victoria, BC, who organized several parties for her friends.

12. Solutions for Schools

1 "Greening America's Schools — Costs and Benefits," by Gregory Kats, *Capital R Report,* October 2006.

2 Ibid.

13. Solutions for Higher Education

1 travelwise.utah.gov/employers_8.php.

16. Start a Climate Action Group

1 "Although the Institute for Cultural Studies [founded by Margaret Mead] has received many inquiries about this famous admonition by Margaret Mead, we have been unable to locate when and where it was first cited. We believe it probably came into circulation through a newspaper report of something said spontaneously and informally. We know, however, that it was firmly rooted in her professional work and that it reflected a conviction that she expressed often, in different contexts and phrasings." interculturalstudies.org/faq.html.

17. Take the Initiative

1 The Goethe couplet referred to here is from an extremely loose translation of Goethe's *Faust* lines 214–30 made by John Anster in 1835.

20. Build a Nation-Wide Movement

1 January 2008 data from The Climate Project

TEN SOLUTIONS FOR COMMUNITIES

25. Encourage More Walking, Cycling and Transit

1 "Copenhagen: City of Cyclists, Bicycle Account 2006," vejpark2.kk.dk/publikationer/pdf/464_Cykelregnskab_UK.%202006.pdf.

2 "No Hassle Transit? Try Hasselt," The Tyee, July 9, 2007. thetyee.ca/Views/2007/07/09/NoFares3

3 "C40 Cities Transport Best Practices," c40cities.org/bestpractices/transport/bogota_bus.jsp

26. Become a City of Green Buildings

1 "How white roofs shine bright green," *Christian Science Monitor,* October 3, 2008.

27. Go for Green Energy

1 "Dead-end Austrian town blossoms with green energy," *Bloomberg News,* August 28, 2007.

2 "New York City has large potential for solar energy," *Renewable Energy Focus,* February 21, 2007.

28. Worship Your Wastes

1 The incineration of 1 tonne of MSW will release approximately 0.7 to 1.2 tonnes of CO_2. Of this, 33%

to 50% is biogenic (part of the natural carbon cycle) while 50 to 67% is non-biogenic (man made). Therefore incineration of 1 tonne of MSW releases between 0.35 and 0.8 tonnes of non-biogenic CO_2. Source: *Emissions from Waste Incineration,* by Bert Johnke. IPCC Good Practice Guidance and Uncertainty Management in National Greenhouse Gas Inventories. ipcc-nggip.iges.or.jp/public/gp/bgp/5_3_Waste_ Incineration.pdf.

2 Energie-Cites case study, energie-cites.org/db/stockholm_113_en.pdf.

3 Energie-Cites case study, energie-cites.org/db/linkoping_113_en.pdf.

4 "Inspiration from Sweden," by Stephen Salter, P.Eng. georgiastrait.org/?q=node/359.

5 "Experimental Determination of Energy Content of Unknown Organics in Municipal Wastewater Streams," by Ioannis Shizas and David M. Bagley, *Journal of Energy Engineering,* Vol. 130, No 2, August 1, 2004.

29. Grow More Food, Plant More Trees

1 *Farming in the City,* by Lester Brown, Earth Policy Institute, 2007.

2 "Seeing Green: Study Finds Greening is a Good Investment," Wharton School of the University of Pennsylvania, 2005, pennsylvaniahorticulturalsociety.org/ phlgreen/see-inggreen.htm.

3 "Tree Ordinances Protect Canopy, Lower A/C Bills, UF Study Shows," July 6, 2000. news.ufl.edu/2000/07/06/tree.

4 Data from the Arbor Day Foundation.

30. Plan Ahead for a Green, Stormy Future

1 "Local Warming," by Christopher Swope, governing.com, December 2007.

TEN SOLUTIONS FOR BUSINESSES

31. Join the Challenge

1 These scores will hopefully change as businesses start to take action. For an immediate score, text "cc" followed by the name of the company to 30644.

2 Carbon Trust, 2006, carbontrust.co.uk/news/ press-centre/2006/061206_Carbonfootprint.htm.

32. Four Solutions for Every Business

1 Almost all of these examples come from "Beyond Neutrality — Moving Your Company Towards Climate Leadership." See bsr.org.

34. Solutions for Retail Stores

1 "Wal-Mart's mixed 'green' bag," *Fortune,* Nov. 17, 2006.

2 "Keep Your Eyes on the Size. The impossibility of a green Wal-Mart," by Stacey Mitchell, *Grist,* March 28, 2007.

35. Solutions for Architects, Builders and Developers

1 "Building a zero-carbon world," *The Observer,* August 13, 2006.

2 *Natural Capitalism,* by Paul Hawkins, Amory Lovins and L. Hunter Lovins, Back Bay Books, 2008, p. 92.

36. Solutions for the Cement Industry

1 The 2006 "Global Cement Report" reported global production of 2.6 GT (billion tonnes), rising to 2.75 GT in 2007 and growing by 9% a year, which suggests 3 GT in 2008. If the average intensity of CO_2 emissions from total global cement production is 900 kg per tonne, the 2007 total emissions from cement were 2.5 GT. Global CO_2 emissions in 2007 (including land-use and forest changes) were 43 GT, of which 2.5 GT is 5.8%. In *Carbon Dioxide Emissions from the Global Cement Industry* (Annual Review of Energy and the Environment, Vol. 6, 2001), Ernst Worrell et al. state 8.14 kg per tonne. Britain's cement industry data says it reduced its emissions from 924 kg to 822 kg per tonne between 1990 and 2006, so a global average of 900 kg seems more likely.

2 A bag of cement weighs 42.5 kg (94 lbs). At 0.822 kg of CO_2 per tonne of cement, this produces 35 kg of CO_2.

3 "Climate Change, Cement and the EU," Cembureau, 1998.

4 "Cement Sustainability Initiative Agenda for Action," July 2002, cement.ca.

5 Ernst Worrel.

6 "Concrete Facts about Sustainability," from the "Cement Sustainability Initiative."

7 "Carbon Strategy Targets and Current Developments," British Cement Association. Website viewed March 2008.

8 "Hope builds for eco-concrete," *New Scientist,* January 26, 2008.

9 "Green Foundations," *New Scientist,* July 13, 2002, and information from tececo.com.

37. Solutions for Industry and Manufacturing

1 *Tracking Industrial Energy Efficiency and CO2 Emissions,* IEA, Paris, 2007.

2 *Natural Capitalism,* p. 245.

3 Ibid.

4 The DuPont information comes from DuPont's 2015 Sustainability Goals; The Climate Group's case study on DuPont; *Beyond Neutrality — Moving Your Company Toward Climate Leadership,* by Business for Social Responsibility, 2007; and DuPont's Economic, Environmental and Social Performance Data, June 2007 Update.

5 "How corporations can save the climate," WWF, 2007.

6 "Accelerating the move to a low carbon economy," National Audit Office, 2007. See also "Industry slow to act on carbon-saving advice, report finds," *Guardian,* Nov. 23, 2007.

7 Data from WWF Climate Savers and *Carbon Down, Profits Up,* The Climate Group, 3rd Edition, 2007.

38. Solutions for Forest Companies

1 Personal communication with Jens Wieting, Coastal Forest Campaigner, Sierra Club of Canada.

2 "Effects on Carbon Storage of Conversion of Old-Growth Forests to Young Forests," by Mark E. Harmon et al., *Science,* Vol. 247, February 9, 1990, pp.699–702.

3 Ibid.

4 "Modeling historical patterns of tree utilization in the Pacific Northwest," by Mark Harmon et al., *Ecological Applications* 6, pp. 641–652. Quoted in "Climate of Destruction: Sierra Pacific Industries' Impact on Global Warming," forestethics.org.

5 "From Theory to Practice — Increasing Carbon Stores through Forest Management," by Laurie Wayburn. *Ecoforestry* 15(2), 2000, pp. 40–42.

39. Solutions for the Media

1 Consensus About Climate Change? Pielke and Oreskes, Science 13 May 2005: 952-954.

2 New York Times, Washington Post, LA Times, Wall St Journal. 18% sample. Source: Al Gore, *An Inconvenient Truth.*

FIVE SOLUTIONS FOR FARMERS

41. Become a Carbon Farmer

1 "Global scale climate–crop yield relationships and the impacts of recent warming," by David B. Lobell and Christopher B. Field, Environmental Research Letters, March 2007.

2 "Global Warming and Agriculture: New Country Estimates Show Developing Countries Face Declines in Agriculture Productivity," by William Cline, Center for Global Development, 2007, cgdev.org.

3 "Modeling the Recent Evolution of Global Drought and Projections for the 21st Century," by E.J. Burke et al., *Journal of Hydrometeorology,* October 2006.

4 "Exposed: the great GM crops myth," *The Independent,* April 20, 2008.

5 "Soil may spoil UK's climate effort," *New Scientist,* Sept. 7, 2005.

6 "Farming without a plough?" *New Scientist,* January 15, 2001.

7 "Reduced tillage helps reduce carbon dioxide levels," Soil Conservation Council of Canada, 2004-02.

8 "Soil Carbon Sequestration Impacts on Global Climate Change and Food Security," by Rattan Lal, *Science,* June 11, 2004.

9 "Bio-Char Sequestration in terrestrial Ecosystems — a Review," by Johannes Lehmann et al., *Mitigation and Adaptation Strategies for Global Change,* Springer Netherlands, 2006, 11: 403–427.

10 For further discussion, see "Biochar, climate change and soil: A review to guide future research," by Saran Sohi et al, CSIRO Land and Water Science Report 2009, csiro.au/files/files/poei.pdf, and "Biochar, reducing and removing CO2 while improving soils: A

significant and sustainable response to climate change?" by Professor Stuart Haszeldine et al, School of GeoSciences, University of Edinburgh, geos.ed.ac.uk/sccs/biochar/Biochar1page.pdf.

42. Become an Organic Farmer

1 See, for example, "Use of Agricultural Pesticides and Prostate Cancer Risk in the Agricultural Health Study Cohort," by M.C.R. Alavanja, et al., *American Journal of Epidemiology,* 157, 2003, pp. 800–814.
2 "Going back to nature down on the farm," Andy Coghlan, *New Scientist,* June 3, 2000.
3 "Nutritional Quality of Organic versus Conventional Fruits, Vegetables and Grains," by Dr. Virginia Worthington, *Journal of Alternative and Complementary Medicine,* Vol .7, No 2, 2001.
4 Analysis by David Thomas of nutritional values listed in McCance and Widdowson, "The Composition of Foods, UK Standards Agency," comparing the 1940 and 2002 edition. "Meat and dairy — where have the minerals gone?" *Food Magazine* 72, Jan-March 2006.
5 "Organic Farming Sequesters Atmospheric Carbon and Nutrients in Soils," by Paul Hepperly, New Farm Research Manager, Rodale Institute, October 15, 2003.
6 Rodale Institute release, April 17, 2008.
7 "Mitigating Climate Change through Organic Agriculture and Localized Food Systems," by Dr. Mae-Wan Ho and Lim Li Ching, Institute for Science in Society, Jan. 31, 2008.
8 Ibid.

43. Become a Green Rancher

1 "Report from my Farm," by Martha Holdridge, Soil Carbon Coalition, soilcarboncoalition.org/opportunity.
2 "Supplements may reduce cattle methane," by Junichi Takahashi, Obihiro University of Agriculture, *Yomiuri Shimbum,* Jan 22, 2008.
3 John Wallace, Rowett Research Institute, Aberdeen, 2005.
4 "Garlic may cut cow flatulence," Research at University of Wales, Aberystwyth, UK, BBC News, July 10, 2007.
5 "Feeding practices can reduce methane production from cattle operations," by Dinah Boadi and Karin Wittenberg, University of Winnipeg Department of Animal Science, March 4, 2004.
6 Ibid.
7 Ibid.
8 Steve Ragsdale, University of Nebraska Institute of Agriculture and Natural Resources, 2002.
9 Boadi and Wittenberg.
10 "Methane Emissions of Beef Cattle on Forages: Efficiency of Grazing Systems," by Alan DeRamus et al., *J. Environ Qual,* 32, 2003, pp. 269–277.
11 Jamie Newbold, Rowett Research Institute, Aberdeen, 2003.
12 CSIRO Livestock Research Centre, New South Wales.
13 Dr. Julian Lee, AgResearch Grasslands, New Zealand, 2002.

44. Become an Energy Farmer

1 Natural Capitalism, by Paul Hawkins, Amory Lovins and L. Hunter Lovins, Back Bay Books, 2008, p. 199.
2 "Harvesting Clean Energy Biogas," by Patrick Mazza, Climate Solutions, 2002.
3 "Net energy of cellulosic ethanol from switchgrass," by Kenneth Vogel et al., *Proceedings of the National Academy of Sciences,* Jan. 7, 2008.
4 "Switching to Switchgrass Makes Sense," by Mark Leibig, *Agricultural Research,* July 2006.
5 "Bioenergy Village — a Concept for Energy Self-Sufficiency in Rural Areas," by Sabine M. Lieberz, Global Agriculture Information Network, 2007, fas.usda.gov/gainfiles/200705/146291233.pdf.

45. Keep on Farming!

1 "Exposed: The Great GM Crops Myth," by Geoffrey Lean. *The Independent,* April 20, 2008.
2 "Toyota, Menicon Develop Advanced Composting Process," Toyota Press Release, June 16, 2006.
3 "Future Farming: A Return to the Roots?" *Scientific American,* August 2007.
4 "What's Happening to our Farmland?" American Farmland Trust. farmland.org/resources/fote, viewed May 10, 2008.

FIVE SOLUTIONS FOR TRANSPORTATION

46. Solutions for the Auto Industry

1 OICA 2007 Production statistics, oica.net/category/production-statistics.
2 "Toyota North American Environmental Report, Energy and Climate Change," 2007.

47. Solutions for Trucking

1 Each gallon of diesel produces 22.4 lbs of CO_2; 100 gallons = 2240 lbs = 1016 kg = 1 tonne.
2 "Estimation of Fuel Use by Idling Commercial Trucks," by Linda Gaines et al., Argonne National Laboratory, 2006, smartidle.com/TRB06FINAL.pdf.
3 Trucks Deliver a Cleaner Tomorrow, trucksdeliver.org/pressroom/drive-smart.html.
4 "Hydrogen-Enhanced Combustion Engine Could Improve Gasoline Fuel Economy by 20% to 30%," Green Car Congress, Nov. 5, 2005, greencarcongress.com/2005/11/hydrogenenhance.html. A similar technology being touted for cars seems to achieve no results, so the jury is still out on this technology.
5 "Aerodynamic Truck Trailer Cuts Fuel And Emissions By Up To 15 Percent," *Science Daily,* April 15, 2008.
6 Schneider News Release, May 8, 2008, schneider.com/news/FINAL_Schneider_Sustainability_News_Release_050808.html.

48. Solutions for Railways

1 "Rail Transit in America — A Comprehensive Evaluation of Benefits," Todd Litman, Victoria Transport Policy Institute, 2006.
2 "Railways and the Environment," Community of European Railway and Infrastructure Companies/International Union of Railways report, November 2004.
3 Personal email, 2001. Cars and Transportation Treehugger interview, January 24, 2008.
4 Average load factors from EcoTransIT. Referenced in "Railways and the Environment."
5 Ibid.
6 "State aims for zero-carbon rail plan," *The Argus,* Feb. 7, 2008.

49. Solutions for Shipping and Ports

1 International Maritime Organization, 2007. 1 ton of bunker oil No 6 fills 7.3 barrels. Each barrel holds 42 gallons, and each gallon of oil No 6 produces 26 lbs of CO_2. 1 ton = 7972 lbs = 3.6 tonnes of CO_2.
2 green-marine.org.
3 "Record fuel prices place stress on ocean shipping," World Shipping Council, May 2, 2008.
4 "Slower Boats to China as Ship Owners Save Fuel," Reuters, Jan. 20, 2008.
5 "Slippery ships float on thin air," *New Scientist,* February 18, 2006.
6 Wikipedia article on black carbon, en.wikipedia.org/wiki/Black_carbon.
7 C40 Cities — Goteburg, c40cities.org.

50. Solutions for Aviation

1 "Airline emissions 'far higher than previous estimates,'" *The Independent,* May 6, 2008. Burning a gallon of kerosene produces 21.5 lbs of CO_2.
2 750 MT is 2% of the total 36.66 MT. 44% of 2% is 0.9%. Multiply by 2.7 comes to 2.43%.
3 "Uneven surfaces conserve fuel," KTH Royal Institute of Technology, Sweden, Feb. 23, 2006.
4 "Take a night flight to Heatsville," *New Scientist,* June 17, 2006. From Nature, Vol 441, p. 864.
5 "IATA Calls for Air Traffic Management Improvements to Reduce CO_2 Emissions," Greener Cars Congress, Feb. 19, 2007.
6 For a thorough analysis of this and other aviation solutions, see "Strategies for Airlines on Aircraft Emissions and Climate Change: Sustainable, Long-term Solutions," Hodgkinson Group, Aviation Advisors, July 2007.
7 "Green fuel for the airline industry," *New Scientist,* Aug. 3, 2008.

TEN SOLUTIONS FOR ENERGY COMPANIES

51. Invest in a New Apollo Project

1 2004 consumption was 17,350 billion kWh = 17,350 TWh. US Energy Information Administration.
2 *IEO 2007* reference case 2004 to 2030, Chapter 6.
3 US Highway statistics, 2006.

4 Proven gas reserves: 177 trillion cubic meters. 2007 consumption = 2.94 trillion cubic meters a year. Exhaustion at this rate = 60 years. BP Statistical Review of World Energy, 2008.

5 Proven coal reserves: 847 billion tonnes. 2007 consumption = 6.5 billion tonnes a year. Exhaustion at this rate = 130 years. BP Statistical Review of World Energy, 2008.

6 This is purely illustrative, to demonstrate the possibilities. Each method of generation will produce a different share of the power in each country, and new technological breakthroughs may change the entire picture.

7 US Energy Information Agency, 2006.

8 Environment Canada, 2004.

9 2007 data.

10 Center for Global Development, CARMA — Carbon Monitoring for Action, 2008.

52. Maximize Energy Efficiency

1 1 cent — "Profitably Getting off Coal," by Amory Lovins, Rocky Mountain Institute, June 7, 2007. 24 cents — "Energy Matters Update," Oct. 14, 2004, Northwest Energy Coalition. 3 cents — American Council for an Energy Efficient Economy, Energy Efficiency Progress and Potential Fact Sheet. aceee.org/energy/effact.htm. (Viewed June 30, 2008.)

2 See *Compendium of Champions: Chronicling Exemplary Energy Efficiency Programs from Across the US,* American Council for an Energy Efficient Economy, February 2008. Also *Successful Strategies for Energy Efficiency: A Review of Approaches in Other Jurisdictions and Recommendations for Canada,* Pembina Institute, 2006.

3 See "Best Practices Guide: Integrated Resource Planning for Electricity," Tellus Institute, 2000.

4 "Smart Meter success in Woodstock," *Municipal World,* January 2005.

5 "Shock tactics," by Terry Slavin, *Guardian,* UK, Feb. 20, 2008.

6 MagnaDrive's disconnected torque-transfer technology, for instance, can increase electric motor efficiency by up to 70%. See magnadrive.com.

54. Invest in Wind Energy

1 At the best sites, wind produces electricity for 7–8 hours a day. On a larger scale, including less windy sites, this falls to 5.5 hours a day.

2 In Germany, 20 GW of wind energy has created 80,000 jobs, but this includes many export jobs. On a comparable basis, 625 GW could create 2.5 million jobs if the US wind industry captured the export market too.

3 "20% Wind Energy by 2030," Black and Veatch Report, October 2007.

4 "Can We Rely on Wind Power?" by Paul Bonavia, chief operating officer of Xcel Energy, Utility Wind Integration Group paper, 2008.

5 "Review of solutions to global warming, air pollution, and energy security," by Mark Jacobson. Stanford University, *Energy and Environmental Science,* Dec 1, 2008. The land areas needed are scaled up from the paper's calculations for wind energy powering all vehicles (up to 3 sq km footprint, and 65,000 sq km spacing).

6 "New Study Shows Quebec's Wind Energy Potential," Electricity Forum, electricityforum.com/news/apr04 /quebecwind.html. Viewed Jan 6, 2009.

7 These state estimates are from the 1991 estimate, which are half the 2007 estimates.

8 AWEA Report, 2008.

9 "Wind Energy and Climate Change: Proposal for a Strategic Initiative," AWEA, 2007.

55. Invest in Solar Thermal and Geothermal Power

1 "Solar Thermal energy: the forgotten energy source," by Reuel Shinnar and Francesco Citro, *Technology in Society* 29 (2007), pp. 261–270. www1.ccny.cuny.edu/ ci/cleanfuels/upload/Solar-Thermal-TIS.pdf.

2 "Solar thermal electricity as the primary replacement for coal and oil in US generation and transportation," by David Mills and Robert Morgan, Ausra Inc., March 14, 2008.

3 "A solar-powered economy: how solar thermal can replace coal, gas and oil," by David Mills and Robert Morgan, *Renewable Energy World,* July 3, 2008.

4 "The Future of Geothermal Energy," an assessment by an MIT-led interdisciplinary panel, 2007.
5 Assumes 16 hours a day capacity factor.
6 Assumes 23 hours a day capacity factor.

56. Invest in Solar Energy
1 "Residential Solar Embraces Leasing, Power Purchase Agreements," *Renewable Energy World*, May 13, 2008.

57. Learn from Austin, Texas
1 Treehugger interview, June 20, 2008.

58. Build a Smart Grid
1 Patrick Mazza interview with Terry Oliver of Bonneville Power Administration. "Adventures in the Smart Grid No. 3," Gristmill, Sept 21, 2007.
2 "Cost of Power Interruptions to Electricity Consumers in the United States," by K. LaCommare and J. Eto, Lawrence Berkeley National Laboratory, 2006.
3 "Powering Up the Smart Grid: A Northwest Initiative for Job Creation, Energy Security and Clean, Affordable Electricity," by Patrick Mazza, Climate Solutions, 2005.
4 "Adventures in the Smart Grid: Demand Response," by Patrick Mazza, Gristmill, July 27, 2007.
5 "Smarter US power usage could save $120 billion," Reuters, Jan. 9, 2008.

59. Solutions for the Gas Industry
1 These two metrics will help you with all natural gas conversions: (a) 1 GJ of gas produces 53 kg of CO_2. (b) 1000 cubic meters of natural gas (37.69 GJ) produce 2 tonnes of CO_2.
2 North Sea gas 88%; US gas 93.4%; Canadian gas 95%.
3 "Methane Emissions from Natural Gas Production, Oil Production, Coal Mining, and Other Sources," Mark Delucchi, Institute of Transportation Studies, University of California, Davis. 2003.
4 "Low methane leakage from gas pipelines," by J. Lelieveld et al., *Nature*, April 14, 2005.
5 "Tech tools detect gas leaks," *Wired*, Jan. 22, 2006.
6 1.4% of 90% of 3000 billion cubic meters.
7 Fossil fuels produced 31 Gt of CO_2 in 2007.
8 "10 million cubic feet of gas released," *AbbeyNews*,

June 6, 2001.
9 One car driving 15,000 miles at 25 mpg produces 5.4 tonnes; 1,286 cars produce 7,000 tonnes; 770,000 cars driving at 90 kph/55 mph for 28 minutes (40km/25 miles) produce 7,000 tonnes of CO_2.
10 For detailed advice on fugitive emissions reporting, see "Fugitive emissions from Oil and Natural Gas Activities," in *Good Practice Guidance and Uncertainty Management in National Greenhouse Gas Inventories*, IPPC Report, 2001, tinyurl.com/5lvyjd.
11 *Wired*, Jan. 22, 2006.
12 "Leak Detection & Measurement of Fugitive Methane Emissions," EPA Best Management Practice for DI&M Programs, Milton W. Heath III Heath Consultants Incorporated, epa.gov/gasstar/documents/heath.pdf.

60. Solutions for the Oil Industry
1 "Oil chief — my fears for the planet," by David Adam, *Guardian*, June 17, 2004.
2 A barrel of oil contains 42 gallons, or 158 liters. When burned, depending on how it is processed, each liter on average releases 2.5 kg of CO_2, totaling 397 kg of CO_2 per barrel. To this, we must add the production energy involved. In 2007, Shell produced 1.2 billion barrels of oil equivalent, and 92 million tonnes of CO_2e, so each barrel of oil equivalent produced 76 kg of CO_2e. This includes Shell's gas and chemicals division, so the production energy for oil alone may be 60 kg of CO_2e per barrel (annualreview.shell.com). In total, therefore, each barrel of oil may produce some 457 kg of CO_2e.
3 *Apollo's Fire*, by Jay Inslee and Bracken Hendricks, Island Press, 2007, p.74
4 "The Future of Geothermal Energy," MIT, January 2007.
5 "Measures to implement DESERTEC," trecers.net/implementation.html. Viewed August 5, 2008.
6 In 2007 Shell's regular oil exploration and production used 1 GJ of energy per tonne of production (7.3 barrels), while its tar sands operations used 6.7 GJ per tonne, fallen from 13 GJ in 2003. "Shell Sustainability Report," 2007.

7 "Return to the Tar Sands," by Kealan Gell, research paper on Tar Sands GHGs, tothetarsands.ca/2007/09/27/greenhouse-gas-emissions. Viewed August 6, 2008.
8 "Scraping the bottom of the oil barrel a significant new climate risk," WWF, July 29, 2008.
9 "Billions up in Flames as Oil Firms Burn Gas," by Bent Svensson, manager of World Bank gas flaring reduction partnership, IPS, Sept. 4, 2007.
10 "Oil industry flares $40 billion a year in gas," *Der Spiegel,* Sept. 7, 2007.
11 "Global Gas Flaring Reduction: A Twelve Year Record of National and Global Gas. Flaring Volumes Estimated Using Satellite Data," National Geophysical Data Center Final Report to the World Bank, May 30, 2007.
12 *Global Fever: How to Treat Climate Change,* by William H. Calvin, University of Chicago Press, 2008, p. 229.

TWENTY FIVE SOLUTIONS FOR GOVERNMENTS

61. Prepare for the Deluge
1 "Economic Impacts of Climate Change on Illinois," Center for Integrative Environmental Research, University of Maryland, July 2008.
2 "California Climate Risk and Response," Next 10, November 2008.
3 "NASA Study Finds World Warming Edging Ancient Limits," NASA Goddard Institute for Space Studies, Sept. 26, 2006.
4 "California farms, vineyards in peril from warming, US energy secretary warns," *Los Angeles Times,* Feb. 4, 2009.
5 "Adaptation Planning: What US States and Localities are Doing," Pew Centre Working Paper, April 2008.
6 Executive Order S-13-08, November 2008.

62. Plan for a Climate-Friendly World
1 33% below 2007 by 2020.
2 30% below 2005 by 2025.
3 "Learning from State Action on Climate Change," Pew Center on Global Climate Change, December 2007.
4 "Tough climate goals may be easier than feared,"

Reuters, Dec. 22, 2008.
5 See, mipiggs.org/govaction/index.html. Viewed Dec. 17, 2008.

64. Get Everyone Engaged
1 "Green strings: Let's make all jobs greener with 'climate quality standards,'" by Greg LeRoy, *Grist,* May 1, 2008.

66. Build a Super-Efficient Nation
1 "Laying the Foundation for Implementing A Federal Energy Efficiency Resource Standard," ACEEE, March 2009, aceee.org/pubs/e091.htm.
2 "The Size of the US Energy Efficiency Market: Generating a More Complete Picture," by Karen Ehrhardt-Martinez and John "Skip" Laitner, ACEEE Report E083, May 2008.
3 "California's investor-owned utilities efficiency programs cost an average 1.2 cents kWh in 2004, and 83 Pacific Northwest utilities cost 1.3 cents/kWh. The national average is 2 cents, but hundreds of utility programs (mainly for businesses) cost less than 1 cent." *Nuclear Power: Climate Fix or Folly?* by Amory Lovins et al, Rocky Mountain Institute Solutions, Updated December 31, 2008.

67. Make All Your Buildings Zero Carbon
1 Data from US Energy Information Administration, collated by Architecture 2030.

69. Plan for 100% Renewable Electricity
1 For updated data, see Paul Gipe's website wind-works.org, the best global inventory of its kind.
2 "Feed-in Tariffs Cheaper than Trading System," Wind-works.org, October 7, 2008.
3 European Commission Table, January 2008.
4 The previous goal of 20% was increased to 27% in 2008, with a further goal of 45% by 2030.

70. Build a Supergrid
1 Personal correspondence.
2 "From AC to DC: Going green with supergrids," by David Strachan. *New Scientist,* March 11, 2009. Also "Invisible, Underground HVDC Power Costs No More Than Ugly Towers," by Thomas Blakeslee, Renewable Energy World, March 10, 2009.

3 "Realizable Scenarios for a Future Electricity Supply based 100% on Renewable Energies," by Gregor Czisch, Institute for Electrical Engineering, University of Kassel, Germany. See "A 'super grid' for Europe?" UPI, Nov. 2, 2007.

4 "The Climate for Change," by Al Gore. *New York Times,* November 9, 2008.

5 "Germany can power itself entirely by renewables," Biopact, Dec 29, 2007, biopact.com.

71. Phase Out All Fossil Fuels

1 *Subsidizing Climate Change,* by Lester Brown, Earth Policy Institute, 2007.

2 "German hard coal production to cease by 2018," *Washington Post,* July 30, 2007.

3 "Government Spending on Canada's Oil and Gas Industry," Pembina Institute, January 2005.

4 "Industry receives $1.4 billion in tax breaks annually while greenhouse gas emissions skyrocket," EcoJustice, June 14, 2006.

5 "Green Scissors" report referenced in *Subsidizing Climate Change,* by Lester Brown, Earth Policy Institute, 2007.

6 "Aiding Oil, Harming the Climate," Oil Change International, 2008.

7 "Subsidies in the US Energy Sector: Magnitude, Causes, and Options for Reform," by Doug Koplow, Earth Track, Nov. 2006.

8 "Subsidizing Unsustainable Development: Undermining the Earth with Public Funds," Earth Council,1997.

72. Stop the Methane Emissions

1 "Oil refineries underestimate release of emissions, study says," *Globe & Mail,* September 6, 2008. The technology used was "differential absorption light detection and ranging." There were also 19 times more benzene emissions (a known cause of cancer) and 15 times more volatile organic compounds (a known cause of smog and suspected cause of cancer).

2 "When burnt, 1 gigajoule of natural gas produces 75 kg of CO_2. If 1.5% of the gas escapes as raw methane, with a GWP of 125 over the life of the methane, that adds 140 kg of CO_2e, for a total 216 kg of CO_2e per gigajoule, increasing its short-term impact by 287%." For data, see "Carbon emissions of different fuels" at biomassenergycentre.org.uk.

3 "Chinese rice farmers reduce methane emissions from paddies by 40%," edie.net/news/news_story.asp?id=6434. Viewed Sept 14, 2008.

73. Reduce from the Impact of Food and Farming

1 "Cool farming: Climate impacts of agriculture and mitigation potential," by Bellarby, Foereid, Hastings and Smith, Greenpeace, 2008.

2 "Environmental impact of products: Analysis of the life cycle environmental impacts related to the final consumption of the EU25," EC Technical Report EUR 22284 EN, May 2006.

3 "Cooking Up a Storm," Food Climate Research Network, September 2008. Britain's production-related food and farming GHGs came to 33 MT of CO_2e. Consumption-related GHGs came to 43.3 MT.

4 "The Food We Waste," Waste and Resources Action Programme, April 2008.

5 "US — Massive food waste & hunger side by side," by Haider Rizvi, Interpress Service, Sept.4, 2004, organicconsumers.org/corp/hunger090604.cfm.

74. Capture Carbon from the Atmosphere

1 "Fifth of world carbon emissions soaked up by extra forest growth, scientists find," *Guardian,* Feb. 18, 2009. The increase of 4.8 billion tonnes is 15% of the 32 billion tonnes that we produce each year by burning fossil fuels.

2 Russia 3,287,243 square miles, Canada 944,294 square miles, USA 872,564 square miles, China 631,200 square miles. Total = 5,735,301 square miles = 1,485,436,140 hectares. ½ = 740 million x 250 Tons = 185,000,000,000 tons over 160 years = 1 Gt a year.

3 "An Assessment of the Potential of Carbon Finance in Rangelands," by Timm Tennigkeit and Andreas Wilkes, ICRAF-SE Asia Working Paper No 68, chinaagroforestry.org.

75. Develop a Sustainable Transportation Strategy

1 "Reducing Freight GHGs: What are the Possibilities?" by Greg Dierkers, Center for Clean Air Policy Freight Solutions Dialogue, June 6, 2005.
2 "Freight Rail Investment Needs," by Steve Winkelman, Center for Clean Air Policy Freight Solutions Dialogue, June 7, 2005.
3 "Clearing the Air: The Myth and Reality of Aviation and Climate Change," Climate Action Network Europe and European Federation for Transport and Environment, 2006.
4 As recommended in "Calculating the Environmental Impact of Aviation Emissions," by Dr. Christian Jardine. Environmental Change Unit, Oxford University Centre for the Environment, 2005.

76. Develop a Sustainable Vehicles Strategy

1 "Andy Grove Calls for Concerted US Effort to Convert Pickups, SUVs and Vans to 40+ Mile PHEVs," Green Car Congress, July 22, 2008. Also "An electric plan for energy resilience," by Andy Grove and Robert Burgelman, *The McKinsey Quarterly*, December 2008.

77. Develop a Sustainable Bioenergy Strategy

1 Starting from 3.75 million barrels a day (mbd). The 20% non-PHEV vehicles need 0.75 mbd, and the 80% PHEVs that use electricity for 80% of their miles need 0.6 mbd. Subtotal = 1.35 mbd. Reduce by 50% for lightweight designs = 0.675 mbd.
2 "Green fuel for the airline industry," *New Scientist*, August 13, 2008.
3 "Farmland is eating up the world's wildernesses," *New Scientist*, Feb. 9, 2008. Twenty-eight million square kilometers are covered in pasture and 15 million square km are used to grow crops.
4 This is an abbreviated version of Version 0.0 (August 2008). The current version is at bioenergywiki.net.

78. Put a Price on Carbon

1 *Stern Review on the Economics of Climate Change*, UK HM Treasury, 2006.
2 "Carbon cost of $150 should stimulate low-carbon innovation," Cambridge University/Tyndall Centre

study by Jonathan Kohler et al., Tyndall Centre, *The Effect*, 2006.
3 Norwegian Pollution Control Authority, sft.no/artikkel____30693.aspx (in Norwegian).

79. Introduce Cap and Trade for Industry

1 "Environmental Defense Fund, Cap and Trade Success Story," edf.org. Viewed October 1, 2008.
2 "Carbon-credit schemes fall 30% short of projections," *Guardian*, June 25, 2008.

80. Launch a Green New Deal

1 "Spend now, reap rewards later" *New Scientist*, December 1, 2007.
2 *Low-Carbon Energy: A Roadmap*, by Christopher Flavin, Worldwatch Institute, November 2008.

82. Build A Zero-Waste Economy

1 *Global Waste Management Market Assessment 2007*, Key Note Publications Ltd.
2 "Municipal Solid Waste Generation, Recycling, and Disposal in the United States: Facts and Figures for 2006," EPA.
3 "Japan wrangling with recycling plan," CNN.com, March 15, 2001. The article says 14%, but by 2008, Japan's annual copper output had fallen to 800,000 tonnes, hence 25%.
4 "Is Recycling Good for California's Economy?" Integrated Waste Management Board, 2008.
5 "Incineration of municipal solid waste: a reasonable energy option?" ICF Consulting, Fact Sheet 3, 2005.

83. Build a Green-Collar Economy

1 "Jobs in Renewable Energy Expanding," by Michael Renner, Worldwatch Institute, July 8, 2008.

84. Ditch Neo-Classical Economics

1 "When the numbers don't add up," *New Scientist*, September 28, 2008.

85. Become a Global Player

1 H.R 1886.
2 "Aiding Oil, Harming the Climate," Oil Change International, December 2007.
3 American University spring commencement speech, June 10, 1963.

FIVE SOLUTIONS FOR DEVELOPING NATIONS

86. Scramble! This is Serious

1 Sajeeda Choudhury, Bangladesh environment minister, January 2000, quoted in "West warned on climate refugees," Alex Kirby, BBC News, Jan. 24, 2000.
2 "Global warming: Tibet's lofty glaciers melt away," *The Independent,* November 17, 2006.
3 "Natural disasters will increase," Reuters, March 19, 2007.
4 "Cities in peril as Andean glaciers melt," *Guardian,* August 29, 2006; also "Town in the Andes face crisis as glaciers melt," *San Francisco Chronicle,* April 24, 2008.
5 "Melting Andean Glaciers Could Leave 30 Million High and Dry," ENS, April 28, 2008.
6 "Climate Change and Food Security in China," Greenpeace, October 2008.
7 "Energy-Hungry Nations also most Wasteful," Inter Service Press, May 30, 2006, reporting on a new UN Foundation report led by the World Bank and UNEP.
8 "Renewable energy can save East Asia two trillion dollars in fuel costs," AFP, Singapore, Aug. 23, 2007, quoting a Greenpeace report.

87. Build a Politics of Change

1 "The Clean Energy Scam," *Time,* March 27, 2008.
2 "World Geothermal Power Generation Nearing Eruption," Earth Policy Institute, Aug. 19, 2008.

88. Build Ecological Cities

1 "Curitiba: A Global Model For Development," by Bill McKibben, CommonDreams.org, Nov. 8, 2005.
2 "Gardening for the Poor: 4,000 vegetable gardens flourish in Caracas," FAO Newsroom, 2008.
3 *Plan B 3.0: Mobilizing to Save Civilization,* by Lester Brown, Chapter 10, "Designing Cities for People," Norton, 2008.

89. Build Solar Villages

1 Solar Electric Light Company data (SELF's affiliate).

90. Solutions for China

1 "A third of China's carbon footprint blamed on exports," *New Scientist,* July 30, 2008. A 2007 study by Tao Wang and Jim Watson, in Tyndall Centre Briefing Note 23, concluded that the number was 23%, using 2004 data.
2 Most of this data comes from "White Paper: China's policies and actions on climate change," published on China.org.cn, Oct 29, 2008; and "Fact Sheet: China Emerging as New Leader in Clean Energy Policies," China Sustainable Energy Programme, efchina.org.
3 "Beijing's Desert Storm," by Ron Gluckman, *Asiaweek,* October 2000, gluckman.com/ChinaDesert.html.
4 See Solution #43 and the Chinese award-winning novel *Wolf Totem,* by Liu Jiamin, under the pseudonym Jiang Rong.

TEN GLOBAL SOLUTIONS

94. Establish a Global Climate Fund

1 "China's Post-Kyoto Roadmap," by Tang Xuepeng, China Dialogue, Nov. 17, 2008.
2 "Towards a global climate fund: principles for Poznan and beyond," December 7, 2008, choike.org.
3 US GDP in 1948 was $258 billion.
4 "Global Green New Deal," UNEP Green Economy Initiative, Oct. 22, 2008.
5 "Executive excess 2008: How average taxpayers subsidize runaway pay," 15th Annual CEO Compensation Survey, Institute of Policy Studies, Aug. 25, 2008. The $588 million number is quoted from the business trade magazine *Alpha,* April 22, 2008.
6 "Sovereign Wealth Funds 2008," IFSL Research/UK Trade and Investment, ifsl.org.uk.
7 James Tobin first proposed the levy in his Janeway Lectures at Princeton University in 1972; first published in the *Eastern Economic Journal,* 4, 1978.

95. Protect the World's Forests and Grasslands

1 "Untouched forests store 3 times more carbon," Reuters, Aug. 4, 2008.
2 "Fifth of world carbon emissions soaked up by extra forest growth, scientists find," *Guardian,* Feb. 19, 2009.
3 "Climate Change: Financing Global Forests," The Eliasch Review, October 2008. occ.gov.uk/activities/eliasch.htm

4 "The 'win-win' solution failing the rainforests,"
Guardian, UK, October 20, 2008.
5 The Eliasch Review.

96. Slow Global Population Growth
1 Population Reference Bureau, Washington.
2 "UNFPA Population Issues Overview," 2008.
3 "The Population Story ... so far," by Danielle
Nierenberg and Mia MacDonald, *World Watch
Magazine*, Sep./Oct. 2004.
4 Ibid.

97. Phase out the F Gases
1 AFEAS data 225,000 tonnes + 20,000 tonnes from
China. "HFC 134a export expansion in great
urgency," China Chemical Reporter, Oct. 26, 2007.
2 The IPCC does not assign numbers for GWPs less
than 20 years, so this is an estimate.
3 5,000 x 1 kilogram = 5 tonnes of CO_2e.
4 "HFC containment has already failed," by Eric
Johnson, Atlantic Consulting, February 2004.
5 "Growing markets of MAC with HFC-134a in
China," by Jianxin Hu and Chunmei Li, College of
Environmental Sciences, Peking University, 2004.
6 "In 2004, the accumulated amount of produced
HFC-134a was equivalent to nearly 2000 million
metric tonnes (of CO_2e). In the same year the total
amount of HFC-134a produced until 2000 — around
1,000 million metric tons of CO_2e — was already
completely released to the atmosphere." This indicates a five-year time-lag before complete release.
"HFC-134a, no sustainable refrigerant," R744.com
report, 24 July 2007.
7 1.25 Gt is 3.3% of the global 36.6 GT of CO_2 released
in 2006.
8 "Act Now to Address the HFC Explosion,"
Environmental Investigation Agency, December
2008. This assumes an 8.8% growth rate, as predicted
in the IPCC (2005) "Special report on safeguarding
the ozone layer and the global climate system: Issues
related to hydrofluorocarbons and perfluorocarbons."
9 Fact sheet No. 46:Industrial greenhouse gases: HFCs,
PFCs and SF6 mst.dk/English/Chemicals/
Legislation/Fact_sheets/Fachtsheet_no_46.htm.

10 "Fluorinated Greenhouse Gases in Products and
Processes — Technical Climate Protection Measures,"
Report of the Federal Environmental Agency, Germany,
Feb. 20, 2004.

98. Decarbonize Global Financing
1 "A Race to the Bottom: Creating Risk, Generating
Debt, and Guaranteeing Environmental
Destruction", ECA NGO Campaign, 1999.
2 "The World Bank and the G-7: Still Changing the
Earth's Climate for Business," 1998, Sustainable
Energy and Economy Network and International
Trade Information Service.
3 In 2004 the World Bank's Extractive Industries
Review recommended that the Bank phase out all
support for oil, gas and mining projects by 2008.
4 "Dirty is the new clean: A critique of the World
Bank's strategic framework for development and climate change," Institute for Policy Studies et al.,
October 2008.
5 "The World Bank and its carbon footprint: why the
World Bank is still far from being an environmental
bank," WWF-UK, June 23, 2008.
6 "How the World Bank's energy framework sells the
climate and poor people short," Bank Information
Center et al., September 2006.
7 "Abettors of destruction," by Robert Goodland,
Guardian, Nov. 2, 2007.
8 "Aiding oil, harming the climate," Oil Change
International, 2007.
9 "The World Bank's Juggernaut: The Coal-Fired
Industrial Colonization of the Indian State of Orissa
and the G-7, the World Bank, and Climate Change,"
Institute for Policy Studies, 1997.
10 "Public Finance Mechanisms to Mobilize Investment
in Climate Change Mitigation," UNEP, 2008.

100. Build a Global Movement
1 Commencement Address at American University,
Washington, D.C., June 10, 1963.

Index

About the Author

Sustainability enables the present generation of humans and other species to enjoy social well-being, a vibrant economy, and a healthy environment, without compromising the ability of future generations to enjoy the same.

— Guy Dauncey

Guy Dauncey is a writer, speaker and social visionary who works to develop a positive vision of a sustainable future, and to translate that vision into action. He lives in Victoria, British Columbia, Canada.

He is President of the BC Sustainable Energy Association, Executive Director of The Solutions Project, and publisher of EcoNews. His website is www.earthfuture.com.

Also by Guy Dauncey

The Unemployment Handbook (1981)

After the Crash: The Emergence of the Rainbow Economy (1987)

Earthfuture: Stories from a Sustainable World (1999)

Stormy Weather: 101 Solutions to Global Climate Change (2001)

Enough Blood Shed: 101 Solutions to Violence, Terror and War (2006) (by Dr. Mary-Wynne Ashford, with Guy Dauncey)

Cancer: 101 Solutions to a Preventable Epidemic (2007) (with Liz Armstrong and Anne Wordsworth)

Building an Ark: 101 Solutions to Animal Suffering (2007) (by Ethan Smith, with Guy Dauncey)

The Solutions Project

The Solutions Project aims to address the world's major problems one by one, informing people about the multitude of solutions that are being embraced around the world, and promoting their adoption.

It seeks to inspire people, businesses and governments with optimism about the achievability of the solutions, and shift the global attitude from pessimism and despair to determination and engagement.

www.earthfuture.com/solutionsproject

If you have enjoyed *The Climate Challenge* you might also enjoy other

BOOKS TO BUILD A NEW SOCIETY

Our books provide positive solutions for people who want to
make a difference. We specialize in:

Sustainable Living • Green Building • Peak Oil • Renewable Energy
Environment & Economy • Natural Building & Appropriate Technology
Progressive Leadership • Resistance and Community
Educational and Parenting Resources

New Society Publishers

ENVIRONMENTAL BENEFITS STATEMENT

New Society Publishers has chosen to produce this book on Enviro 100, recycled
paper made with **100% post consumer waste**, processed chlorine free, and old
growth free.

For every 5,000 books printed, New Society saves the following resources:[1]

40	Trees
3,603	Pounds of Solid Waste
3,965	Gallons of Water
5,171	Kilowatt Hours of Electricity
6,550	Pounds of Greenhouse Gases
28	Pounds of HAPs, VOCs, and AOX Combined
10	Cubic Yards of Landfill Space

[1]Environmental benefits are calculated based on research done by the Environmental Defense Fund and
other members of the Paper Task Force who study the environmental impacts of the paper industry.

For a full list of NSP's titles, please call **1-800-567-6772** *or check out our website at:*

www.newsociety.com

NEW SOCIETY PUBLISHERS